Bolyai Society Mathematical Studies

Volume 31

This series will typically consist of survey volumes. It may also include conference proceedings and monographs.

János Pach · Géza Tóth
Editors

Courses in Discrete and Computational Geometry

BOLYAI SOCIETY

Springer

Editors
János Pach
Rényi Institute
Budapest, Hungary

Géza Tóth
Rényi Institute
Budapest, Hungary

ISSN 1217-4696 ISSN 2947-9460 (electronic)
Bolyai Society Mathematical Studies
ISBN 978-3-032-10502-8 ISBN 978-3-032-10503-5 (eBook)
https://doi.org/10.1007/978-3-032-10503-5

This Springer imprint is published by the registered company Springer Nature Switzerland AG
The registered company address is: Gewerbestrasse 11, 6330 Cham, Switzerland

Bolyai Society Mathematical Studies

Volume 31

This series will typically consist of survey volumes. It may also include conference proceedings and monographs.

János Pach · Géza Tóth
Editors

Courses in Discrete and Computational Geometry

BOLYAI SOCIETY

Springer

Editors
János Pach
Rényi Institute
Budapest, Hungary

Géza Tóth
Rényi Institute
Budapest, Hungary

ISSN 1217-4696 ISSN 2947-9460 (electronic)
Bolyai Society Mathematical Studies
ISBN 978-3-032-10502-8 ISBN 978-3-032-10503-5 (eBook)
https://doi.org/10.1007/978-3-032-10503-5

This Springer imprint is published by the registered company Springer Nature Switzerland AG
The registered company address is: Gewerbestrasse 11, 6330 Cham, Switzerland

Preface

In the Fall of 2023, the Erdős Center of the Rényi Institute in Budapest hosted a special thematic semester on "Discrete Geometry and Convexity." It brought together some of the leading experts in the field along with many outstanding students and young researchers. The first major event of the semester was a Summer School held from August 28 to September 1, which featured four intensive mini-courses, attended by more than 70 participants from all over the world. The core of the present volume consists of material from three of these courses, complemented by the previously unpublished lecture notes of one the last courses given by the late Jirka Matoušek.

Thanks to the monumental work of Robertson and Seymour at the turn of the millenium, we have a vast generalization of the Kuratowski-Wagner theorem on planar graphs: every minor-closed class of graphs can be characterized by a finite number of excluded minors. In particular, every element of such a class can be constructed from smaller graphs that can be almost embedded in a fixed surface, by gluing them together in a tree-like fashion. But what about planar graphs? Can they also be constructed from even simpler, more elementary parts? In her course, **Vida Dujmović** presents a product structure theory for planar graphs that can be combined with the Robertson-Seymour theory and has many applications in algorithmic, topological, and coloring problems.

Suppose we randomly and independently select n points in a Euclidean space, according to some probability distribution. What is the expected measure of their convex hull? The mini-course by **Apostolos Giannopoulos** concentrates on this question, in the fairly general setting where the probability measure is log-concave. It turns out that under some mild conditions one can prove that the measure of the convex hull is sharply concentrated around its expected value. In the special case $n = 3$, the problem was already considered by Sylvester, and the first detailed analysis for the uniform distribution in a polygon or a smooth convex body in the plane–as the number of points tends to infinity–was given by Rényi and Sulanke in the 1960s.

How well can a convex body be approximated by an ellipsoid or by a polytope with a given number of vertices or facets? These questions are not only fascinating in their own right, but any nontrivial answer has meaningful applications in geometric algorithms, computer graphics, discrete geometry, and the theory of normed spaces.

Márton Naszódi's course presents several useful general techniques for approaching these problems, including cap coverings and epsilon-nets. The methods are illustrated through a classical example: deriving quantitative Helly-type theorems.

Part II of this volume is dedicated to–primarily geometric–applications of the existential theory of reals, usually denoted by $\exists\mathbb{R}$. This is an important subdiscipline of mathematical logic concerned with testing whether a given first-order statement over the real numbers, using only polynomials and existential quantifiers, is true. It is equivalent to deciding whether a semialgebraic set defined by a system of polynomials is empty or not. **Jirka Matoušek**'s lecture notes offer a crystal-clear case study: a detailed proof that recognizing whether a graph can be realized as the intersection graph of segments in the plane is $\exists\mathbb{R}$-complete, meaning that it is equivalent under polynomial time reductions to the existential theory of reals. In the process, it is shown that many similar problems in discrete geometry share the same property.

Our volume concludes with the work of **Marcus Schaefer**, with contributions by **Jean Cardinal**, and **Tillmann Miltzow**, which is essentially a self-contained monograph on the complexity class $\exists\mathbb{R}$. This class lies between the classes NP and PSPACE, but it is entirely possible that it differs from both. In recent years, it has turned out that many fundamental problems in computational geometry, graph drawing, game theory, and machine learning are $\exists\mathbb{R}$-complete. In addition to tracing the history of the field, the work includes a comprehensive list of computational problems that are $\exists\mathbb{R}$-complete or $\exists\mathbb{R}$-hard, along with an extensive "compendium" of exciting open problems. There is little doubt that it will serve as an indispensable resource for research in this field for years to come. We are grateful to Márton Beke for his invaluable help as a technical editor. We would also like to thank Yulia Rylova, who designed the covers for both this book and its companion volume. Additionally, she created the logo for our special semester as well as the posters for its individual programs.

We wish our readers to enjoy and make good use of this book and–if they are engaged in research–to successfully apply the methods and ideas presented here.

Budapest, Hungary
July 2025

János Pach
Géza Tóth

Contents

Part I
Lecture Notes

Chapter 1
Graph Product Structure Theory with Applications

Vida Dujmović

Graph product structure theory describes planar and other complex graph classes as subgraphs of products of much simpler graphs: a tree-like graph and a path. This discovery opened a new line of research and has played a key role in resolving several long-standing open problems. These lecture notes provide an introduction to graph product structure theory.

1.1 Introduction

Graph product structure theory describes how graphs in some complex classes can be represented as subgraphs of products of simpler, tree-like graphs. Since 2019, there has been a surge of interest in this field following the discovery that every planar graph is contained in the product of a tree-like graph and a path. This result opened up a new research direction and has been the key tool in solving several longstanding open problems. These lecture notes will introduce graph product structure theory and provide the tools necessary to understand the theory and the proof for planar graphs.

The lecture notes will begin with a geometric graph theory problem. The nearly twenty-year pursuit of its solution led to the discovery of the product structure theorem for planar graphs. Sometimes, the tools developed to solve a problem become more important than the problem itself. By the end of these lectures, we hope to convince you that this is one such case.

We begin with some graph theory preliminaries in Sect. 1.2. In Sect. 1.3, we introduce and study the geometric graph theory problem that was the original motivation

---

Research supported by NSERC, Australian Research Council and a University of Ottawa Research Chair.

V. Dujmović (✉)
School of Computer Science and Electrical Engineering, University of Ottawa, Ottawa, Canada
e-mail: vida.dujmovic@uottawa.ca

J. Pach and G. Tóth (eds.), *Courses in Discrete and Computational Geometry*,
Bolyai Society Mathematical Studies 31,
https://doi.org/10.1007/978-3-032-10503-5_1

behind this discovery. Tree-like graphs, specifically graphs of constant treewidth, play an important role in the product structure theorem. In Sect. 1.4, we introduce treewidth and key scholarship on treewidth that is important for understanding the product structure theorem and its proof. Readers already familiar with treewidth may skip this section, except for Sect. 1.4.3, where we introduce a new class of graphs, plane (k, s)-trees. This is a class of bounded treewidth graphs that appears in the product structure theorem.

In Sect. 1.5, we present the product structure theorem of planar graphs and its proof in Sect. 1.5.1. We then discuss variants of the theorem in Sect. 1.5.2 and extensions to classes of graphs beyond planar graphs in Sect. 1.5.3. Several problems resolved or advanced with the help of the product structure theorem are detailed in Sect. 1.6. The final section, Sect. 1.7, contains exercises.

The sections are fairly independent, allowing readers to choose their preferred order. For example, those primarily interested in the product structure theorem (and less in its applications and original motivation) may skip directly to Sect. 1.5.

1.2 Preliminaries

A graph is a pair $G = (V, E)$ where V is a finite set of *vertices* (also called *nodes*) and E is a set of *edges* where an *edge* is an unordered pair of vertices, that is, $E = \{x, y, |, x, y \in V \text{ and } x \neq y\}$. When no confusion may arise, we sometimes use the notation xy for an edge x, y. In the edge x, y, x and y are the *endpoints* of the edge and the edge is *incident* to x and y. The *degree* of a vertex is the number of edges incident to that vertex, and the maximum degree over all vertices of a graph G is denoted by $\Delta(G)$. In this manuscript, we often denote by n the number of vertices in a graph G, and call such a graph an n-vertex graph. All the graphs studied in this manuscript are *simple* and *undirected*.

A *drawing* of a graph G is a representation of G where each vertex v of G is represented by a distinct point p_v in the plane (that is, $\mathbb{R}^2$) and where each edge vw is represented by a simple curve, c_{vw}, in the plane such that the endpoints of c_{vw} are p_v and p_w, and c_{vw} intersects no vertex of $V(G)$ other than v and w. If in the previous definition we replace the plane (that is, $\mathbb{R}^2$) by $\mathbb{R}^3$, we arrive at the definition of a *3D drawing*. We say that two edges *cross* in a drawing (or 3D drawing) if they intersect at some point other than a common endpoint. A drawing (or a 3D drawing) is *crossing-free* if no pair of edges cross. A *planar drawing* is a crossing-free drawing. A graph is *planar* if it has a planar drawing. For a planar drawing $D(G)$ of a graph G, the regions of $\mathbb{R}^2 \setminus D(G)$ are called *faces* of $D(G)$. We consider only finite graphs G, so $D(G)$ has exactly one unbounded face, known as the *outerface*. The other faces of $D(G)$ are *inner faces*. The *boundary* of a face is the subgraph of G that includes all the edges and vertices adjacent to that face in $D(G)$. The *size* of a face is defined as the number of edges encountered during a closed walk around its boundary. For 2-connected graphs, this is simply the number of edges (or vertices) on the boundary of the face. A *plane graph* (also called *embedded planar graph*) is a planar graph together with a planar drawing. An edge maximal planar graph (on at least 3 vertices) is a *triangulation*. The name comes from the fact that every face in a plane triangulation is bounded by a 3-cycle (a triangle). One can argue that planar graphs are the most studied class of graphs. They are the key class of graphs in these lecture notes.

Edge contraction is an operation that removes an edge, vw, from a graph and merges its endpoints v and w into one vertex. A graph H is called a *minor* of the graph G if H can be obtained from G by applying the following operations: deleting edges, deleting vertices and contracting edges. A class $\mathcal{G}$ of graphs is *minor-closed* if for every graph $G \in \mathcal{G}$, every minor of G is in $\mathcal{G}$. A minor-closed class is *proper* if it is not the class of all graphs.

In a connected graph G, the *distance* between vertices v and w of G, denoted by $\text{dist}_G(v, w)$, is the number of edges on a shortest path between v and w in G. The *diameter* of a connected graph G is $\max\{\text{dist}_G(v, w) : v, w \in V(G)\}$.

If r is a vertex in a connected graph G and $V_i := \{v \in V(G) : \text{dist}_G(r, v) = i\}$ for all $i \geq 0$, then $(V_0, V_1, \dots)$ is called a *BFS layering* of G. Associated with a BFS layering is a *BFS spanning tree* T obtained by choosing, for each non-root vertex $v \in V_i$ with $i \geq 1$, a neighbour w in V_{i-1}, and adding the edge vw to T. Thus $\text{dist}_T(r, v) = \text{dist}_G(r, v)$ for each vertex v of G.

In this manuscript, P_ℓ denotes the path on ℓ vertices. K_n denotes the complete graph on n vertices and $K_{n,n}$ denotes the complete bipartite graph with parts of size n.

We will often make statements like "some family of graphs has some parameter bounded". For those unfamiliar with this terminology, we will define it formally. A *graph parameter* is a function α that assigns a non-negative integer $\alpha(G)$ to every graph G. For a family of graphs, $\mathcal{G}$, let $\alpha(\mathcal{G})$ denote the function $f : \mathbb{N} \to \mathbb{N}$, where $f(n)$ is the maximum of $\alpha(G)$, taken over all n-vertex graphs $G \in \mathcal{G}$. We say $\mathcal{G}$ has *bounded* α if $\alpha(\mathcal{G}) \in O(1)$. A graph parameter α is *bounded by* a graph parameter β (for some graph family $\mathcal{G}$), if there exists a function g such that $\alpha(G) \leq g(\beta(G))$ for every graph G (in $\mathcal{G}$).

Undefined terms and notation can be found in the textbook by Diestel [Die18].[1]

1.3 Motivational Problem

A *3D straight-line crossing-free grid drawing* of a graph G, or *3D grid drawing* for short, is a 3D crossing-free drawing of G that represents the vertices of G by distinct points with integer coordinates in 3D (that is, with points in $\mathbb{Z}^3$, called *grid-points*), and represents each edge as a line-segment between its endpoints, as illustrated in Fig. 1.1. Replacing $\mathbb{Z}^3$ with $\mathbb{Z}^2$ defines *2D grid drawings*. Note that having line-

[1] The textbook is freely available here: https://diestel-graph-theory.com/.

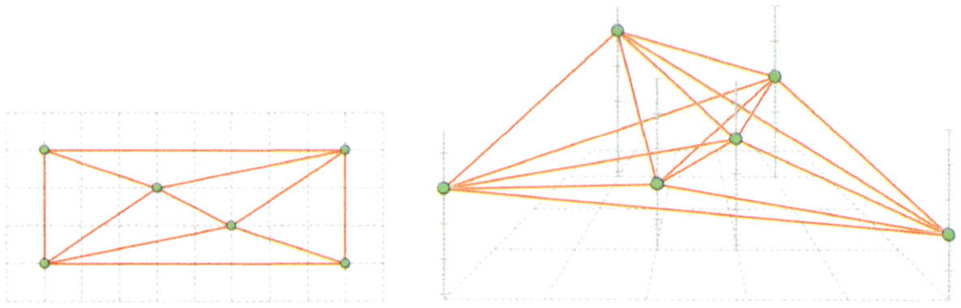

Fig. 1.1 2D grid drawing and 3D grid drawing. Figure by David R. Wood

segments for edges implies that once the vertices are assigned to grid-points, the rest of the drawing is determined.

It is well known that not every graph has a crossing-free drawing in the plane (equivalently, not every graph is planar). In contrast, a folklore result states that every graph has a 3D grid drawing, as we argue next. First, consider the following simple observation: if two line-segments cross, then their 4 endpoints lie in the same plane. Therefore, any set of n points in 3D where no four points are coplanar will suffice–we just have to find such a pointset with integer coordinates. In particular, a 3D grid drawing of a graph can be constructed using the 'moment curve' algorithm as follows. Given an n-vertex graph G with the vertex set $\{v_i : 1 \leq i \leq n\}$, represent each vertex v_i by the grid-point (i, i^2, i^3). By the earlier observation, to conclude that this defines a 3D grid drawing of G, it suffices to show that no four points in such a pointset are coplanar. If some four distinct points v_i, v_j, v_k, v_ℓ are coplanar, then the determinant of the following matrix is zero: $\begin{pmatrix} 1 & i & i^2 & i^3 \\ 1 & j & j^2 & j^3 \\ 1 & k & k^2 & k^3 \\ 1 & \ell & \ell^2 & \ell^3 \end{pmatrix}$. This, however, is a well-known Vandermonde matrix whose determinant is non-zero if and only if i, j, k, ℓ are distinct. This concludes the exposition of this folklore construction and the arguments.

Since every graph has a 3D grid drawing, one is interested in optimizing certain measures of the aesthetic quality of a 3D grid drawing. If a 3D grid drawing is contained in an axis-aligned box with side lengths $X - 1$, $Y - 1$, and $Z - 1$, then we speak of an $X \times Y \times Z$ grid drawing with *volume* $X \cdot Y \cdot Z$. This definition is formulated so that 2D grid drawings have positive volume (that is, a positive area). Observe that the drawings produced by the above folklore algorithm have $O(n^6)$ volume for any n-vertex graph. Cohen, Eades, Lin, and Ruskey [CELR96] improved this bound by proving that if p is a prime with $n < p \leq 2n$, and each vertex v_i is represented by the grid-point $(i, i^2 \bmod p, i^3 \bmod p)$, then the resulting pointset still has no four coplanar points, and thus it defines a 3D grid drawing of G in $O(n^3)$ volume. This construction is a generalization of an analogous two-dimensional technique due to Erdős [Erd51]. Furthermore, as observed by Cohen et al. [CELR96], it is easy to argue that the $O(n^3)$ volume bound is asymptotically optimal in the case of the complete graph on n vertices, K_n. First, recall that K_5 is not a planar graph (in fact, it is the smallest non-planar graph). Thus, any drawing of

K_5 in the plane has at least one crossing. That fact implies that no five vertices of K_n can be assigned the same x-coordinate, as that would imply that those five vertices lie in the same plane, which is not possible since they form a K_5 subgraph in K_n. The same argument applies to the y-coordinate and the z-coordinate of any five vertices of K_n. Therefore, every $X \times Y \times Z$ grid drawing of K_n has $X \geq \lceil n/5 \rceil$, $Y \geq \lceil n/5 \rceil$, and $Z \geq \lceil n/5 \rceil$, resulting in the $\Omega(n^3)$ volume lower bound.

It is therefore of interest to identify classes of graphs that allow for 3D grid drawings with small volume. Pach, Thiele, and Tóth [PTT99] proved that every graph with bounded chromatic number has a 3D grid drawing with volume $O(n^2)$, and this volume bound is optimal for $K_{n,n}$. By the 4-color theorem, planar graphs have chromatic number 4, thus the aforementioned result implies that planar graphs have a 3D grid drawing with volume $O(n^2)$. More strongly, a classical result by de Fraysseix, Pach, and Pollack [dPP88, dPP90] states that planar graphs have 2D grid drawings in $O(n^2)$ volume (area) and that the bound is best possible for 2D grid drawings. A major open problem in the area was to determine if the $O(n^2)$ volume can be improved in 3D. More specifically, Felsner, Liotta, and Wismath [FLW03] asked the following problem. It remained open for almost 20 years and is a motivational problem behind these lecture notes.

Problem 1.3.1 ([FLW03]) Does every n-vertex planar graph have a 3D grid drawing with $O(n)$ volume?

1.3.1 *From a Geometric to a Topological Problem*

Our goal is to solve Problem 1.3.1, that is, to prove that every n-vertex planar graph has a 3D grid drawing in $O(n)$ volume. Among other possibilities, one may aim to find $O(1) \times O(1) \times O(n)$ grid drawings, or $O(\sqrt{n}) \times O(\sqrt{n}) \times O(1)$ grid drawings, or

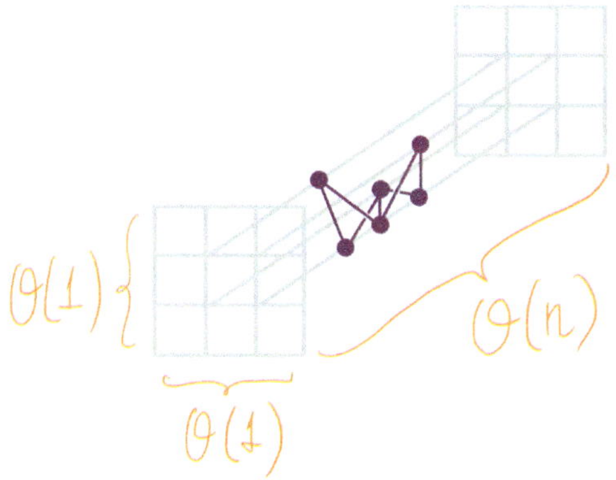

Fig. 1.2 A $O(1) \times O(1) \times O(n)$ axis aligned box and partial 3D grid drawing of a graph (in purple)

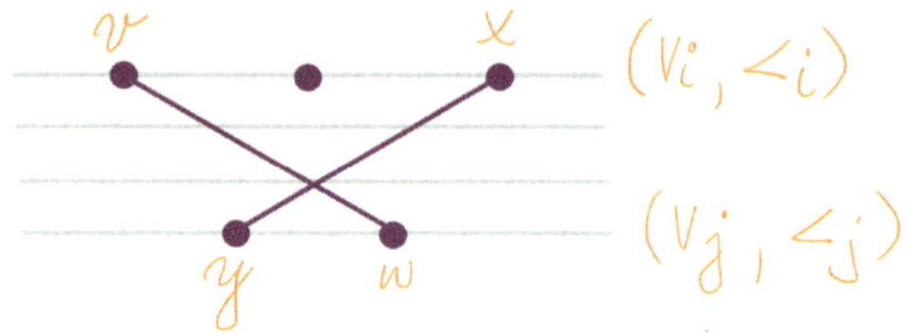

Fig. 1.3 An example of an X-crossing in a track assignment

$O(n^{1/3}) \times O(n^{1/3}) \times O(n^{1/3})$ grid drawings. $O(1) \times O(1) \times O(n)$ grid drawings have important properties that make them more suitable for converting Problem 1.3.1 to a topological/combinatorial problem. Specifically, notice that if a graph G has a $O(1) \times O(1) \times O(n)$ grid drawing, D, then all the vertices of G in D lie on $O(1)$ parallel lines (parallel to z-axis)—see Fig. 1.2. Moreover, each of these lines, L_i, defines a total order $<_i$ of the vertices on L_i (by increasing z-coordinates). The following key observation holds for these total orders: for every pair of edges vw and xy of G, if v and x are on L_i with $v <_i x$; and if y and w on L_j, then $w <_j y$—see Fig. 1.3. Otherwise, (if $y <_j w$), vw and xy would cross in D for any assignment of z-coordinates to these four points, unless that assignment changes the order of v and x on L_i, or changes the order of y and w on L_j. This observation motivates the introduction of the following graph layouts (void of geometry), called *track layouts*. As we will see, there is a tight relationship between $O(n)$ volume 3D grid drawings (more specifically, $O(1) \times O(1) \times O(n)$ grid drawings) and track layouts. We define them next.

A *track assignment* of a graph G is a partition $\{V_1, \ldots, V_t\}$ of $V(G)$ into independent sets where each V_i is equipped with a total order $<_i$. Each pair $(V_i, <_i)$ is called a *track*. When it is clear from the context, it will be convenient to also refer to V_i as a track. An *X-crossing* in a track assignment consists of two edges vw and xy such that $v <_i x$ and $y <_j w$, for distinct tracks V_i and V_j, as illustrated in Fig. 1.3. A track assignment with no X-crossing is called a *track layout*. A *t-track assignment/layout* is a track assignment/layout with t tracks. The minimum k such that G has a k-track layout is called the *track-number* of G, denoted by $\operatorname{tn} G$.

Let $\{(V_i, <_i) : i \in I\}$ be a t-track layout of a graph G. The *span* of an edge vw of G, with respect to a numbering of the tracks $I = \{1, 2, \ldots, t\}$, is defined to be $|i - j|$ where $v \in V_i$ and $w \in V_j$.

Going back to the discussion at the beginning of the section, consider a $O(1) \times O(1) \times O(n)$ grid drawings, and note that the vertices on each line L_i almost form a track except that they do not form an independent set, but instead, the consecutive vertices on a line may be adjacent. That leads to a notion of *improper track* and associated notions of improper track assignment and improper track layout that were implicit in earlier literature on the subject (Felsner et al. [FLW03], Heath, Leighton, and Rosenberg [HLR92], Rengarajan and Veni Madhavan [RVM95]). Allowing these *intratrack* edges (that is, the edges between consecutive vertices in a track) does not buy us much as every improper t-track layout can be turned easily into a $2t$-track layout by doubling the number of tracks and moving every second vertex of an edge on track V_i to the new track V_i'.

We now formalize the connection between 3D grid drawings and track layouts, starting with the easy direction, which was discussed above.

Lemma 1.3.2 ([DMW05]) *If a graph G has an $A \times B \times C$ grid drawing, then* $\operatorname{tn} G \leq 2AB$.

The proof of the converse statement is inspired by the moment curve algorithm discussed earlier (at the beginning of Sect. 1.3). Loosely speaking, the proof by Cohen et al. [CELR96] (that gives a $O(n^3)$ volume 3D grid drawing of K_n) allows for all three dimensions of the axis-aligned box to grow with n. The proof by Pach et al. [PTT99] (that gives a $O(n^2)$ volume 3D grid drawing of $K_{n,n}$) uses the assignment of vertices to colour classes to 'fix' one dimension and allow two dimensions of the axis-aligned box to grow with n. The proof of the converse of Lemma 1.3.2, that is, the proof of the next lemma by Dujmović et al. [DMW05], uses an assignment of vertices to tracks to fix two dimensions and allow one dimension of the axis-aligned box to grow with n. Specifically, for a graph G with a t-track layout $\{(V_i, <_i) : 1 \leq i \leq t\}$ and the smallest prime $p > t$ (and thus $p \leq 2t$), they assign the following grid-points to the vertices of G. For each i, $1 \leq i \leq t$, represent the vertices in V_i by the grid-points

$$\{(i, i^2 \bmod p, r) : 1 \leq r \leq p \cdot |V_i|,\ r \equiv i^3 \pmod{p}\},$$

such that the z-coordinates respect the given total order $<_i$. The edges are represented by line-segments between its endpoints. They then show that no pair of (non-incident) edges lies in the same plane and thus cannot cross. Specifically, they show that given any two edges e and e' with four distinct endpoints $(i_\alpha, i_\alpha^2 \bmod p, r_\alpha)$, $1 \leq \alpha \leq 4$, the determinant of $\begin{pmatrix} 1 & i_1 & i_1^2 \bmod p & r_1 \\ 1 & i_2 & i_2^2 \bmod p & r_2 \\ 1 & i_3 & i_3^2 \bmod p & r_3 \\ 1 & i_4 & i_4^2 \bmod p & r_4 \end{pmatrix}$ is non-zero and thus the two edges do not cross, leading to the following result.

Lemma 1.3.3 ([DMW05]) *For any graph G, if* $\operatorname{tn} G \leq t$, *then G has a $t \times 2t \times 2tn$ grid drawing (of $4t^3n$ volume).*

Lemmas 1.3.2 and 1.3.3 imply the following theorem.

Theorem 1.3.4 ([DMW05]) *A family of graphs has an $O(1)$ track-number if and only if every n-vertex graph in the family has an $O(1) \times O(1) \times O(n)$ 3D grid drawing.*

1.3.2 Track Layouts of Some Graph Classes

Recall that our goal is to answer Problem 1.3.1. Theorem 1.3.4 implies that Problem 1.3.1 has an affirmative answer if planar graphs have a bounded track-number, meaning there exists a constant c such that every planar graph has a track-number at most c. Before tackling this problem directly, we will first develop some intuition about constructing track-layouts of simple graphs. It is easy to see that every path has a 2-track layout, so we turn our attention to trees first. The following result is implicit in the work of Felsner et al. [FLW03].

Lemma 1.3.5 ([FLW03]) *Every tree T has a 3-track layout.*

Proof Root T at an arbitrary vertex r and consider a natural crossing-free drawing of T based on a BFS layering of T. Specifically, let $D(T)$ be a drawing of T where vertices at distance d from the root are drawn on the line $y = d$ and where edges are drawn between two consecutive lines, as depicted in Fig. 1.4a. The resulting drawing defines a track-layout, L, on (possibly many) tracks $\{V_1, V_2, V_3 \dots\}$. To get a 3-track layout of T, "wrap" the drawing $D(T)$ onto 3 tracks as indicated in Fig. 1.4b. More formally, let σ be the total order of the vertices of T obtained from $D(T)$ by taking first the vertices of V_1 (that is, by taking r), followed by the vertices of V_2 in order in which they appear on line $y = 2$, followed by the vertices of V_3 in order in which they appear on line $y = 3$, and so on. We now "wrap" track layout L into 3-track layout L', as follows. For $i \in \{1, 2, 3\}$, let V_i' be the set of vertices of T whose distance d from r is congruent to $(i - 1) \pmod 3$. With each V_i' ordered by σ, we obtain a 3-track assignment L' of T. Assume for the sake of contradiction that there is an X-crossing in L' between edges vw and xy with $v <_\sigma x$ in V_i' and $y <_\sigma w$ in V_j'. Assume first that v and x are on the same track V_ℓ in L. If w and y are on distinct tracks $V_{\ell-1}$ and $V_{\ell+1}$ in L, then vw and xy cannot form an X-crossing in L' as w and y are not on the same track in L'. If w and y are on the same track $V_{\ell-1}$ or $V_{\ell+1}$ in L then the endpoints of edges vw and xy are in the same pair of tracks in L and do not form an X-crossing in L. Therefore, if $v <_\sigma x$ in L then $w <_\sigma y$ in L. Since the same order is used in L', $w <_\sigma y$ in L', and vw and xy cannot form an X-crossing. Assume now that v and x are on distinct tracks V_p and V_q in T. Then $p \neq q$ and $p \equiv q \pmod 3$. That implies that v and w both appear before x and y in σ (or the other way around), which contradicts the assumption that vw and xy form an X-crossing in L'. □

In the previous proof, we successfully modified the track layout L, which had many tracks, to a track layout L', with a constant number of tracks (three). This was possible because, in L, all the edges had endpoints in two consecutive tracks, meaning the edges had a span of 1. This concept of span, when combined with the wrapping strategy, can be formalized and generalized, as described next. It is a common technique used to obtain bounded track-number results for many graph classes.

Let $\{(V_i, <_i) : i \in I\}$ be a track layout of a graph G. The *span* of an edge vw of G, with respect to a numbering of the tracks $I = \{1, 2, \dots, t\}$, is defined to be $|i - j|$ where $v \in V_i$ and $w \in V_j$. The proof of the following lemma is an easy generalization of the wrapping argument from Lemma 1.3.5. Its proof is left as an exercise (see Exercise 1 in Sect. 1.7).

Lemma 1.3.6 ([DMW05]) *If a graph G has a track layout with maximum edge span s, then G has a $(2s + 1)$-track layout.*

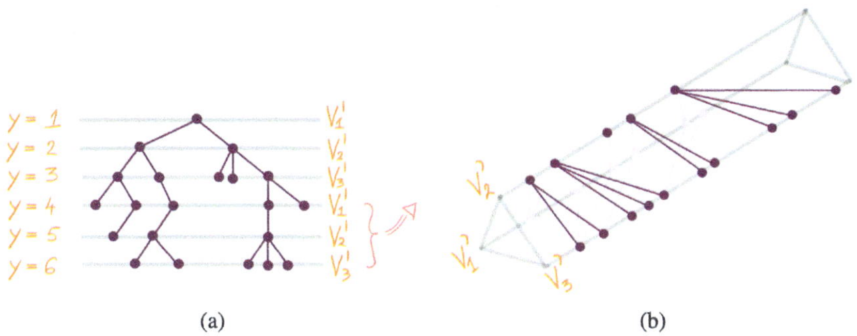

Fig. 1.4 **a** From a BFS layering-based crossing-free drawing of a tree **b** to a 3-track layout of that tree

The next class of graphs we consider are outerplanar graphs. A graph is *outerplanar* if it has a crossing-free drawing with all vertices on one face (for example, the outerface). Felsner et al. [FLW03] proved that every outerplanar graph G has a crossing-free drawing based on BFS layering, similar to that of trees. Specifically, they showed that every outerplanar graph has a crossing free straight-line drawing with vertices on parallel lines (as per BFS layering) where the endpoints of every edge are either on two consecutive lines or they are two consecutive vertices on the same line, as depicted in Fig. 1.5a. Such a drawing thus defines an improper track layout with span one. As before, such improper track layout can be wrapped onto 3 tracks to obtain an improper 3-track layout. As discussed earlier, every improper t-track layout can be turned into $2t$-track layout by doubling the number of tracks and moving every second vertex of each track i to a new track i'. Thus these improper 3-track layouts of outerplanar graphs can be turned into 6-track layouts. Dujmović, Pór, and Wood [DPW04] show that 6 can be improved to 5, which is optimal by the recent result of Pupyrev [Pup20]. Namely, a straightforward inductive proof [DPW04] shows that every outer planar graph has a crossing-free straight-line drawing where the endpoints of every edge have y-coordinates that are integers differing by 1 or 2 (see Exercise 1.7 in Sect. 1.7). Thus every outerplanar graph has a track-layout with span two (Fig. 1.5b), and thus Lemma 1.3.6 implies the following result.

Lemma 1.3.7 ([DPW04]) *Every outerplanar graph has a 5-track layout.*

What trees and outerplanar graphs have in common is that they are, vaguely speaking, tree-like. More specifically, they have bounded *treewidth*. Trees have treewidth 1, and outerplanar graphs have treewidth 2. Treewidth, defined independently by Halin [Hal76], Robertson and Seymour [RS86] and Arnborg and Proskurowski [AP89], is a measure of the similarity of a graph to a tree (see Sect. 1.4 for the definition). Treewidth will be formally defined and studied Sect. 1.4. Treewidth plays a critical role in Robertson and Seymour's graph minor theory and in graph algorithms. Many problems that are NP complete on general graphs have polynomial-time solutions on graphs of bounded treewidth.

The importance of treewidth in the resolution of Problem 1.3.1 will become evident later. At this point, we state the following result on the track-number of bounded treewidth graphs, known since 2005.

Theorem 1.3.8 ([DMW05]) *Graphs of bounded treewidth have bounded track-number.*[2]

Theorem 1.3.8 does not resolve Problem 1.3.1, as there are planar graphs with unbounded treewidth. For instance, it is known that the $n \times n$ planar grid graph has a treewidth n (as will be discussed in Sect. 1.4). However, having unbounded treewidth does not preclude a graph from having a constant track-number. For example, while $n \times n$ planar grid graphs have unbounded treewidth, they have track layouts with span one, as depicted in Fig. 1.5c for the 4×4 planar grid graph. Therefore, $n \times n$ planar grid graphs have the track number at most 3 by Lemma 1.3.6.

Over the years, some partial progress has been made on Problem 1.3.1, from an $O(\log^8 n)$ bound by Battista, Frati, and Pach [BFP10], an $O(\log n)$ bound by Dujmović [Duj15], an $O(1)$ bound for bounded degree planar graphs by Bekos, Förster, Gronemann, Mchedlidze, Montecchiani, Raftopoulou, and Ueckerdt [BFG+19].

1.3.3 Track Layouts of Graph Products

Consider the following question. Suppose you want to find a good bound on the track-number for a certain class of graphs. Further, suppose that you know that every graph in this class is "composed" of simpler graphs for which a good track-number

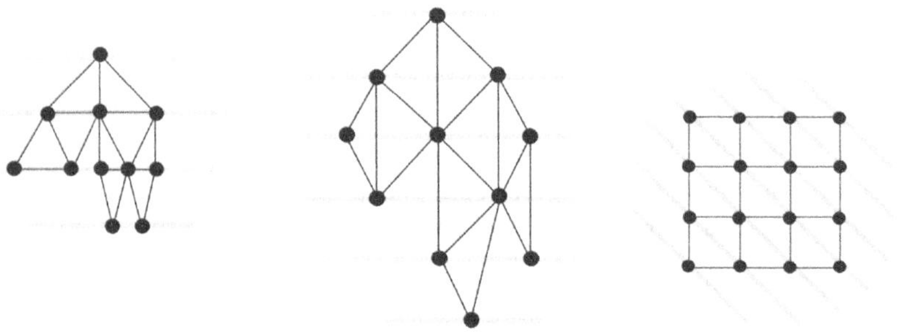

(a) (b) (c)

Fig. 1.5 An outerplanar graph with **a** an improper track layout with span one, **b** a track layout with span two. **c** A track layout with span one of a 4×4 planar grid graph. Tracks are depicted diagonally

[2] More precisely, Dujmović et al. [DMW05] proved that every graph of treewidth k has track-number at most $3^k \cdot 6^{4^k}$. That doubly exponential bound in k, was much improved, to $2^{O(k^2)}$, by Wiechert [Wie17].

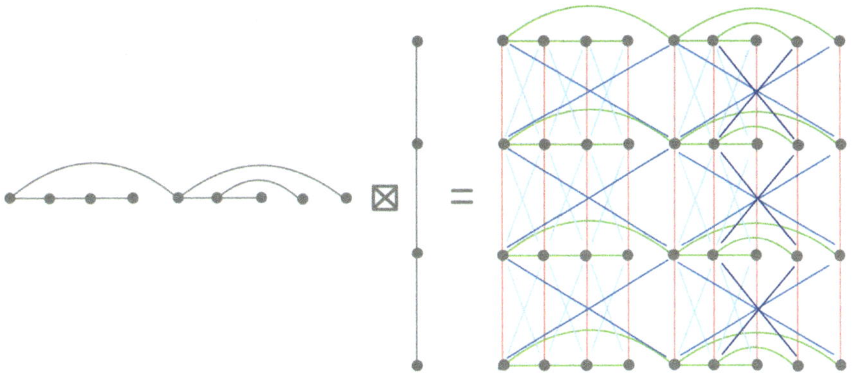

Fig. 1.6 The strong product of a tree T and a path P. The T-edges are depicted in green, the P-edges in pink, and the direct edges in different shades of blue

bound is already known. Is that helpful? In this section (see Lemma 1.3.9), we show that the answer is yes when "composed" means "graph products".

Let G_1 and G_2 be graphs. The *strong product* of graphs G_1 and G_2, denoted by $G_1 \boxtimes G_2$, is the graph whose vertex set is $V(G_1) \times V(G_2) = \{(v, x) : v \in V(G_1), x \in V(G_2)\}$. The edge set of $G_1 \boxtimes G_2$ is defined as follows. Two vertices $(v, x), (w, y)$ of $G_1 \boxtimes G_2$ are adjacent if:

- $x = y$ and $vw \in E(G_1)$ (such an edge is called a G_1*-edge*), or
- $v = w$ and $xy \in E(G_2)$, (such an edge is called a G_2*-edge*), or
- $vw \in E(G_1)$ and $xy \in E(G_2)$ (such an edge is called a *direct-edge*).

See Fig. 1.6 for an illustration.

Intermission: Another well-studied graph product is the *Cartesian product*, denoted by $G_1 \square G_2$ for two graphs G_1 and G_2. It is defined similarly to the strong product but omits direct edges. While Cartesian products will not play a significant role in these lecture notes, some well-known graphs derived from Cartesian products will be mentioned on several occasions. For example, the *$a \times b$ planar grid graph* is $P_a \square P_b$.

A version of the following result (Lemma 1.3.9) was originally proved by Wood [Woo05] for the queue number of graph products, see Exercise 1.7 in Sect. 1.7 for the definitions. Since the track-number is known to be bounded by the queue-number [Woo02], Wood's result also applies to track-numbers. We present a direct proof here.

Lemma 1.3.9 *For every graph H and every path P,* $\operatorname{tn} H \boxtimes P \leq 4 \operatorname{tn} H - 1$.

Proof Let $G := H \boxtimes P$. Let $p = |V(P)|$ and $p_1, p_2 \dots p_p$ be the vertices of P in order of their appearance from one end of P to the other. Let $L(H)$ denote a track layout of H on $t := \operatorname{tn} H$ layers. We start with a straightforward idea for the track assignment, $L(G)$, of G on $t \cdot p$ layers. Place all the vertices $V_1 := \{(v, p_1) : v \in$

$V(H)\}$ of G on the first t layers exactly as in $L(H)$ track layout of H. Recall that $G[V_1]$ is isomorphic to H. Repeat that on each set of t subsequent layers, with vertices $V_i := \{(v, p_i) : v \in V(H)\}$ to obtain the track assignment $L(G)$ of G on $t \cdot p$ layers. We call $G[V_i]$ the i-th *copy* of H and denote the layers $G[V_i]$ is placed on by $L_i = (L_{i,1}, L_{i,2}, \ldots, L_{i,t})$. Figure 1.7a, b depict this construction. Figure 1.7a depicts a tree H and a 3-track layout $L(H)$ of H. The track assignment of $H \boxtimes P_4$ is (partially) depicted in Fig. 1.7b. In particular, Fig. 1.7b depict the copies of H and the H-edges of G on layers L_1 and L_2. On layers L_3 and L_4 some H-edges, P-edges and direct-edges are depicted.

It is clear that the maximum span of an edge in $L(G)$ is $2t - 1$. Thus Lemma 1.3.6 would imply the result if $L(G)$ does not have X-crossings. We prove that next.

Each edge in G is an H-edge, a P-edge or a direct-edge (see Fig. 1.7b). The assignment of the vertices of G to the layers of $L(G)$ in the above construction implies that the following properties hold for the endpoints of the (H-, P-, direct-) edges of G.

For each H-edge e, there is an $i \in [p]$, such that the endpoints of e are assigned to the tracks in the same set of layers L_i.

For each P-edge e, there is an $i \in [p]$ and $j \in [t]$, such that one endpoint of e is assigned to the track $L_{i,j}$ and the other to the track $L_{i+1,j}$. Figure 1.7c, with $i = 1$ and $j = 2$, depicts two such P-edges in red.

Consider now a direct-edge e of G. By the definition of the strong product, e is generated by some edge ab of H and some edge $p_i p_{i+1}$ of P. Thus in $L(G)$, the endpoints of the H-edge ab in the i-th copy of H are assigned to two distinct layers, $L_{i,j}$ and $L_{i,k}$ in the set of layers L_i. Similarly, the endpoints of the H-edge ab in $(i+1)$-st copy of H are assigned to two distinct layers, $L_{i+1,j}$ and $L_{i+1,k}$ in the set of layers L_{i+1}. That implies that the endpoints of two direct-edges (one of them being e) generated by ab and $p_i p_{i+1}$ have the following layer assignments in $L(G)$. One direct-edge has one endpoint in $L_{i,j}$ and the other in $L_{i+1,k}$ where $j \neq k$. The other direct-edge has one endpoint in $L_{i,k}$ and the other in $L_{i+1,j}$ where $j \neq k$.

Since the four endpoints of any pair of edges that form an X-crossing must lie on just two layers, the structural properties established above imply that an X-crossing can occur only between two edges of the same type–that is, either two H-edges, two P-edges, or two direct-edges. We now show that no such pair of same-type edges can actually form an X-crossing.

If the endpoints of two H-edges belong to two tracks of $L(G)$, then they all belong to the same copy of H in $L(G)$. Such a pair of H-edges $(v, p_i)(w, p_i)$ and $(x, p_i)(y, p_i)$ cannot form an X-crossing in $L(G)$ since the corresponding pair of edges (v, w) and (x, y) of H do not form an X-crossing in $L(H)$.

Consider now two P-edges $(v, p_i)(v, p_{i+1})$ and $(x, p_i)(x, p_{i+1})$ with endpoints in two layers. Then there is $j \in [t]$, such that (v, p_i) and (x, p_i) are in layer $L_{i,j}$; and, (v, p_{i+1}) and (x, p_{i+1}) are in layer $L_{i+1,j}$. In $L(H)$, v and w are in the same layer. This case (with $j = 2$ and $i = 1$) is depicted in Fig. 1.7c with the two P-edges depicted in red.

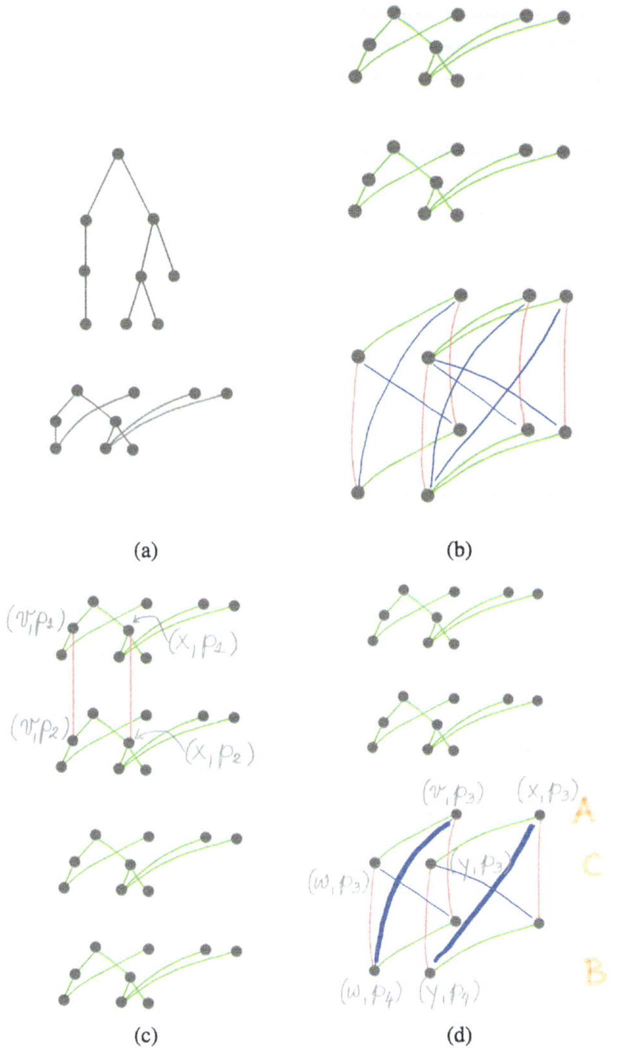

Fig. 1.7 Illustration for the proof of Lemma 1.3.9

Assume, without loss of generality, that $v < w$ in that layer of $L(H)$. Then by construction, (v, p_i) comes before (x, p_i) in layer $L_{i,j}$ and (v, p_{i+1}) comes before (x, p_{i+1}) in layer $L_{i+1,j}$. Thus they do not form an X-crossing in $L(G)$.

Consider now two direct-edges forming an X-crossing: $e_1 = (v, p_i)(w, p_{i+1})$ and $e_2 = (x, p_i)(y, p_{i+1})$ with $(v, p_i) < (x, p_i)$ in some layer A of L_i and $(w, p_{i+1}) > (y, p_{i+1})$ in some layer B of L_{i+1}. The two think blue edges in Fig. 1.7d, depict e_1 and e_2. Then, by the properties of the strong product, vw and xy are edges of H. That further implies that $(v, p_i)(w, p_i)$, $(x, p_i)(y, p_i)$, $(v, p_{i+1})(w, p_{i+1})$,

$(x, p_{i+1})(y, p_{i+1})$ are H-edges of G (as illustrated by the four green edges in layers L_3 and L_4 in Fig. 1.7d). These edges do not form an X-crossing in $L(G)$, as proved above. Furthermore, the fact that $(v, p_i)(w, p_i)$ and $(x, p_i)(y, p_i)$ do not form an X-crossing, implies that $(w, p_i) <_C (y, p_i)$ in some layer C that contains (w, p_i) and (y, p_i) in the set of layers L_i. However, the corresponding vertices (w, p_{i+1}) and (y, p_{i+1}) in the set of layers L_{i+1} must have the same ordering by construction and thus $(w, p_{i+1}) <_B (y, p_{i+1})$. This contradicts the starting assumption, and thus the direct-edges e_1 and e_2 do not form an X-crossing. □

Lemma 1.3.9 tells us that if every planar graph is a subgraph of a strong product of a path and some graph of bounded track-number, then Problem 1.3.1 would have an affirmative answer. For example, by Theorem 1.3.8, every graph of bounded treewidth has bounded track-number. Thus, Lemma 1.3.9 implies that if every planar graph is a subgraph of a strong product of a path and a graph of bounded treewidth, then Problem 1.3.1 would be resolved affirmatively. However, no such statement was known about planar graphs despite decades of extensive studies of that graph class, and there was no particularly strong reason to believe it were true.

The statement was proved in 2019 [DJM+20] and later named "the product structure theorem of planar graphs". This result, combined with everything presented in the previous sections, not only provided an affirmative answer to Problem 1.3.1 but also resolved the 27-year-old conjecture by Heath, Leighton, and Rosenberg [HLR92] on queue layouts and several other open problems that we have not mentioned yet.

Section 1.5 will be dedicated to proving the product structure theorem of planar graphs. Given the critical importance of treewidth in this result, the next section is dedicated to introducing treewidth and a generalization of planar k-trees (that we call plane (k, s)-trees) which also play a role in the product structure theorem.

1.4 Treewidth

Providing an extensive lecture on treewidth is beyond the scope of these notes. However, we will introduce the relevant concepts, background, and ideas necessary to understand the material and follow the proof of the product structure theorem in Sect. 1.5.

1.4.1 Definition and Examples

Recall that a *tree* is a connected graph without cycles and that a *subtree* is a connected subgraph of a tree. Let G be a graph and T a tree. We refer to the elements of $V(T)$ as *nodes*, and to the elements of $V(G)$ as *vertices*. Let $\mathcal{B} := \{B_x \subseteq V(G) : x \in V(T)\}$ be a collection of subsets of $V(G)$, called *bags*, indexed by the nodes of T. A pair $(T, \mathcal{B})$ is a *tree-decomposition* of G if

(T1) for every vertex v of G, the set $\{x \in V(T) : v \in B_x\}$ induces a subtree of T, and

(T2) for every edge vw of G, there is a vertex $x \in V(T)$ for which $v, w \in B_x$.

For a graph G and its tree-decomposition $(T, \mathcal{B})$, we will sometimes say that T is a tree-decomposition of G when the collection of subsets $\mathcal{B}$ is clear from the context.

One way to depict a tree-decomposition $(T, \mathcal{B})$ is to imagine placing the vertices of each bag B_x in the node x of T, as illustrated in later examples (Figs. 1.8b and 1.9b). With this depiction in mind, it makes sense to say that a vertex of G is in a bag/node of T.

The first condition, (T1), ensures that every vertex v of G is in some bag/node of T and that the set of bags/nodes of T containing v form a connected subgraph of T. The second condition, (T2), ensures that the endpoints of every edge are in some bag. It is easy to see that every graph has a tree-decomposition: take T to be a tree with just one node and place all vertices of G in the corresponding bag. Clearly, conditions (T1) and (T2) are satisfied. However, such a tree-decomposition, with $n := |V(G)|$ vertices all in one bag, is not very useful. The goal is typically to find a tree-decomposition where the maximum number of vertices in any bag is minimized. This number is considered a measure of how similar a given graph G is to a tree. We now define this measure formally.

The *width* of a tree-decomposition $(T, \mathcal{B})$ is the maximum size of a bag in $\mathcal{B}$, that is, $\max\{|B_x| : x \in V(T)\} - 1$. The *treewidth* of a graph G is the minimum width of a tree-decomposition of G. If T is a path, then the decomposition is also called a *path-decomposition* and the associated measure is called the *pathwidth*.

As mentioned earlier, treewidth can be viewed as a measure of how similar a graph G is to a tree. One useful way to view this similarity is through the concept of balanced separators. For a graph G, a subset S of vertices of G is called a *balanced separator* of G if removing S from G results in a graph in which every connected component contains at most half of the vertices of G. The *size of a separator* is the number of vertices in S. It is well known (and not difficult to prove—see Exercise 10) that every tree has a balanced separator of size 1. In fact, one can characterize forests using this property: a graph G is a forest if and only if every non-empty subgraph of G has a balanced separator of size 1. This extends to graphs of higher treewidth. In one direction, every graph of treewidth k has a balanced separator of size at most $k + 1$ (see Exercise 12). The converse, while more challenging to prove, also holds: if every non-empty subgraph of a graph G has a balanced separator of size at most k, then the treewidth of G is at most $O(k)$ [DN19, HMM25]. In this way, the smaller the treewidth of a graph, the closer it is to a tree. This connection has important algorithmic consequences: many problems that are NP-complete on general graphs can be solved efficiently–in polynomial time–on trees and, more broadly, on graphs of bounded treewidth.

One particularly appealing and simple way to view tree-decompositions is as an *intersection graph of subtrees*. From a tree-decomposition of G, one can derive a subtree intersection representation of G. Specifically, condition (T1) can be restated as requiring that each vertex v of G is represented by a non-empty subtree T_v of

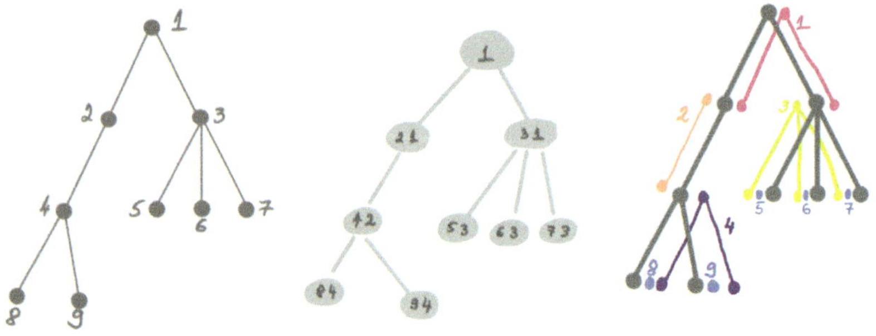

(a) (b) (c)

Fig. 1.8 **a** a tree G **b** a tree-decomposition of G **c** a subtree intersection representation of G

T. Condition (T2) can be restated as requiring that for every edge vw of G, T_v and T_w must have at least one vertex in common. Conversely, one can obtain a tree-decomposition $(T, \mathcal{B})$ of G from a subtree intersection representation of G. The tree T is the union of all the subtrees in the intersection representation. The set of bags $\mathcal{B}$ can be obtained as follows: for a node x of T, the bag B_x contains all the vertices of G whose subtrees include x.

We consider some examples next.

Example 1 Figure 1.8: Every tree has treewidth at most 1.

Indeed, let G be a tree as shown in Fig. 1.8a. Root G at an arbitrary vertex. Then each vertex of G (except for the root) has one parent. To construct a tree-decomposition of G, we will use a tree T isomorphic to G. Place each vertex v of G and its parent vertex in the node v of the decomposition T, as depicted in Fig. 1.8b. In terms of the intersection graph of subtrees, this is equivalent to representing each vertex v of G as a subtree of T composed of v and all its children, as shown in Fig. 1.8c. It is straightforward to verify that conditions (T1) and (T2) are satisfied. Since each bag contains at most 2 vertices, the treewidth is at most 1. In fact, the treewidth of G is exactly 1 unless G is a trivial tree with only one vertex.

The motivation for the "-1" in the definition of treewidth was to ensure that all (non-trivial) trees have a treewidth of 1, not 2.

Example 2 Consider the graph G depicted in Fig. 1.9a. Figure 1.9b depicts a tree-decomposition of G of width 2. Figure 1.9c depicts the equivalent intersection representation of G. Thus G has treewidth at most 2. In fact, G has treewidth equal to 2 owing to the following.

Helly property holds for subtrees of a tree: given a tree T and a set of subtrees $S := \{T_1, T_2, \ldots, T_p\}$ of a tree T, if every pair of subtrees in S share at least one vertex, then some vertex of T is contained in all subtrees in S (see Exercise 1.7 in Sect. 1.7). Combined with the subtree intersection representation view of tree-decompositions, the Helly property of subtrees implies the following lemma.

Lemma 1.4.1 *Let $(T, \mathcal{B})$ be a tree-decomposition of a graph G. For every clique C of G there is a node x of T such that all the vertices of C are in bag B_x.*

Since the graph G in Fig. 1.9a has a clique of size 3, Lemma 1.4.1 implies that every tree-decomposition of G has a bag that contains all the vertices of the clique. Thus the treewidth of G is at least 2, and the tree-decomposition of G depicted in Fig. 1.9b is optimal.

Example 3 Grids play an important role in structural graph theory and treewidth in particular. It can be proved that an $n \times n$ planar grid graph has treewidth (and, in fact, pathwidth) equal to n [Die18]. Consider the 3×3 grid graph in Fig. 1.10a. Its tree (path) decomposition is depicted in Fig. 1.10c. Here is an intuition on how it is obtained.

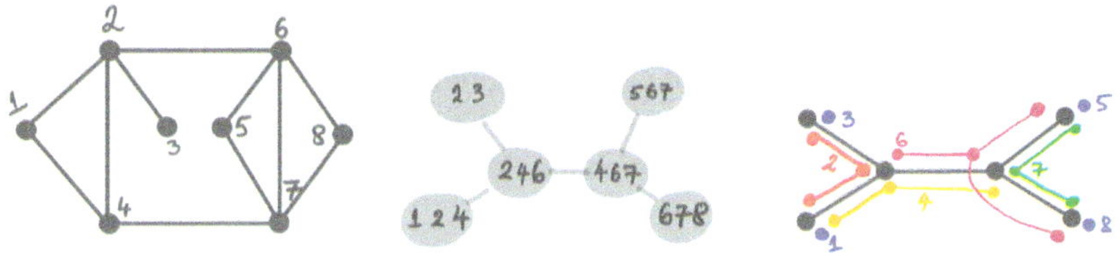

(a) (b) (c)

Fig. 1.9 **a** A graph, **b** a tree-decomposition of the graph, and **c** a subtree intersection representation corresponding to the decomposition

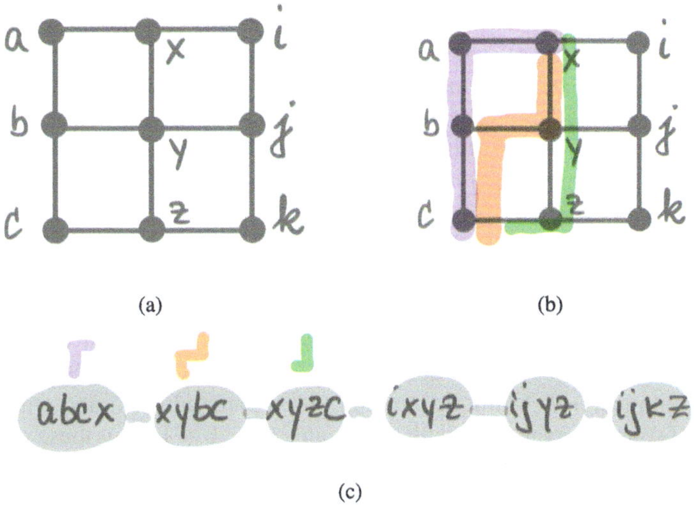

Fig. 1.10 **a** 3×3 planar grid graph G, **b** the first, second, and third bags of the path decomposition of G are shown in pink, orange, and green, respectively, and **c** the final path decomposition of G

We want to build a path decomposition $(P, \mathcal{B})$ of 3×3 planar grid graph, where P is a path. Start by placing all the vertices in the first column and the first vertex of the second column in the first node/bag of P (the content of the first bag is highlighted in purple in Fig. 1.10b). This bag contains the top-left vertex of the grid and all its neighbours. Therefore, it no longer needs to appear in any subsequent bags/nodes of P. We can safely omit it from the second bag/node of P and add a new vertex without increasing the width of the decomposition.

The new vertex added in the second bag is the second vertex of the second column (the content of the second bag is highlighted in orange in Fig. 1.10b). At this point, the second vertex of the first column is contained in the first and second bags along with all of its neighbours. Therefore, it no longer needs to appear in any subsequent bags/nodes of P. We can thus safely omit it from the third bag/node of P. We replace it in the third bag with the next vertex in the second column (the content of the third bag is highlighted in green in Fig. 1.10b). We continue this process until the full path decomposition is obtained.

For an $p \times q$ planar grid graph G, this process gives a decomposition where each vertex of G appears in some set of consecutive nodes of P, thus meeting condition (T1). Each bag in the decomposition has $p + 1$ vertices of G, making the width of the path decomposition p. Thus, the pathwidth (and treewidth) of an $p \times q$ planar grid graph is at most $\min\{p, q\}$.

1.4.2 Edge-Maximal Graphs of Treewidth k

To better understand treewidth, it is essential to develop an intuition for what edge-maximal graphs of treewidth k look like. These graphs are also closely related to k-trees, which are in turn related to the specific bounded treewidth graphs that appear in the product structure theorem of planar graphs. Thus in this subsection, we will discuss edge-maximal graphs of treewidth k and their relatives.

We first revisit the notion of intersection graphs of subtrees. Consider a family $\mathcal{K}$ of nonempty sets. An *intersection graph* of $\mathcal{K}$ is a graph G where each set in $\mathcal{K}$ is a vertex of G and two vertices of G are adjacent if and only if the two corresponding sets have a nonempty intersection. A *subtree intersection graph* is an intersection graph of a family of subtrees of a tree (see two examples in Fig. 1.11). An *interval graph* is an intersection graph of a family of subpaths of a path.

From the discussion in the previous section, we can see that a graph G has treewidth k if and only if G is a subgraph of some subtree intersection graph with maximum clique $k + 1$. Similarly, a graph G has pathwidth k if and only if G is a subgraph of some interval graph with maximum clique $k + 1$. Thus these intersection graphs are some form of edge maximal graphs of a given treewidth.

A graph is *chordal* if every cycle of length greater than 3 has an edge connecting two vertices that are not consecutive on the cycle. Note that this implies that in chordal graphs every induced subgraph has either a clique-cutest or is a clique. Gavril [Gav74]

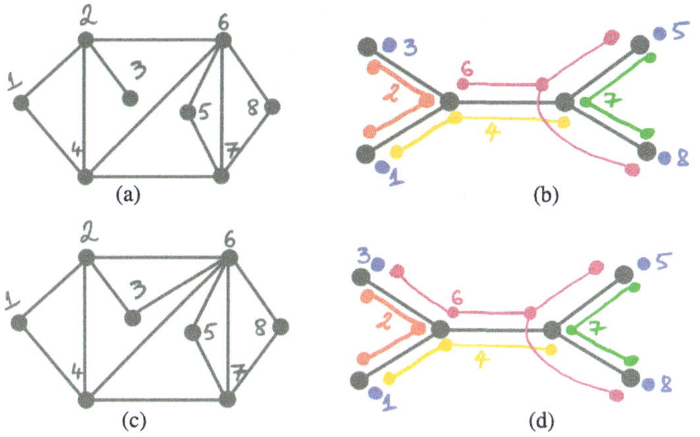

Fig. 1.11 **a** The graph G is a subtree intersection graph, a chordal graph, and a 2-tree. **b** A subtree intersection representation of G. **c** The graph G' is also a subtree intersection graph and a chordal graph, and in addition, it is a *strict* 2-tree. **d** A subtree intersection representation of G'

showed in 1974 that a graph is a subtree intersection graph if and only if it is a chordal graph. The graphs in Fig. 1.11a, c are examples of chordal graphs.

Another concept closely related to treewidth is that of k-trees. The definition of k-trees varies slightly in the literature, depending on whether the term "at most" appears before the size of the maximum cliques involved. We will use the following definition.

For a positive integer k, a *k-tree* is defined recursively as follows: the complete graph on $k+1$ vertices is a k-tree. Any graph obtained from a k-tree by adding a new vertex adjacent to each vertex of a clique of size at most k is also a k-tree. If the clique is required to be of size exactly k, we arrive at a variant of the k-tree definition called a *strict k-tree*. Note that strict k-trees are edge-maximal graphs of treewidth k. The graph in Fig. 1.11a is a 2-tree but not a strict 2-tree, whereas the graph in Fig. 1.11c is a strict 2-tree and therefore a 2-tree as well.

A graph is a *partial k-tree* if it is a subgraph of a k-tree. The following statement is well established: a graph has treewidth at most k if and only if it is a partial k-tree.

To summarize, the following notions are known to be equivalent. For more information, see the classical survey on treewidth by Bodlaender [Bod98] or a more recent one by Harvey and Wood [HW17].

Theorem 1.4.2 *Let G be a graph. The following statements are equivalent.*

1. *G has treewidth at most k.*
2. *G is a partial k-tree.*
3. *G is a subgraph of a subtree intersection graph whose maximum clique size is at most $k+1$.*
4. *G is a subgraph of a chordal graph whose maximum clique size is at most $k+1$.*

1-trees are trees and thus planar graphs. It is not difficult to see that all 2-trees are also planar. Not all 3-trees are planar. For example, consider the graph obtained by starting with K_3 and adding three new vertices, each adjacent to all three vertices of the starting K_3. The resulting graph is a 3-tree that contains $K_{3,3}$ as a subgraph, and is therefore non-planar. For any $k \geq 4$, no k-tree is planar.

Planar k-trees, $k \in \{1, 2, 3\}$, are well-studied and well-understood families of graphs. 1-trees, being trees, are ubiquitous. 2-trees are also known as (edge-maximal) series-parallel graphs. Edge maximal outerplanar graphs are also 2-trees. Planar 3-trees commonly appear in the literature under various names such as *stacked triangulations* or *Appolonian network*.

1.4.3 Plane (k, s)-Trees

Planar 3-trees and their generalization, *plane* (k, s)*-trees*, which we now introduce, will play an important role in these lecture notes, particularly in the product structure theorem. This notion of plane (k, s)-trees is new to these notes. We introduce it to simplify the understanding of the product structure theorem and its proof.

For integers $k \geq 3$ and $s \geq 1$, we define a *plane* (k, s)*-tree* as an embedded planar graph defined recursively as follows. The smallest plane (k, s)-tree is a cycle on at most k vertices, drawn crossing-free in the plane. Every embedded planar graph that can be obtained from a plane (k, s)-tree G by the following operations is also a plane (k, s)-tree. Pick a face F of G, and add a set S of at most s new vertices to F. Add edges with both endpoints in S or with one endpoint in S and one in $V(F)$, drawing them crossing-free in F such that the resulting graph is a 2-connected embedded planar graph with faces of size at most k.

Note that every plane (3, 1)-tree is a planar 3-tree. Plane (6, 3)-trees will appear in the product structure theorem. Importantly, they have bounded treewidth (in fact, treewidth at most 8). We now provide a simple proof of this fact (Fig. 1.12).

Lemma 1.4.3 *For all integers $k \geq 3$, $s \geq 1$, every plane (k, s)-tree G has treewidth at most $k + s - 1$. More specifically, G has a tree decomposition of width $k + s - 1$, such that for every face F of G there is a bag of the tree-decomposition that contains $V(F)$.*

Proof The proof is by induction on the number of vertices of G. If $|V(G)| \leq k + s$, then the statement is immediate. T is a tree with one node, and all the vertices of $|V(G)|$ are placed in the same bag indexed by that node of T.

If $|V(G)| > k + s$, then by the definition of plane (k, s)-trees there is a set S, $1 \leq |S| \leq s$, of vertices of G such that $G' := G - S$ is a smaller plane (k, s)-tree and the vertices of S are drawn in the same face F of G'. By induction, G' has a tree-decomposition T' of width $k + s - 1$, such that $V(F)$ is contained in some bag indexed by a node x of T'. To construct the desired tree-decomposition T of

G add a new leaf y to T' adjacent to x and place $V(F) \cup S$ to the bag indexed by y. Since G' is a plane (k, s)-tree, the size of F is at most k, thus the width of the resulting decomposition is at most $k + s - 1$, as claimed. It remains to prove that the resulting decomposition is indeed a tree-decomposition. Recall the two conditions, labelled (T1) and (T2) in the definition of tree-decompositions. Every edge that is in G and not in G' has endpoints in $V(F) \cup S$. Since $V(F) \cup S$ is in one bag of T, the condition (T2) is satisfied for new edges. It is satisfied trivially for the edges in G' by induction and by the construction of T. Similarly, vertices not in $V(F) \cup S$ satisfy condition (T1). Vertices of S belong to only one bag (the one indexed by y) and thus meet condition (T1). Consider a vertex $v \in V(F)$. All the bags containing v in T' form a subtree T_v' of T'. Since v is in the bag indexed by x and is added to the bag indexed by y, the bags of T that contain v form a subtree T_v' plus the leaf y adjacent to x of T_v'. This is a subtree of T and thus the condition (T1) is satisfied. □

An improvement to the product structure theorem by Ueckerdt, Wood, and Yi [UWY22a] relies on obtaining "better" plane (k, s)-trees. Specifically, a plane (k, s)-tree is *flexible* if the definition of plane (k, s)-trees is modified as follows. In the recursive step, every time a set S is added to a face F, either $|F| + |S| < k + s$, or there exists a vertex $x \in V(F)$ and a vertex $y \in S$ such that no face in the resulting embedded planar graph contains both x and y. A stronger treewidth bound can be obtained for such flexible plane (k, s)-trees, as stated in the next lemma. The proof of this lemma is left as an exercise (see Exercise 2 in Sect. 1.7).

Lemma 1.4.4 *For all integers $k \geq 3$, $s \geq 2$, every flexible plane (k, s)-tree G has treewidth at most $k + s - 2$. More specifically, G has a tree decomposition of width $k + s - 2$, such that for every face F of G there is a bag of the tree-decomposition that contains $V(F)$.*

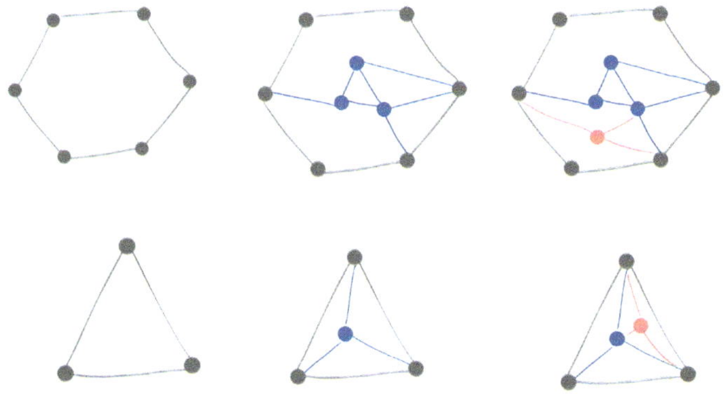

Fig. 1.12 An example of a plane (6, 3)-tree construction and a 3-tree construction

1.5 Product Structure of Planar Graphs

This section focuses on the central tool of these lecture notes, the *product structure theorem of planar graphs*, stated as follows.

Theorem 1.5.1 ([DJM+20]) *Every planar graph is a subgraph of $H \boxtimes P$ for some path P and some graph H of treewidth at most* 8. *In addition, H is a minor of G.*

Section 1.5.1 presents a proof of this theorem. Section 1.5.2 discusses improvements (in terms of treewidth) and variations. Section 1.5.3 presents extensions to other graph classes beyond planar graphs.

Before presenting the formal proof, we first outline the underlying intuition and strategy behind the proof. Consider any graph G and a BFS layering of G. Now, consider a partition of the vertex set of G into parts $\mathcal{P} := \{P_1, P_2, \dots\}$, where each part $P_i \in \mathcal{P}$ induces a path in G and has at most one vertex from each BFS layer. Let H be the graph obtained by contracting each path $P_i \in \mathcal{P}$ into a single vertex. Observe that every graph trivially admits such a partition–for instance, by placing each vertex of G into its own singleton part. In that case, H is simply the same graph as G.

Given such a partition of any graph G one can show (see Proposition 1.5.2) that G is a subgraph of the strong product $H \boxtimes P$, where P is a path (on h vertices with h corresponding to the number of BFS layers). Therefore, to prove Theorem 1.5.1, it suffices to show that planar graphs admit such a partition where the graph H has treewidth at most 8. This is proved in Theorem 1.5.3.

1.5.1 Proof of the Product Structure Theorem of Planar Graphs

The following definitions are central notions in the proof of the product structure theorem. A *vertex-partition*, or simply *partition*, of a graph G is a set $\mathcal{P}$ of non-empty sets of vertices in G such that each vertex of G is in exactly one element of $\mathcal{P}$. Each element of $\mathcal{P}$ is called a *part*. The *quotient* of $\mathcal{P}$ is the graph, denoted by $G/\mathcal{P}$, with vertex set $\mathcal{P}$ where distinct parts $A, B \in \mathcal{P}$ are adjacent in $G/\mathcal{P}$ if and only if some vertex in A is adjacent in G to some vertex in B. If T is a tree rooted at a vertex r, then a non-empty path $(x_p, \dots, x_p)$ in T is *vertical* if for there is some integer $d \geq 0$ such that $\mathrm{dist}_T(x_i, r) = d + i$ for all $i \in \{0, \dots, p\}$.

The following proposition explains a key connection between these notions and the strong product of a certain graph and a path. Examining Fig. 1.13 should provide an intuitive understanding of the proposition, even before considering the formal proof below.

Proposition 1.5.2 *Let G be a connected graph, T a rooted BFS tree of G, $\mathcal{P}$ a partition of G into vertical paths and $H := G/\mathcal{P}$ the quotient graph. Then G is a subgraph of $H \boxtimes P$ for some path P.*

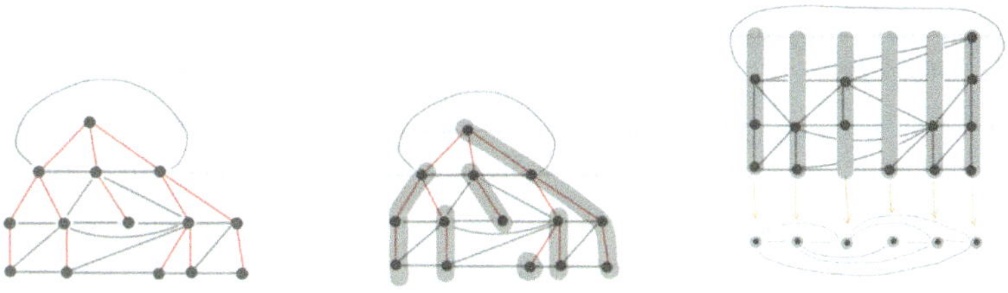

(a) (b) (c)

Fig. 1.13 Illustration for Proposition 1.5.2. **a** A BFS layering of a planar graph G, with the BFS tree drawn in pink. **b** A vertical path partition $\mathcal{P}$ of G, where each part is shown as a shaded gray path. **c** The parts of $\mathcal{P}$ are depicted vertically to illustrate the origin of the $\times P$ factor in the strong product. The lower graph shows the quotient graph obtained by contracting each vertical path in the partition

Proof The *height* of a rooted tree is the number of vertices on the longest path from the root to a leaf. If h denotes the height of T, then let $P := P_h$ be the path $(p_0, \dots, p_{h-1})$. We will prove that G is a subgraph of $H \boxtimes P$. First, we map the vertices of G to distinct vertices of $H \boxtimes P$. Then, we show that for every edge vw in G, there is an edge in $H \boxtimes P$ between the vertices to which v and w are mapped.

Let v be a vertex of G at distance d from the root of T, and let v be in the vertical path Z of $\mathcal{P}$. Recall that Z is a vertex of H. Map v to (Z, p_d). This maps the vertices of G to distinct vertices of $H \boxtimes P$.

Consider an edge vw in G. Without loss of generality, suppose that v is the closer to, or at the same distance from, the root of T as w. Suppose v is mapped to (Z, p_d) and w to $(Z', p_{d'})$. Since T is a BFS tree, $d' = d$ or $d' = d + 1$.

First, consider the case where $Z \neq Z'$. In this case, $ZZ' \in E(H)$ by the definition of the quotient graph $H := G/\mathcal{P}$ since $v \in Z$ and $w \in Z'$ in $\mathcal{P}$, and $vw \in E(G)$. If $d' = d$, then $(Z, p_d)(Z', p_d)$ is an H-edge in $H \boxtimes P$. If $d' = d + 1$, then $(Z, p_d)(Z', p_{d+1})$ is a direct edge in $H \boxtimes P$ since $p_d p_{d+1}$ is an edge in P.

Now consider the case where $Z = Z'$. Here, w is mapped to (Z, p_{d+1}), so $(Z, p_d)(Z, p_{d+1})$ is a P-edge in $H \boxtimes P$ since $p_d p_{d+1}$ is an edge in P. □

A partition $\mathcal{P}$ of G is *connected* if the subgraph of G induced by each part is connected. In this case, the quotient $G/\mathcal{P}$ is the minor of G obtained by contracting each part into a single vertex. This fact makes connected partitions particularly useful in some applications. We will not pay particular attention to minors, but all the partitions that we will construct in this section will be connected and thus minors of the original graph G.

We now set out to prove the product structure theorem of planar graphs, that is Theorem 1.5.1. The proof is inspired by the following elegant result of Pilipczuk and Siebertz [PS19]: Every planar graph G has a partition $\mathcal{P}$ into geodesics such that $G/\mathcal{P}$ has treewidth at most 8. Here, a *geodesic* is a shortest path between its endpoints. Note that every vertical path in a BFS spanning tree of a graph G is a geodesic in G. Thus the next theorem strengthens the result of Pilipczuk and Siebertz [PS19].

Theorem 1.5.3 *Let T_0 be a rooted spanning tree in a connected planar graph G_0. Then G_0 has a partition $\mathcal{P}_0$ into vertical paths in T_0 such that $G_0/\mathcal{P}_0$ has treewidth at most* 8.

The remainder of this section is dedicated to proving Theorem 1.5.3. Together with Proposition 1.5.2, it implies the product structure theorem of planar graphs as shown next.

Proof of Theorem 1.5.1 *assuming Theorem* 1.5.3 We may assume that G is connected, since if each component C_i of G is a subgraph of $H_i \boxtimes P_i$ for some treewidth-8 graph H_i that is a minor of C_i, then the graph H obtained from the disjoin union of all H_i has treewidth at most 8 and is a minor of G.

Let G be connected and let T be a BFS spanning tree of G. By Theorem 1.5.3, G has a connected partition $\mathcal{P}$ into vertical paths in T such that $H := G/\mathcal{P}$ has treewidth at most 8. Proposition 1.5.2 then implies that G is a subgraph of $H \boxtimes P$ for some path P. Since $\mathcal{P}$ is a connected partition, H is a minor of G. □

Note that Theorem 1.5.3 does not require that T_0 be a BFS tree. The theorem holds for *all* spanning trees T_0 of G_0. BFS assumption is critical in Proposition 1.5.2, that is, in concluding that the vertical path partition gives a strong product between a path and $G_0/\mathcal{P}_0$. Proposition 1.5.2 is not true for all spanning trees[3] but it is true for all BFS trees.

We now turn to the proof of Theorem 1.5.3. We start with a couple of useful definitions. A *near-triangulation* is a plane graph, where the outerface is a simple cycle, and every internal face is a triangle. For a cycle C, we write $C = [P_1, \dots, P_k]$ if $P_1, \dots, P_k$ are pairwise disjoint non-empty paths in C, and the endpoints of each path P_i can be labelled x_i and y_i so that $y_i x_{i+1} \in E(C)$ for $i \in \{1, \dots, k\}$, where x_{k+1} means x_1. This implies that $V(C) = \bigcup_{i=1}^{k} V(P_i)$.

Consider a vertical path $P = (x_1, \dots, x_p)$ in T. The vertex x_1 is called the *upper endpoint* of the path and x_p is its *lower endpoint*. For every vertical path that does not contain the root of T, the parent of x_1 in T is referred to as the *parent of P*.

We now present (the key ideas behind) the proof of Theorem 1.5.3 while trying to avoid tedious special cases. We will point out, though, where those special cases occur in the full proof.

Description of the Proof of Theorem 1.5.3 It is often helpful to work with edge-maximal planar graphs, specifically planar triangulations, rather than arbitrary planar graphs. More precisely, the proof will proceed by induction on graphs that are near-triangulations.

First, we demonstrate that G_0 can be expanded into a (near) triangulation G such that the resulting partition of G is the claimed partition of G_0. Let v be the root of T_0, see Fig. 1.14a. Let G be a triangulation containing G_0 as a spanning subgraph, with v on the outerface of G, as illustrated in Fig. 1.14 (b). It is useful to assume that the root of the spanning tree of our near-triangulation G is not in G but in the outerface

[3] For example, let G be K_3 with vertex set $\{v, w, u\}$. Let T be the depth-first search tree rooted at v. Let $\mathcal{P}$ be comprised of one vertical path $\{v, w, u\}$. Then $H := G/\mathcal{P}$ is K_1. However, G is not a subgraph of $H \boxtimes P$ for any path P.

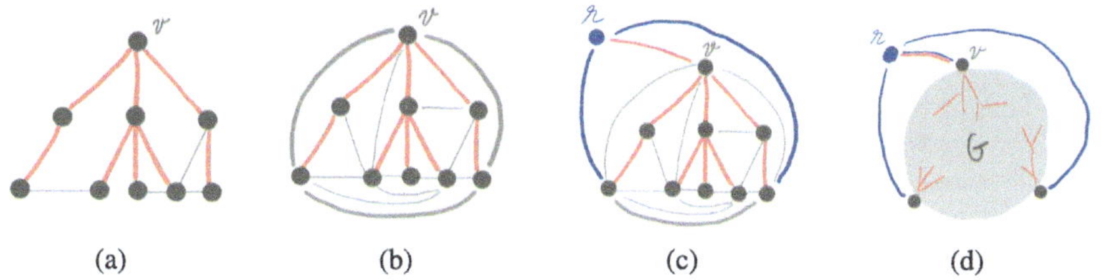

Fig. 1.14 **a** A graph G_0 with the edges of its spanning tree T_0 drawn in pink. **b** The graph G. **c** The augmented graph G^+, with the edges of its spanning tree T shown in pink. **d** A general depiction of G and G^+

of G. To ensure this from the beginning, we define G^+ as the triangulation obtained from G by adding a new vertex r in the outerface of G and making it adjacent to every vertex on the boundary of the outerface of G, as illustrated in Fig. 1.14c. Let T be the spanning tree of G^+ obtained from T_0 by adding r and the edge rv. Consider T to be rooted at r. Note that all edges of T are present in G_0 (specifically, in T_0) except rv.

Suppose that G has a partition $\mathcal{P}$ into vertical paths in T. Then all vertical paths in $\mathcal{P}$ are also vertical paths in T_0. Thus, $\mathcal{P}_0 := \mathcal{P}$ is also a partition of G_0 into vertical paths. Since G_0 is a subgraph of G, $G_0/\mathcal{P}_0$ is a subgraph of $G/\mathcal{P}$. Therefore, if $G/\mathcal{P}$ has treewidth of at most 8, so does $G_0/\mathcal{P}_0$, as Theorem 1.5.3 claims.

We now explain the key ideas of the proof. Our goal is to prove the following statement:

(*) G has a partition $\mathcal{P}$ into vertical paths in T such that $G/\mathcal{P}$ is a plane $(6, 3)$-tree.

Statement (*) implies Theorem 1.5.3 since plane $(6, 3)$-trees have treewidth at most 8 by Lemma 1.4.3. We should note here, however, that statement (*) is slightly incorrect. The statement could be made correct with small yet tedious adjustments to the definition of plane (k, s)-trees (allowing for loops and parallel edges or by introducing clique-sums). For the sake of simplicity, we focus on presenting the main cases only, which lead to plane $(6, 3)$-trees as defined in the previous section.

Now let's try to build the desired partition $\mathcal{P}$ of G into vertical paths in T. Let v, w and u be the three vertices on the outerface of G, as illustrated in Fig. 1.15a. Suppose we begin constructing $\mathcal{P}$ by adding to $\mathcal{P}$ three vertical paths, each containing exactly one of v, w or u, as illustrated in Fig. 1.15b. Specifically, we add $P_1 = (v)$, $P_2 = (w)$, and $P_3 = (u)$ to $\mathcal{P}$. If $\mathcal{P}$ partitions G (equivalently if v, w, and u are the only vertices of G) then we would be done, as $G/\mathcal{P}$ is K_3, which is a plane $(3, 1)$-tree and thus a plane $(6, 3)$-tree. More generally, if $\mathcal{P}$ does not partition G and thus G contains more vertices, let G' be the subgraph induced in G by the vertices that are currently in $\mathcal{P}$. Then the quotient graph $G'/\mathcal{P}$ is K_3 as illustrated in Fig. 1.15c. Additionally, if G contains more vertices, then it is a triangulation (and thus a near-triangulation) that is a subgraph of G^+, where the outerface F of G consists of the three vertical paths, that is $F = [P_1, P_2, P_3]$, with the vertices inside of F not yet assigned to $\mathcal{P}$.

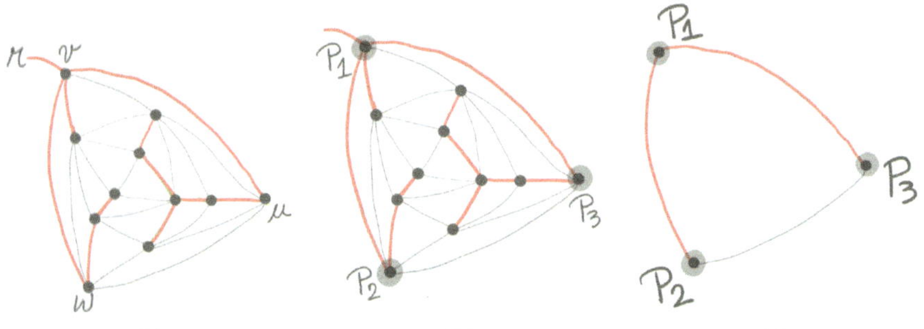

Fig. 1.15 **a** A near triangulation G with edges of T and T_0 in pink. **b** The parts P_1, P_2 and P_3 of $\mathcal{P}$ in shaded gray. **c** The quotient graph $G'/\mathcal{P}$

The following generalization of this setup leads to a stronger statement that we aim to prove by induction. This stronger statement is more conducive to an inductive proof than Statement (*).

(**) Given a near-planar triangulation G that is a subgraph of G^+, given k disjoint vertical paths $\{P_1, \ldots, P_k\}$ in T that comprise the outerface F of G (that is, $F = [P_1, \ldots, P_k]$), G has a partition $\mathcal{P}$ into vertical paths such that $\{P_1, \ldots, P_k\} \subseteq \mathcal{P}$ and such that $G/\mathcal{P}$ is a plane (k, s)-trees, with $k \leq 6$ and $s \leq 3$.

As already noted, at the very beginning, we start with G with its outerface F partitioned into vertical paths $P_1 = (v)$, $P_2 = (w)$, and $P_3 = (u)$. The remaining vertices of G are all inside of F but have not yet been assigned to any vertical paths, as illustrated in Fig. 1.15b. To prove (**), we aim to proceed by induction. Our strategy involves finding a small number s of vertical paths strictly inside F such that these new vertical paths, together with those on F, decompose G into new smaller near-triangulations. Each of these new near-triangulations and their partitioned outerfaces lead to a smaller instance of the problem (**), to which we can then apply induction.

Recall that our goal is for $G/\mathcal{P}$ to be plane (k, s)-trees. In the above strategy, s will be determined by the number of vertical paths inside of F, and thus we aim to keep s at most 3. The parameter k will come from the number of vertical paths on the boundary of the outer face of each new near-triangulation, thus we aim to keep k at most 6.

How do we go about finding $s \leq 3$ new vertical paths inside F that align with the strategy described earlier? Suppose we can find an inner face τ in G with vertices v_1, v_2, v_3 on its boundary, as illustrated in Fig. 1.16a. Note that the path in T from v_i to the root r has to include a vertex of F, given that r lies outside of F. Assume the path in T from v_1 to r contains v; denote this path excluding v by P_1'. Similarly, assume the path in T from v_2 to r contains w, and from v_3 to r contains u, denote the corresponding paths by P_2' and P_3', as illustrated in Fig. 1.16b shaded in gray. Assume now that each of these three paths is non-empty (that is, each has at least one vertex) and that the three paths are pairwise vertex-disjoint.

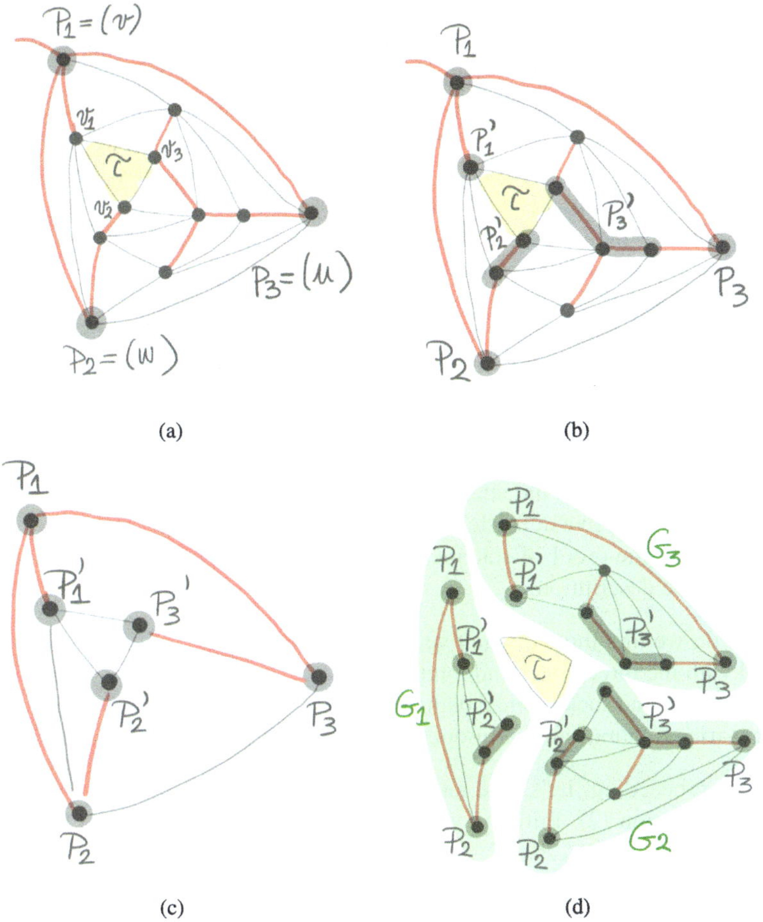

Fig. 1.16 **a** A near-triangulation G with three vertical paths partitioning the boundary of the outerface, shaded in gray. **b** Three new vertical paths P_1', P_2' and P_3', shaded in gray. **c** The quotient graph $G'/\mathcal{P}$. **d** Three new near-triangulations G_1, G_2, and G_3, shaded in green

P_1', P_2', P_3' are vertical paths in T. Since all their vertices are strictly inside of F, and since we assume they are vertex disjoint and non-empty, we can add them to the partition $\mathcal{P}$. The assumption of each of P_1', P_2', P_3' being non-empty allows us to avoid the tedious special cases in this proof description.

Imagine that G does not have any other vertices, i.e. that $\mathcal{P}$ partitions G. Then $G/\mathcal{P}$ begins with face F of size 3 (the outerface of G, comprised of P_1, P_2, P_3). Contracting, in G, the three new vertical paths, P_1', P_2', P_3', effectively introduces three new vertices inside of F. In the resulting graph $G/\mathcal{P}$, the maximum face size is at most 4. According to the definition of plane (k, s)-trees, s is 3 in this scenario, and k is 4. Therefore, if $\mathcal{P}$ partitions G, $G/\mathcal{P}$ is a plane $(4, 3)$-tree, hence a plane

(6, 3)-tree. More generally, if $\mathcal{P}$ does not partition G (equivalently, if G contains more vertices), let G' be the subgraph induced in G by the vertices that are currently in $\mathcal{P}$. Then $G'/\mathcal{P}$ is a plane (4, 3)-tree, as illustrated in Fig. 1.16c.

Note the increase in the size of the outer face in the subproblems (from $k = 3$ initially to $k = 4$ in the inductive subproblem). To uphold the requirement that $G/\mathcal{P}$ remains a plane (6, 3)-tree, we need to prevent k from growing beyond 6 (especially since the treewidth of plane (k, s)-trees increases with k and s).

In the previous paragraph, we considered the special case when $\mathcal{P}$ partitions G and we are done. In general, however, we are not done as G has more vertices. In that case, we have the following setup. After adding P_1', P_2', P_3' to $\mathcal{P}$, we obtain three new near-triangulations G_1, G_2, G_3, each with fewer vertices than G, and each having at most 4 vertical paths partitioning their outerfaces: $F_1 = [P_1', v, w, P_2']$, $F_2 = [P_2', w, u, P_3']$, and $F_3 = [P_3', u, v, P_1']$, as illustrated in Fig. 1.16d. By ensuring that P_1', P_2', P_3' each have a parent in F, we ensure that these new vertical paths indeed define three new near-triangulations. Thus if each G_i has a partition $\mathcal{P}_i$ such that: the vertical paths partitioning F_i are in $\mathcal{P}_i$ and such that $G_i/\mathcal{P}_i$ is a plane (6, 3)-tree, then we would be done by the recursive definition of plane (6, 3)-trees.

To complete the proof of statement (**), we must resolve two unresolved steps from the above description:

(1) How do we find three disjoint vertical paths inside F, each with a parent in F, and with the three lower endpoints of these paths on an internal face of G?
(2) how do we find such vertical paths so that the outerfaces of each of the three new near-triangulations are partitioned into at most $k = 6$ vertical paths?

Before proceeding, let's take an intermission to understand the significance of $k = 6$ in condition (2). Initially, we began with a face F composed of 3 vertical paths. After adding P_1', P_2', P_3', we ended up with three subproblems, each with associated near-triangulations having 4 vertical paths on their outerfaces. We need to ensure that the number of vertical paths on the outerfaces does not grow indefinitely. Specifically, we want to prevent it from exceeding $k \leq 6$, as our goal is to obtain plane (6, 3)-trees for the quotient graph. More importantly, we aim for a bounded treewidth quotient graph, and recall that the treewidth of plane (k, s)-trees depends on both k and s, see Lemma 1.4.3.[4]

While we cannot maintain only $k \leq 3$ vertical paths on the outerface, it turns out that, with some care, we can maintain $k \leq 6$. Here's how we achieve this and resolve points (1) and (2).

The input to our problem (**) is a near-triangulation G with $k \in \{3, 4, 5, 6\}$ vertical paths on its outerface F. Our goal is to find three new vertical paths in F as described in (1). To achieve (2), we must be careful about where these new vertical paths end on F, that is, where their parents are located. Otherwise, the outerfaces of the new instances might grow beyond 6.

[4] In fact, the treewidth of plane (k, s)-trees grows with k and s as $n \times n$ grid graph is a plane (n, n)-tree.

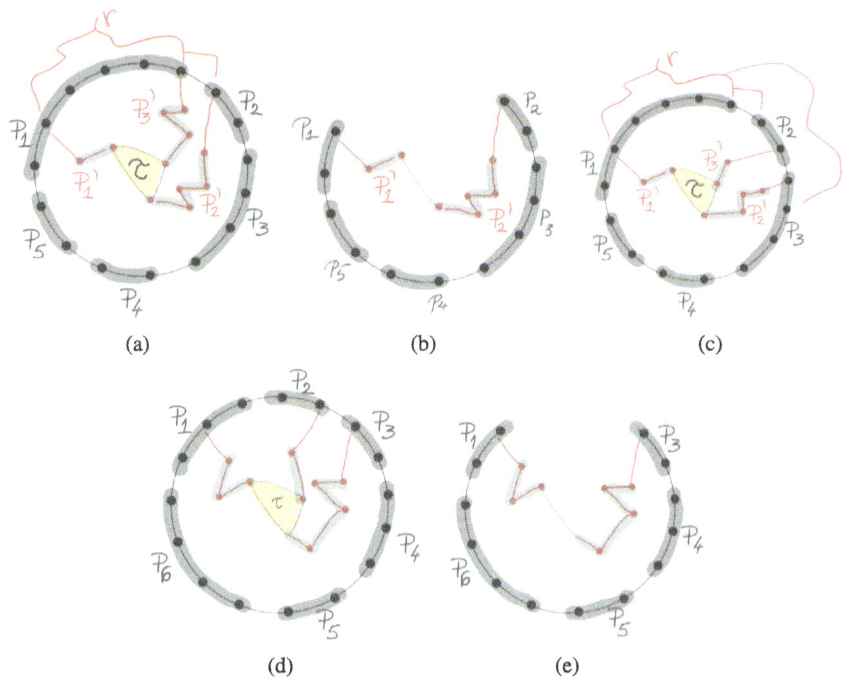

Fig. 1.17 **a–b** An example illustrating a poor choice of vertical paths. **c** A contrasting example with a well-chosen set of vertical paths. **d–e** Another example demonstrating a poor choice of vertical paths

Consider the example in Fig. 1.17a, where starting with 5 vertical paths on F, we ended up creating a subproblem with an outerface of size 7, as illustrated in Fig. 1.17b. In this example, two of the three new vertical paths ended up in the same vertical path of F, namely the vertical path P_1. If we can ensure that each of the three new vertical paths has parents in distinct vertical paths of F, we can achieve (2) as long as we start with at most $k \leq 5$ vertical paths on F, as illustrated in Fig. 1.17c. However, if $k = 6$, we can still create a face of size 7 even if the three parents are in three distinct vertical paths on F, as illustrated in Fig. 1.17d, e. This happens when the parents are in three consecutive paths on F. In this case, we need not only to have parents in distinct vertical paths in F but also need to ensure that they are not all in three consecutive paths.

All of this, (1) and (2) to be precise, can be achieved with a help of a well-known variation of Sperner's Lemma (see [AZ10]):

Lemma 1.5.4 (Sperner's Lemma) *Let G be a near-triangulation whose vertices are coloured* 1, 2, 3, *with the outerface $F = [R_1, R_2, R_3]$ where each vertex in R_i is coloured i. Then G contains an internal face whose vertices are coloured* 1, 2, 3.

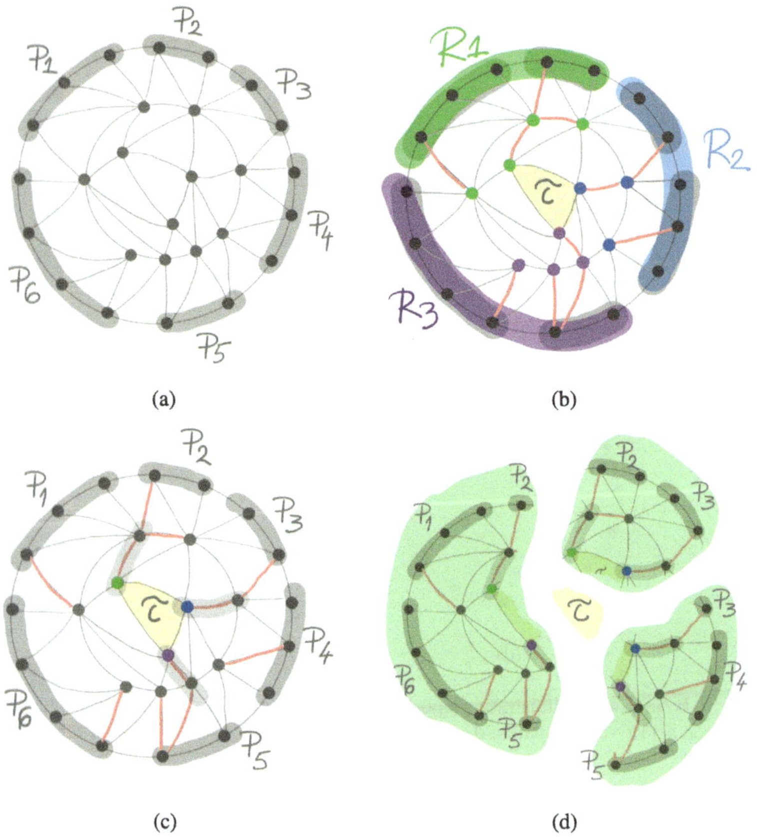

Fig. 1.18 **a** A near-triangulation G with six vertical paths that partition the boundary of the outer face, shaded in gray. **b** Sperner's Lemma application. **c** Addition of three new vertical paths, shaded in gray **d** Formation of three new near-triangulations G_1, G_2, and G_3, each shaded in green

Recall that in the general case, we have up to $k = 6$ vertical paths forming the outer face F of G, that is, $F = [P_1, \ldots, P_k]$. To apply Sperner's Lemma and ensure that the parents of the three new vertical paths are in the desired vertical paths on F, we first group consecutive pairs of vertical paths on F (before colouring vertices and applying Sperner's Lemma).

Specifically, we make three disjoint groups R_1, R_2, and R_3 of vertical paths on F such that each group has one or two consecutive vertical paths. For example, when $k = 6$, we set $R_1 := [P_1, P_2]$, $R_2 := [P_3, P_4]$ and $R_3 := [P_5, P_6]$. More generally, let $R_1 := [P_1, \ldots, P_{\lfloor k/3 \rfloor}]$, $R_2 := [P_{\lfloor k/3 \rfloor + 1}, \ldots, P_{\lfloor 2k/3 \rfloor}]$ and $R_3 := [P_{\lfloor 2k/3 \rfloor + 1}, \ldots, P_k]$, as illustrated in Fig. 1.18a, b.

For $i \in \{1, 2, 3\}$, colour each vertex in R_i by i, as illustrated in Fig. 1.18b. Now, for each remaining vertex x in G, consider the path P_x from x to the root of T. Since r is on the outerface of G^+, P_x contains at least one vertex of F. If the first vertex of

P_x that is on the boundary of F is in R_i then assign the colour i to x, as illustrated in Fig. 1.18b. In this way, we obtain a 3-colouring of the vertices of G that satisfies the conditions of Sperner's Lemma. Therefore, by Sperner's Lemma, there exists a triangular face $\tau = v_1, v_2, v_3$ of G whose vertices are coloured $1, 2, 3$ respectively, as illustrated in Fig. 1.18c. For each $i \in \{1, 2, 3\}$, consider the path in T from v_i to the root r and let v_i' be the first vertex on that path that is contained in F. Let P_i' be the path in T from v_i to v_i' not including v_i', as illustrated in Fig. 1.18c.

Observe that P_1', P_2', and P_3' are vertex disjoint since P_i' consists only of vertices coloured i. Assuming that each P_i' is non-empty, we obtain the desired vertical paths. The full proof handles cases where one or more of the paths P_1', P_2', P_3' are empty. This can occur if some vertices of τ are on F. Such faces τ may lead to subproblems where the number of vertical paths on the outer faces of the new near-triangulations is 1 or 2. These cases, while not difficult to handle, are tedious and would make the definition of plane (k, s)-trees cumbersome without adding much to the understanding of the solution.

Finally, consider the subgraph of G that consists of the edges and vertices of F, the edges and vertices of τ, and the edges and vertices of $P_1' \cup P_2' \cup P_3'$. Each inner face of this graph is bounded by two vertical paths from among $P_1' \cup P_2' \cup P_3'$ and by up to four vertical paths from among $P_1, P_2, P_3, P_4, P_5, P_6$, as illustrated in Fig. 1.18d. Thus, the number of vertical paths on the outerface of each of the subproblem near-triangulations is maintained at at most $k = 6$. □

1.5.2 Improvements and Variations

Recall the definition of flexible plane (k, s)-trees from Sect. 1.4.3. It is not difficult to see that the plane $(6, 3)$-trees obtained in Theorem 1.5.3 are indeed flexible and thus have treewidth at most 7 by Lemma 1.4.4. This observation was not made in the original publication by [DJM+20]. It is made here, but thanks to the insights from the improved results by Ueckerdt, Wood, and Yi [UWY22b], who, in fact, prove the following strengthening of Theorem 1.5.3.

Theorem 1.5.5 ([UWY22b]) *Let T_0 be a rooted spanning tree in a connected planar graph G_0. Then G_0 has a partition $\mathcal{P}_0$ into vertical paths in T_0 such that $G_0/\mathcal{P}_0$ has treewidth at most* 6.

The main idea behind their proof is an improved statement (*) and (**) where plane $(6, 3)$-trees are replaced by flexible plane $(5, 3)$-trees, and thus treewidth 6 graphs by Lemma 1.4.4. Ueckerdt et al. [UWY22b] obtain plane $(5, 3)$-trees by circumventing Sperner's Lemma. Starting with the outer face $F = [P_1, \dots, P_k]$ where $k \leq 5$, they show that one can always find three vertical paths inside of F whose parents are not in three consecutive vertical paths of F. That is sufficient to ensure that the outerfaces in the subproblems remain at most $k \leq 5$.

It seems, however, that we have reached the limits of this method in reducing the treewidth of $G_0/\mathcal{P}_0$ and thus the treewidth of H in the product structure theorem,

Theorem 1.5.1. The best-known lower bound is 3 [DJM+20], leading to the following open problem.

Open Problem 1.5.6 *Is every planar graph G a subgraph of $H \boxtimes P$ for some path P and some graph of H of treewidth 4 or even treewidth 3?*

The following theorem provides a better treewidth bound, albeit at the cost of adding a complete graph K_3 to the product. The upper bound of 3 on treewidth in the next theorem is best possible, even if K_3 is replaced by any complete graph of constant size, as proved in [DJM+20]. For many applications of the product structure theorem of planar graphs, this version has led to the best results because the best-known bounds on many graph parameters are exponential in the treewidth.

Theorem 1.5.7 ([DJM+20]) *Every planar graph G is a subgraph of $H \boxtimes P \boxtimes K_3$ for some path P and some planar 3-tree H (thus a graph of treewidth 3).*

Having seen the main ideas behind the proof of Theorem 1.5.3, and thus the proof of the product structure theorem, it is fairly easy to understand the origins and proof of the above theorem. Let's discuss that now. For a graph G and a spanning tree T, a *tripod* in G is a union of up to 3 vertical paths whose lower endpoints form a clique in G, as illustrated in Fig. 1.19a. We refer to the vertical paths in a tripod as the *legs* of the tripod. Note that vertical paths and tripods with one leg are equivalent notions. Proposition 1.5.2 can easily be generalized to the following statement.

Proposition 1.5.8 *Let G be a connected graph, and T a rooted BFS tree of G. If, for some integer $c \in \{1, 2, 3\}$, G has a partition $\mathcal{P}$ into tripods with c legs, then G is a subgraph of $H \boxtimes P \boxtimes K_c$ for some path P and $H := G/\mathcal{P}$.*

This proposition, along with the following theorem, implies Theorem 1.5.7.

Theorem 1.5.9 *Let T be a rooted spanning tree of a planar triangulation G. Then G has a partition $\mathcal{P}$ into tripods such that $G/\mathcal{P}$ is a planar 3-tree (thus a graph of treewidth 3).*

To understand why this theorem is true, recall the proof of Theorem 1.5.3 and how we obtained a partition $\mathcal{P}$ there. We started with three vertical paths (thus three tripods) on the outer face F of G. Specifically, we added the vertical paths (tripods) $T_1 = (v)$, $T_2 = (w)$, and $T_3 = (u)$ to $\mathcal{P}$, as illustrated in Fig. 1.19b. Together with the partition $F = [T_1, T_2, T_3]$ of F, they decompose G into three near-triangulations, each of which still has at most three tripods on its boundary. This process can continue as follows:

(i) In the general case, we start with a near-triangulation whose outerface F is partitioned into three tripods $F = [T_1, T_2, T_3]$. Then we find a tripod T' inside of F such that the parents of each leg of T' are on distinct tripods T_1, T_2, T_3 of F, as illustrated in Fig. 1.19c. The outerfaces of the resulting near-triangulations are then also bounded by three tripods. This maintains the size of the outerface in the near-triangulations of the subproblems, as illustrated in Fig. 1.19c, allowing us to obtain

the desired partition $\mathcal{P}$ of G into tripods.
(ii) Contracting tripods in the obtained partition $\mathcal{P}$ results in a planar 3-tree. The process starts by contracting the three tripods on the outer face F of G, resulting in a triangle outerface in the quotient graph, as illustrated in Fig. 1.19d. Then, contracting the tripod T' inside of that face of G is equivalent to adding a 3-simplicial vertex to that face of the quotient graph, resulting in another planar 3-tree, as illustrated in Fig. 1.19e. This process continues until all the tripods in $\mathcal{P}$ are contracted, as illustrated in Fig. 1.20.

Instead of maintaining three 3-legged tripods on the outer face of near-triangulations, which led to the previous theorem (Theorem 1.5.7), we can opt to maintain four 2-legged tripods. That strategy leads to the following (slightly incorrect) statement akin to the statement (*) in the previous section. Let T be a rooted spanning tree of a planar triangulation G, then G has a partition $\mathcal{P}$ into 2-legged tripods such that $G/\mathcal{P}$ is a plane (4, 1)-tree (thus a graph of treewidth 4 by Lemma 1.4.3). Together with Theorem 1.5.8, this implies the following final version of the product structure theorem of planar graphs (see Exercise 6 in Sect. 1.7).

Theorem 1.5.10 ([UWY22b]) *Every planar graph G is a subgraph of $H \boxtimes P \boxtimes K_2$ for some path P and some planar graph H of treewidth* 4.

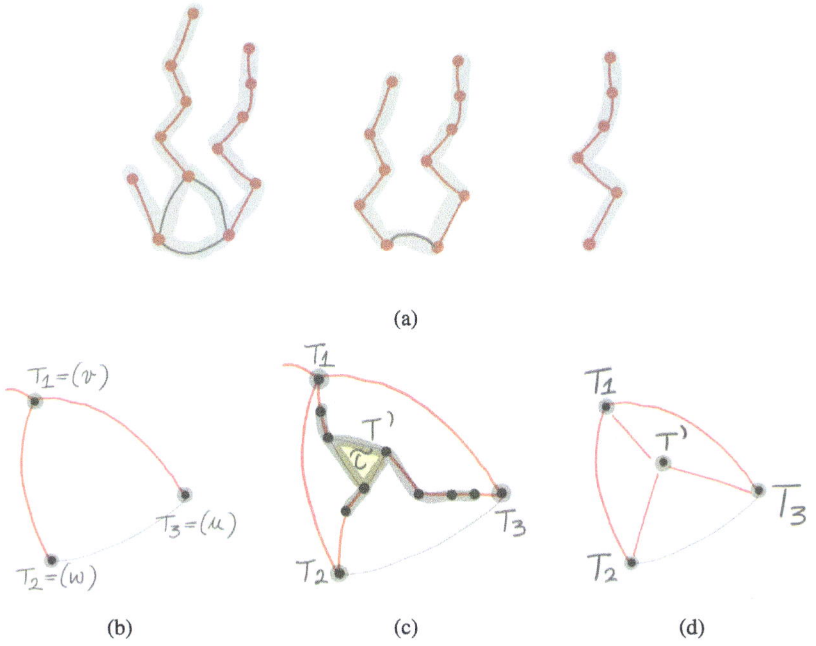

Fig. 1.19 **a** Tripods with 3, 2, and 1 legs, respectively. **b–d** Illustrations conveying the key ideas behind the proof of Theorem 1.5.9

1.5.3 Product Structure Theorem: Beyond Planar Graphs

We say that a class of graphs $\mathcal{G}$ *admits product structure* if there exist integers k and c such that for every graph G in $\mathcal{G}$, there exists a graph H of treewidth at most k such that G is a subgraph of $H \boxtimes P \boxtimes K_c$.

The work in the previous section can be summarized by saying that planar graphs admit product structure. This statement (and the utility of product structure theorems described in the next section) leads to the following natural question: what classes of graphs admit product structure? We mention several results (without proofs), starting with minor-closed classes of graphs.

To appreciate the differences, or lack thereof, between the variants of the product structure theorem that will follow, we first present a useful lemma. Its proof is left as an exercise (see Exercise 4 in Sect. 1.7). Loosely speaking, the lemma states that products of the form $H \boxtimes P \boxtimes K_c$ (for a constant c) are not significantly different from products of the form $H \boxtimes P$, up to a multiplicative constant c.

Lemma 1.5.11 ([DJM+20]) *If a graph G is a subgraph of $H \boxtimes P \boxtimes K_c$ for some graph H of treewidth k and a path P, then G is a subgraph of $H' \boxtimes P$ for some graph H' of treewidth at most $(k + 1) \cdot c - 1$.*

We start with graphs of bounded Euler genus.[5] Planar graphs have Euler genus zero, thus graphs of bounded genus generalize planar graphs. Dujmović [DJM+20] proved that such graphs indeed admit product structure. In particular, they proved the following theorem. Dujmović [DJM+20] proved that such graphs indeed admit product structures.

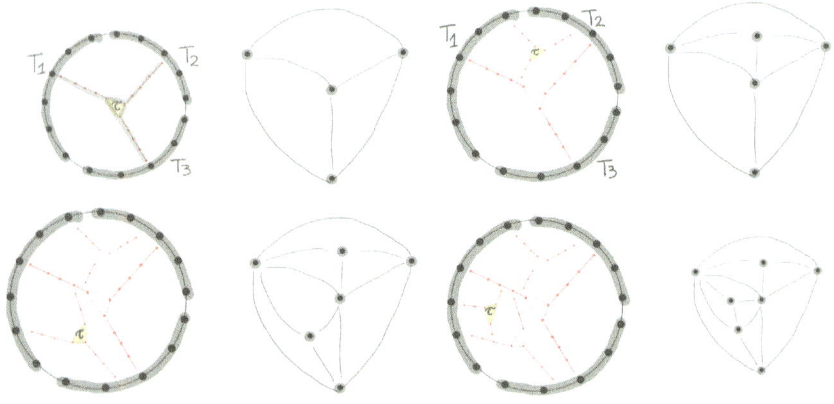

Fig. 1.20 Illustrations conveying the key ideas behind the proof of Theorem 1.5.9

[5] The *Euler genus* of the orientable surface with h handles is $2h$. The *Euler genus* of the non-orientable surface with x cross-caps is x. The *Euler genus* of a graph G is the minimum integer k such that G has a crossing-free drawing in a surface of Euler genus k. Of course, a graph is planar if and only if it has Euler genus 0; see [MT01] for more on graphs on surfaces.

Theorem 1.5.12 ([DJM+20]) *Every graph G of Euler genus g is a subgraph of both:*

(i) $H \boxtimes P$ *for some path P and some graph H such that the treewidth of H is at most* $2g + 8$ *(and H is a minor of G); and,*
(ii) $H \boxtimes P \boxtimes K_{\max\{2g,1\}}$ *for some path P and some graph H such that the treewidth of H is at most* 4.

Part (i) of the above theorem has been improved by Ueckerdt et al. [UWY22b] to $H \boxtimes P$ where H is a graph of treewidth at most $2g + 6$. Part (ii) of the above theorem has been improved by Distel, Hickingbotham, Huynh, and Wood [DHHW22] to $H \boxtimes P \boxtimes K_{\max\{2g,3\}}$ where H is a planar graph of treewidth 3.

Planar graphs and graphs of Euler genus g are closed under taking minors, thus one might be tempted to conjecture that every proper minor-closed family of graphs admits product structure. In particular, consider the following question: is it true that for every proper minor-closed family of graphs $\mathcal{G}$ there exists a constant h such that every graph in $\mathcal{G}$ is a subgraph of $H \boxtimes P$ such that the treewidth of H is at most k. The answer is no. First, consider the following two straightforward lemmas, the proofs of which are left as exercises (see Exercises 5 in Sect. 1.7).

Lemma 1.5.13 *Let G be a graph that is a subgraph of $H \boxtimes P_\ell$ for some graph H of treewidth k. Then the treewidth of G is at most $(k + 1) \cdot \ell - 1$. More generally, let G be a graph that is a subgraph of $H \boxtimes P_\ell \boxtimes K_c$ for some graph H of treewidth k. Then the treewidth of G is at most $(k + 1) \cdot c \cdot \ell - 1$*

Lemma 1.5.14 *Let G be a connected graph that is a subgraph of $H \boxtimes P$ for some graph H and a path P. Let ℓ be the minimum integer such that G is a subgraph of $H \boxtimes P_\ell$, then the diameter of G is at least $\ell - 1$.*

Now, consider the graph obtained from an $n \times n$ planar grid graph by adding one more vertex adjacent to all the vertices of the grid. Let $\mathcal{G}$ be the class of such graphs for every integer $n \geq 1$. The diameter of every graph in $\mathcal{G}$ is at most 2. Supposed $G \in \mathcal{G}$ is a subgraph of $H \boxtimes P$ for some graph H and let ℓ be the minumum integer such that G is a subgraph of $H \boxtimes P_\ell$. Let the treewidth of G be denoted by t. By Lemma 1.5.14, G is a subgraph of $H \boxtimes P_3$. Then by Lemma 1.5.13, the treewidth of H is at least $(t + 1)/3 - 1 \in \Omega(k)$. Since G contains an $n \times n$ planar grid graph as a subgraph, G has treewidth at least n and thus the treewidth of H is $\Omega(n)$. Thus there exists no constant k such that every graph G in $\mathcal{G}$ is a subgraph of $H \boxtimes P$, for some graph H of treewidth at most k and some path P. By Lemma 1.5.11, $\mathcal{G}$ does not admit product structure.

Every graph in $\mathcal{G}$ is a so-called apex graph. A graph G is an *apex* graph if there exists a vertex v in G such that $G - v$ is planar. By Kuratowski's theorem planar graphs have no K_5 as a minor. Since apex graphs are comprised of planar graphs plus at most one vertex, every apex graph has no K_6 as a minor. Thus, $\mathcal{G}$ is a class of K_6-minor free graphs (thus a proper minor-closed class) that does not admit product structure.

The above arguments demonstrate that apex graphs present an obstruction to proper minor-closed families of graphs admitting product structure. This turns out to provide a characterization of minor-closed classes that admit product structure, as detailed in the following theorem.

Theorem 1.5.15 ([DJM+20]) *Let $\mathcal{G}$ be a minor-closed class of graphs. There exist integers k and c such that for every $G \in \mathcal{G}$, G is a subgraph of $H \boxtimes P \boxtimes K_c$ where H is some graph of treewidth at most k and P is a path, if and only if there is an apex graph not in $\mathcal{G}$.*

The next natural question to consider is: what about non-minor-closed families of graphs? It turns out that the usefulness of product structure extends beyond minor-closed classes. The following theorem, by Dujmović, Morin, and Wood [DMW23], exemplifies a product structure theorem of a well-studied non-minor-closed class of graphs. A graph is *k-planar* if it has a drawing such that each edge has at most k crossings. (Thus, 0-planar graphs are planar graphs.) k-planar graphs are not closed under taking minors, even for $k = 1$ (see Exercise 1.7 in Sect. 1.7).

Theorem 1.5.16 ([DMW23]) *Every k-planar graph G is a subgraph of $H \boxtimes P$ for some graph H of treewidth $O(k^5)$ and a path P.*

Impressively, Distel, Hickingbotham, Seweryn, and Wood [DHSW23] show that in the above theorem, the treewidth of H can be made an absolute constant independent of k, at the cost of adding a complete graph to the product. Specifically, they prove that there exists a function f and a constant c (with $c \leq 15288899$) such that every k-planar graph G is a subgraph of $H \boxtimes P \boxtimes K_{f(k)}$ for some graph H of treewidth c and a path P.

For other non-minor-free classes of graphs that admit product structure, see [DMW23, DHSW23, HW24].

1.6 Applications

1.6.1 Solution to Problem 1.3.1 and Related Applications

(A) *3D Grid Drawings.* At the end of Sect. 1.3.3, we have already argued that if planar graphs admit product structure, then Problem 1.3.1 has an affirmative answer. We formalize this below.

Corollary 1.6.1 ([DJM+20]) *Problem 1.3.1 has an affirmative answer. Specifically, every n-vertex planar graph G has a 3D grid drawing with $O(n)$ volume.*

Proof Theorem 1.5.1 implies that G is a subgraph of $H \boxtimes P$ for some H of treewidth at most 8. By Theorem 1.3.8, H has $O(1)$ track-number. By Lemma 1.3.9, G also has $O(1)$ track-number. Finally, Lemma 1.3.3 implies that G has a 3D grid drawing with $O(n)$ volume. □

(B) *Track-number*. The previous proof already argued the following result on track-number, answering an open problem posed by Felsner et al. [FLW03] affirmatively.

Corollary 1.6.2 ([DJM+20]) *Every planar graph has a bounded track-number.*

Currently, the best-known bound on the track-number of planar graphs, due to Pupyrev [Pup20], is 225. This was achieved with the help of Theorem 1.5.7, that is, the version of the product structure theorem H is a planar 3-tree. Pupyrev [Pup20] showed that planar 3-trees have track-number of at most 25. He then improved Lemma 1.3.9 by showing that if G is a subgraph of $H \boxtimes P \boxtimes K_\ell$, then $\operatorname{tn} G \leq 3 \cdot \ell \cdot \operatorname{tn}(H)$. Thus, Theorem 1.5.7 implies that every planar graph has track-number at most $3 * 3 * 25 = 225$.

(C) *Queue-number*. In the early 1990s, Heath et al. [HLR92] introduced and studied queue layouts of graphs and the associated graph parameter queue-number. They conjectured that the queue-number of planar graphs is bounded. Since the early 2000s, it has been known that minor-closed families of graphs have $O(1)$ queue-number if and only if they have $O(1)$ track-number [DMW05]; and, in fact that queue-number and track-number are tied [DPW04].

Therefore, Corollary 1.6.2 resolved the 27-year-old problem on the queue-number of planar graphs posed by Heath et al. [HLR92].

Corollary 1.6.3 ([DJM+20]) *Every planar graph has a bounded queue-number.*

1.6.2 Other Applications

It is beyond the scope of these lecture notes to detail all the applications of the product structure theorem of planar and other graphs. However, we briefly list some applications here and provide references for interested readers who wish to explore further.

(D) *Nonrepetitive colouring*. Soon after the preprint publication of [DJM+20] appeared, Dujmović, Esperet, Joret, Walczak, and Wood [DEJ+20] used the product structure theorem to prove that planar graphs have bounded non-repetitive chromatic number. This solves a problem posed by Alon, Grytczuk, Hałuszczak, and Riordan [AGHR02] in 2002.

(E) *p-Centered Colouring*. Dębski, Felsner, Micek, and Schröder [DFMS20] used the product structure theorem to show that planar graphs have p-centered colourings using $O(p^3 \log p)$ colours. This dramatically reduces the previous $O(p^{18})$ bound and gives the current best result on a problem posed by Nešetřil and Ossona de Mendez [NO06] in 2006.

(F) *Adjacency labelling.* Dujmović, Esperet, Gavoille, Joret, Micek, and Morin [DEG+21] used the product structure theorem to show that planar graphs have $(1 + o(1)) \log n$-bit adjacency labelling schemes. This settles a problem posed by Kannan, Naor, and Rudich [KNR88] in 1988.

(G) *Sparse universal graphs.* Esperet, Joret, and Morin [EJM23] used the product structure theorem to show that for every $n \geq 1$, there exists a graph G_n with $(1 + o(1))n$ vertices and $n^{1+o(1)}$ edges such that every n-vertex planar graph is isomorphic to a subgraph of G_n, essentially solving the problem of universal graphs for planarity. The previous best bound of $O(n^{3/2})$ was proved by Babai, Chung, Erdős, Graham, and Spencer [BCEGS+82] in 1982.

(H) *ℓ-Vertex Ranking.* Bose, Dujmovic, Javarsineh, and Morin [BDJM20] used the product structure theorem to show that planar graphs have ℓ-vertex rankings using $O(\log n / \log\log\log n)$ colours, and this is tight. This settles a problem posed by Karpas, Neiman, and Smorodinsky [KNS15] in 2015.

(I) *Twin-width.* Jacob and Pilipczuk [JP22] used a version of the product structure theorem explicitly to give concrete upper bounds on the twin-width of planar graphs. Twin-width is a graph parameter recently introduced by Bonnet, Kim, Thomassé, and Watrigant [BKTW22].

(J) *The Clustered Hadwiger Conjecture.* Dujmović, Esperet, Morin, and Wood [DEMW23] combined the product structure theorem with the Robertson-Seymour Graph Minor Structure Theorem to obtain a new structural decomposition of K_h-minor-free graphs and used it to prove that such graphs have $(h-1)$-colourings with monochromatic components of bounded size. This solves an open problem of Edwards, Kang, Kim, Oum, and Seymour [EKK+15] posed in 2015, and concludes a line of research initiated in 2007.

1.7 Exercises

Omitted Proofs:

1. Prove Lemma 1.3.6 (wrapping of track layouts).
2. Prove Lemma 1.4.4 (treewidth of flexible plane (k, s)-tree).
3. Prove Theorem 1.5.8 (from partition into tripods to graph products).
4. Prove Lemma 1.5.11 (from $H \boxtimes P \boxtimes K_c$ to $H' \boxtimes P$).
5. Prove Lemma 1.5.13 (from graph products to treewidth) and Lemma 1.5.14 (diameter and graph products).
6. Prove Theorem 1.5.10 (partition into tripods with two legs).

Track Layouts:

7. Prove that every outerplanar graph G has a track layout with the maximum edge span at most two. More strongly, prove (by induction) that G has a straight-line crossing-free drawing where the endpoints of every edge have y-coordinates that are integers differing by 1 or 2.

Minors:

8. Prove that the class of 1-planar graphs is not closed under taking minors. Hint: prove it for the class of $n \times n \times 2$ grid graphs, that is, $P_n \square P_n \square 2$ graph.

Treewidth:

9. Prove Helly property of trees (see Sect. 1.4) and with that prove Lemma 1.4.1 on cliques in tree-decompositions.

Separators:
Let S be a subset of the vertices of an n-vertex graph G. S is a $1/c$-*separator* of G if removing S from G (along with the edges incident to S) results in a graph in which every connected component has at most n/c vertices. $|S|$ is the *size of the separator*.

10. Prove that every n-vertex tree has a 1/2-separator of size 1.
11. (a) Prove that every n-vertex outerplanar graph has 1/2-separator of size 3.
    (b) Prove that every n-vertex outerplanar graph has 2/3-separator of size 2.
12. Prove that every n-vertex graph G of treewidth k has 1/2-separator of size $k + 1$.

Lipton Tarjan Theorem:

12. With the help of Exercise 1.7 and the product structure theorem of planar graphs, Theorem 1.5.1, and Lemma 1.5.11 try to prove the Lipton-Tarjan's theorem (it states that every n-vertex planar graph has a 1/2-separator of size $O(\sqrt{n})$.

Applications of the Product Structure Theorem of Planar Graphs:

14. Let $p \geq 1$ be an integer. A vertex colouring ϕ of a graph G is p-*centered* if for every connected subgraph C of G either ϕ uses more than p colors on C or there is a color that appears exactly once on C. The p-*centered chromatic number*, $\chi_p(G)$, of G is the minimum integer k such that there is a p-centered colouring of G using k colors. Pilipczuk and Siebertz [PS19] showed that for every $p \geq 1$, and every graph G, $\chi_p(G) \leq \binom{p+t}{t}$ where t is the treewidth of G. Using that theorem and Theorem 1.5.1, prove that every planar graph G has $\chi_p(G) \leq O(p^9)$.
15. Dębski et al. [DFMS20] proved that every planar 3-tree G has $\chi_p(G) \in O(p^2 \log p)$. Use that result and Theorem 1.5.7 to improve the $O(p^9)$, bound from Exercise 1.7, to $O(p^3 \log p)$, which is currently the best-known bound on the p-centered chromatic number of planar graphs.
16. A *queue layout* of a graph G consists of a total order σ of $V(G)$ and a partition of $E(G)$ into sets (called *queue*) such that no two edges in the same stack *nest*; that is, there are no edges vw and xy in a single queue with $v <_\sigma x <_\sigma y <_\sigma w$. The minimum number of queues in a queue layout of G is the *queue-number* of G, $\mathrm{qn}(G)$.

    (a) Prove that for every graph H and path P, $\mathrm{qn}(H \boxtimes P) \leq 3 \cdot \mathrm{qn}(H) + 1$.

(b) Wiechert [Wie17] proved that for every graph G, $\text{qn}(G) \leq 2^t - 1$ where t is the treewidth of G. Use that result, part (a) of this question, and the product structure theorem of planar graphs, Theorem 1.5.1, to show that every planar graph has queue-number at most 766.

Acknowledgements I thank the anonymous referees for their detailed and insightful comments, which significantly improved the quality of this manuscript.

References

[AGHR02] Noga Alon, Jarosław Grytczuk, Mariusz Hałuszczak, and Oliver Riordan. Nonrepetitive colorings of graphs. *Random Structures Algorithms*, 21(3-4):336–346, 2002.

[AP89] Stefan Arnborg and Andrzej Proskurowski. Linear time algorithms for NP-hard problems restricted to partial k-trees. *Discrete Appl. Math.*, 23(1):11–24, 1989.

[AZ10] Martin Aigner and Günter M. Ziegler. *Proofs from The Book*. Springer, 4th edition, 2010.

[BCEGS+82] László Babai, Fan R. K. Chung, Paul Erdős, Ron L. Graham, and Joel H. Spencer. On graphs which contain all sparse graphs. In *Theory and practice of combinatorics*, volume 60 of *North-Holland Math. Stud.*, pages 21–26. Elsevier, 1982.

[BDJM20] Prosenjit Bose, Vida Dujmovic, Mehrnoosh Javarsineh, and Pat Morin. Asymptotically optimal vertex ranking of planar graphs. *CoRR*, abs/2007.06455:33, 2020.

[BFG+19] Michael A. Bekos, Henry Förster, Martin Gronemann, Tamara Mchedlidze, Fabrizio Montecchiani, Chrysanthi N. Raftopoulou, and Torsten Ueckerdt. Planar graphs of bounded degree have bounded queue number. *SIAM* J. Comput., 48(5):1487–1502, 2019. Also in *51st Annual ACM SIGACT Symposium on Theory of Computing*, (STOC), pp. 176–184, 2019.

[BFP10] Giuseppe Di Battista, Fabrizio Frati, and János Pach. On the queue number of planar graphs. In *Proc. 51st Annual Symp. on Foundations of Computer Science (FOCS '10)*. IEEE, 2010.

[BKTW22] Édouard Bonnet, Eun Jung Kim, Stéphan Thomassé, and Rémi Watrigant. Twin-width I: tractable FO model checking. *J. ACM*, 69(1):3:1–3:46, 2022.

[Bod98] Hans L. Bodlaender. A partial k-arboretum of graphs with bounded treewidth. *Theoret. Comput. Sci.*, 209(1-2):1–45, 1998.

[CELR96] Robert F. Cohen, Peter Eades, Tao Lin, and Frank Ruskey. Three-dimensional graph drawing. *Algorithmica*, 17(2):199–208, 1996.

[DEG+21] Vida Dujmović, Louis Esperet, Cyril Gavoille, Gwenaël Joret, Piotr Micek, and Pat Morin. Adjacency labelling for planar graphs (and beyond). *J. ACM*, 68(6):42, 2021.

[DEJ+20] Vida Dujmović, Louis Esperet, Gwenaël Joret, Bartosz Walczak, and David R. Wood. Planar graphs have bounded nonrepetitive chromatic number. *Adv. Comb.*, #5:11, 2020.

[DEMW23] Vida Dujmović, Louis Esperet, Pat Morin, and David R. Wood. Proof of the clustered Hadwiger conjecture. In *64th IEEE Annual Symposium on Foundations of Computer Science, FOCS 2023, Santa Cruz, CA, USA, November 6-9, 2023*, pages 1921–1930. IEEE, 2023.

[DFMS20] Michał Dębski, Stefan Felsner, Piotr Micek, and Felix Schröder. Improved bounds for centered colorings. In Shuchi Chawla, editor, *Proc. ACM-SIAM Symp. on Discrete Algorithms (SODA '20)*, pages 2212–2226, 2020.

[DHHW22] Marc Distel, Robert Hickingbotham, Tony Huynh, and David R. Wood. Improved product structure for graphs on surfaces. *Discrete Math. Theor. Comput. Sci.*, 24(2):#6, 2022.

[DHSW23] Marc Distel, Robert Hickingbotham, Michał T. Seweryn, and David R. Wood. Powers of planar graphs, product structure, and blocking partition, 2023.

[Die18] Reinhard Diestel. *Graph theory*, volume 173 of *Graduate Texts in Mathematics*. Springer, 5th edition, 2018.

[DJM+20] Vida Dujmović, Gwenaël Joret, Piotr Micek, Pat Morin, Torsten Ueckerdt, and David R. Wood. Planar graphs have bounded queue-number. *J. ACM*, 67(4):22:1–22:38, 2020. Also in *60th IEEE Annual Symposium on Foundations of Computer Science*, (FOCS), pp. 862–875, 2019.

[DMW05] Vida Dujmović, Pat Morin, and David R. Wood. Layout of graphs with bounded tree-width. *SIAM J. Comput.*, 34(3):553–579, 2005.

[DMW23] Vida Dujmović, Pat Morin, and David R. Wood. Graph product structure for non-minor-closed classes. *J. Comb. Theory, Ser. B*, 162:34–67, 2023.

[DN19] Zdeněk Dvořık and Sergey Norin. Treewidth of graphs with balanced separations. *Journal of Combinatorial Theory, Series B*, 137:137–144, 2019.

[dPP88] Hubert de Fraysseix, János Pach, and Richard Pollack. Small sets supporting Fáry embeddings of planar graphs. In *Proc. 20th Annual ACM Symp. on Theory of Computing (STOC '88)*, pages 426–433. ACM, 1988.

[dPP90] Hubert de Fraysseix, János Pach, and Richard Pollack. How to draw a planar graph on a grid. *Combinatorica*, 10(1):41–51, 1990.

[DPW04] Vida Dujmović, Attila Pór, and David R. Wood. Track layouts of graphs. *Discrete Math. Theor. Comput. Sci.*, 6(2):497–522, 2004.

[Duj15] Vida Dujmović. Graph layouts via layered separators. *J. Comb. Theory, Ser. B*, 110:79–89, 2015.

[EJM23] Louis Esperet, Gwenaël Joret, and Pat Morin. Sparse universal graphs for planarity. *Journal of the London Mathematical Society*, 108(4):1333–1357, 2023.

[EKK+15] Katherine Edwards, Dong Yeap Kang, Jaehoon Kim, Sang-il Oum, and Paul D. Seymour. A relative of hadwiger's conjecture. *SIAM J. Discret. Math.*, 29(4):2385–2388, 2015.

[Erd51] Paul Erdős. Appendix, in KLAUS F. ROTH, On a problem of Heilbronn. *J. London Math. Soc.*, 26:198–204, 1951.

[FLW03] Stefan Felsner, Giuseppe Liotta, and Stephen K. Wismath. Straight-line drawings on restricted integer grids in two and three dimensions. *J. Graph Algorithms Appl.*, 7(4):363–398, 2003. Also in *9th International Symposium on Graph Drawing*, (GD 2001), LNCS 2265, pp. 328–342, 2002.

[Gav74] Fănică Gavril. The intersection graphs of subtrees in trees are exactly the chordal graphs. *Journal of Combinatorial Theory, Series B*, 16(1):47–56, 1974.

[Hal76] Rudolf Halin. S-functions for graphs. *J. Geometry*, 8(1-2):171–186, 1976.

[HLR92] Lenwood S. Heath, F. Thomson Leighton, and Arnold L. Rosenberg. Comparing queues and stacks as mechanisms for laying out graphs. *SIAM J. Discrete Math.*, 5(3):398–412, 1992.

[HMM25] Hussein Houdrouge, Babak Miraftab, and Pat Morin. Separation number and treewidth, revisited, 2025.

[HW17] Daniel J. Harvey and David R. Wood. Parameters tied to treewidth. *J. Graph Theory*, 84(4):364–385, 2017.

[HW24] Robert Hickingbotham and David R. Wood. Shallow minors, graph products, and beyond-planar graphs. *SIAM* J. Discret. Math., 38(1):1057–1089, 2024.

[JP22] Hugo Jacob and Marcin Pilipczuk. Bounding twin-width for bounded-treewidth graphs, planar graphs, and bipartite graphs. In Michael A. Bekos and Michael Kaufmann, editors, *Graph-Theoretic Concepts in Computer Science - 48th International Workshop, WG 2022, Tübingen, Germany, June 22-24, 2022, Revised Selected Papers*, volume 13453 of *Lecture Notes in Computer Science*, pages 287–299. Springer, 2022.

[KNR88] Sampath Kannan, Moni Naor, and Steven Rudich. Implicit representation of graphs. In *Proc. 20th Annual ACM Symp. on Theory of Computing (STOC 1988)*, pages 334–343, 1988.

[KNS15] Ilan Karpas, Ofer Neiman, and Shakhar Smorodinsky. On vertex rankings of graphs and its relatives. *Discret. Math.*, 338(8):1460–1467, 2015.

[MT01] Bojan Mohar and Carsten Thomassen. *Graphs on surfaces*. Johns Hopkins University Press, 2001.

[NO06] Jaroslav Nešetřil and Patrice Ossona de Mendez. Tree-depth, subgraph coloring and homomorphism bounds. *European Journal of Combinatorics*, 27(6):1022–1041, 2006.

[PS19] Michał Pilipczuk and Sebastian Siebertz. Polynomial bounds for centered colorings on proper minor-closed graph classes. In Timothy M. Chan, editor, *Proc. 30th Annual ACM-SIAM Symp. on Discrete Algorithms*, pages 1501–1520, 2019.

[PTT99] János Pach, Torsten Thiele, and Géza Tóth. Three-dimensional grid drawings of graphs. In Bernard Chazelle, Jacob E. Goodman, and Richard Pollack, editors, Advances in discrete and computational geometry, volume 223 of *Contemporary Math.*, pages 251–255. Amer. Math. Soc., 1999.

[Pup20] Sergey Pupyrev. Improved bounds for track numbers of planar graphs. *J. Graph Algorithms Appl.*, 24(3):323–341, 2020.

[RS86] Neil Robertson and Paul D. Seymour. Graph minors. II. algorithmic aspects of tree-width. *J. Algorithms*, 7(3):309–322, 1986.

[RVM95] S. Rengarajan and C. E. Veni Madhavan. Stack and queue number of 2-trees. In Ding-Zhu Du and Ming Li, editors, *Proc. 1st Annual Int'l Conf. on Computing and Combinatorics (COCOON '95)*, volume 959 of *Lecture Notes in Comput. Sci.*, pages 203–212. Springer, 1995.

[UWY22a] Torsten Ueckerdt, David R. Wood, and Wendy Yi. An improved planar graph product structure theorem. *Electron. J. Comb.*, 29(2):P2–51, 2022.

[UWY22b] Torsten Ueckerdt, David R. Wood, and Wendy Yi. An improved planar graph product structure theorem. *Electron. J. Comb.*, 29(2):P2–51, 2022.

[Wie17] Veit Wiechert. On the queue-number of graphs with bounded tree-width. *Electron. J. Combin.*, 24:P1.65, 2017.

[Woo02] David R. Wood. Queue layouts, tree-width, and three-dimensional graph drawing. In Manindra Agrawal and Anil Seth, editors, *Proc. 22nd Foundations of Software Technology and Theoretical Computer Science (FST TCS '02)*, volume 2556 of *Lecture Notes in Comput. Sci.*, pages 348–359. Springer, 2002.

[Woo05] David R. Wood. Queue layouts of graph products and powers. *Discrete Math. Theor. Comput. Sci.*, 7(1):255–268, 2005.

Chapter 2
Threshold for the Measure of Random Polytopes

Apostolos Giannopoulos

Starting with the work of Dyer, Füredi and McDiarmid who established a sharp threshold for the expected volume of random polytopes with independent vertices uniformly distributed in the discrete cube $E_2^n = \{-1, 1\}^n$, in this survey article we focus on a very general variant of the problem. Let μ be a log-concave probability measure on $\mathbb{R}^n$ and for any $N > n$ consider the random polytope $K_N = \mathrm{conv}\{X_1, \ldots, X_N\}$, where $X_1, X_2, \ldots$ are independent random points in $\mathbb{R}^n$ distributed according to μ. We discuss an approach to the question if there exists a threshold for the expected measure $\mathbb{E}_{\mu^N}[\mu(K_N)]$ of K_N, based on joint works with S. Brazitikos and M. Pafis, via the Cramér transform Λ_μ^* of μ. We show that, under some conditions, one has a sharp threshold for the expectation $\mathbb{E}_{\mu^N}[\mu(K_N)]$ of the measure of K_N: it is close to 0 if $\ln N \ll \mathbb{E}_\mu(\Lambda_\mu^*)$ and close to 1 if $\ln N \gg \mathbb{E}_\mu(\Lambda_\mu^*)$. The main condition is that the parameter $\beta(\mu) = \mathrm{Var}_\mu(\Lambda_\mu^*)/(\mathbb{E}_\mu(\Lambda_\mu^*))^2$ should be small.

2.1 The Case of the Discrete Cube

Let X be a random vector in $\mathbb{R}^n$ with independent coordinates that take each of the values ± 1 with probability $\frac{1}{2}$. Given $N > n$, we consider N independent copies $X_1, \ldots, X_N$ of the random vector X and define the random "0/1 polytope"

The author acknowledges support by the Hellenic Foundation for Research and Innovation (H.F.R.I.) under the call "Basic research Financing (Horizontal support of all Sciences)" under the National Recovery and Resilience Plan "Greece 2.0" funded by the European Union—Next Generation EU (Project Number: 15445).

A. Giannopoulos (✉)
School of Applied Mathematical and Physical Sciences, National Technical University of Athens, Department of Mathematics, Athens, Greece
e-mail: apgiannop@math.ntua.gr

J. Pach and G. Tóth (eds.), *Courses in Discrete and Computational Geometry*, Bolyai Society Mathematical Studies 31,
https://doi.org/10.1007/978-3-032-10503-5_2

$$K_N = \mathrm{conv}\{X_1, \ldots, X_N\} \subseteq Q_n := [-1, 1]^n. \tag{2.1}$$

Dyer, Füredi and McDiarmid established in [DFM92] a sharp threshold for the volume of these random 0/1 polytopes.

Theorem 2.1.1 (Dyer-Füredi-McDiarmid) *Let $\kappa = \ln 2 - \frac{1}{2}$ and for any $N > n$ consider the random polytope K_N defined by* (2.1). *For every $\varepsilon \in (0, \kappa)$ we have that*

$$\lim_{n\to\infty} \sup\left\{2^{-n}\mathbb{E}\,|K_N| : N \leqslant \exp((\kappa - \varepsilon)n)\right\} = 0 \tag{2.2}$$

and

$$\lim_{n\to\infty} \inf\left\{2^{-n}\mathbb{E}\,|K_N| : N \geqslant \exp((\kappa + \varepsilon)n)\right\} = 1. \tag{2.3}$$

In the statement of the theorem we denote by $|K_N|$ the n-dimensional volume of K_N; since $|Q_n| = 2^n$, the ratio $2^{-n}\mathbb{E}\,|K_N|$ is the "expected percentage" of Q_n occupied by K_N. In this section we present the main points of the proof of Theorem 2.1.1. Several of the lemmas that are used will be proved in a more general and stronger form later on.

The function $\varphi : (-1, 1)^n \to \mathbb{R}$ defined by

$$\varphi(x) := \inf\left\{\mathrm{Prob}\big(X \in H\big) : x \in H,\ H \text{ is a closed half-space}\right\} \tag{2.4}$$

plays a key role in the argument. Note that the infimum in (2.4) is determined by those half-spaces H for which x lies on the boundary $\partial(H)$ of H. Next, for any origin symmetric convex body $A \subset (-1, 1)^n$ we define

$$\varphi_+(A) = \sup_{x\notin A} \varphi(x) \quad \text{and} \quad \varphi_-(A) = \inf_{x\in A} \varphi(x). \tag{2.5}$$

Lemma 2.1.2 *Let $N > n$ and let A be an origin symmetric convex body contained in $(-1, 1)^n$. Then,*

$$\mathbb{E}(|K_N|) \leqslant |A| + 2^n N\varphi_+(A).$$

Proof We write

$$\mathbb{E}\,(|K_N|) = \mathbb{E}\,(|K_N \cap A|) + \mathbb{E}\,(|K_N \setminus A|) \leqslant |A| + \mathbb{E}\,(|K_N \setminus A|). \tag{2.6}$$

Note that if H is a closed half-space containing x, and if $x \in K_N$, then we may find $i \leqslant N$ such that $X_i \in H$ (otherwise, we would have $x \in K_N \subseteq H'$, where H' is the complementary half-space of H). It follows that $\mathrm{Prob}\left(x \in K_N\right) \leqslant N\,\varphi(x)$. Using Fubini's theorem we see that

$$\mathbb{E}\,(|K_N \setminus A|) = \int_{Q_n\setminus A} \mathrm{Prob}(x \in K_N)\,dx \leqslant \int_{Q_n\setminus A} N\varphi(x)\,dx \leqslant N\,\varphi_+(A)\,|Q_n \setminus A|.$$

where in the last inequality we use the fact that $\varphi(x) \leqslant \varphi_+(A)$ for every $x \notin A$. Going back to (2.6) we get the lemma. □

Lemma 2.1.2 will be useful for the proof of (2.2). It remains to choose, if possible, suitable A (depending on N and n) such that for all $N \leqslant \exp((\kappa - \varepsilon)n)$ we will have simultaneously $|A|/2^n \to 0$ and $N\varphi_+(A) \to 0$ as $n \to \infty$.

A second basic observation is given by the following lemma.

Lemma 2.1.3 *Let A be an origin symmetric convex body contained in $(-1, 1)^n$. Then,*

$$1 - \mathrm{Prob}\big(K_N \supseteq A\big) \leqslant \binom{N}{n} 2^{-(N-n)} + 2\binom{N}{n}\big(1 - \varphi_-(A)\big)^{N-n}.$$

We skip the proof since we shall discuss a more general version of Lemma 2.1.3 in Sect. 2.3 (see Lemma 2.3.8). What is important is that Lemma 2.1.3 allows us to use the function φ in order to prove (2.3). Roughly speaking, it remains to choose, if possible, suitable A (depending on N and n) so that for all $N \geqslant \exp((\kappa + \varepsilon)n)$ we will have simultaneously $|A|/2^n \to 1$ and $1 - \mathrm{Prob}\big(K_N \supseteq A\big) \to 0$ as $n \to \infty$.

Given a bounded random variable X, consider the moment generating function $M(s) := \mathbb{E}\left(e^{sX}\right)$ and the logarithmic moment generating function $\Lambda(s) := \ln M(s)$ of X. Since X is bounded, we see that $M(s) < \infty$ for every $s \in \mathbb{R}$. By the symmetry of X it also follows that M and Λ are even. Using Hölder's inequality we easily check that Λ, and hence also M, is convex. Finally, M is C^∞ on $\mathbb{R}$; the n-th derivative of M is the function $M^{(n)}(s) = \mathbb{E}\left(X^n e^{sX}\right)$.

Returning to our case, where X takes the values ± 1 with probability $\frac{1}{2}$, direct computation shows that

$$M(s) = \cosh(s) \quad \text{and} \quad \Lambda(s) = \ln\cosh(s).$$

Consider the Legendre transform of Λ: this is the function

$$f(x) := \sup\{sx - \Lambda(s) : s \in \mathbb{R}\}, \qquad x \in (-1, 1).$$

Direct computation shows that

$$f(x) = \tfrac{1}{2}(1 + x)\ln(1 + x) + \tfrac{1}{2}(1 - x)\ln(1 - x)$$

for every $x \in (-1, 1)$. Note that f is an even convex function and $\lim\limits_{x \to \pm 1} f(x) = \ln 2$.

From the definition of f and Markov's inequality we get the next upper bound for $\varphi(x)$ in terms of $\sum_{i=1}^n f(x_i)$, for every $x \in (-1, 1)^n$.

Lemma 2.1.4 *For every $x \in (-1, 1)^n$ we have that $\varphi(x) \leqslant \exp\left(-\sum_{i=1}^n f(x_i)\right)$.*

Proof Let H be a closed half-space such that $x \in \partial(H)$. Then, there exists $s = (s_1, \dots, s_n) \in \mathbb{R}^n$ such that $H = H(s) = \{y : \langle s, y - x\rangle \geqslant 0\}$. From Markov's inequality,

$$\mathrm{Prob}\big(X \in H(s)\big) = \mathrm{Prob}\left(\sum_{i=1}^n s_i(X_i - x_i) \geqslant 0\right) \leqslant \mathbb{E}\left(\exp\Big(\sum_{i=1}^n s_i(X_i - x_i)\Big)\right)$$
$$= \prod_{i=1}^n \mathbb{E}\left(\exp(s_i(X_i - x_i)\right) = \prod_{i=1}^n \exp(\Lambda(s_i) - s_i x_i).$$

By the definition of $\varphi(x)$ we have

$$\varphi(x) \leqslant \inf_{s\in\mathbb{R}^n} \prod_{i=1}^n \exp(\Lambda(s_i) - s_i x_i) = \prod_{i=1}^n \exp(-\sup\{s x_i - \Lambda(s) : s \in \mathbb{R}\})$$
$$= \prod_{i=1}^n \exp(-f(x_i)).$$

This proves the lemma. □

We extend f continuously on $[-1, 1]$ setting $f(\pm 1) = \ln 2$ and for every $x = (x_1, \dots, x_n) \in Q_n$ we set

$$F(x) = \frac{1}{n}\sum_{i=1}^n f(x_i).$$

For every $0 < t < \ln 2$, we define

$$F_t = \left\{x \in (-1, 1)^n : F(x) \leqslant t\right\}.$$

Since f is even and convex on $(-1, 1)$, the set F_t is an origin symmetric convex body contained in $(-1, 1)^n$. From the definition of F_t we see that $\sum_{i=1}^n f(x_i) = nF(x) = tn$ for all $x \in \partial(F_t)$. Therefore, Lemma 2.1.4 and the definition of φ_+ give us the next fact.

Lemma 2.1.5 *Let $0 < t < \ln 2$. For every $x \in (-1, 1)^n$ we have $\varphi(x) \leqslant \exp(-nF(x))$. In particular,*

$$\varphi_+(F_t) \leqslant \exp(-tn).$$

Let $U_1, \dots, U_n$ be independent random variables, uniformly distributed in $(-1, 1)$. Then, for every $0 < t < \ln 2$,

$$2^{-n}\,|F_t| = \mathrm{Prob}((U_1, \dots, U_n) \in F_t) = \mathrm{Prob}\left(\frac{1}{n}\sum_{i=1}^n f(U_i) \leqslant t\right).$$

Note that

$$\kappa = \mathbb{E}\,(f(U_i)) = \tfrac{1}{2}\int_{-1}^1 f(x)dx = \ln 2 - \tfrac{1}{2}.$$

By the law of large numbers we conclude the following.

Lemma 2.1.6 *For every* $t \in (0, \kappa)$ *we have* $\lim_{n\to\infty} 2^{-n} |F_t| = 0$ *and, similarly, for every* $t \in (\kappa, \ln 2)$ *we have* $\lim_{n\to\infty} 2^{-n} |F_t| = 1$.

Now, we can prove the first part of the theorem of Dyer, Füredi and McDiarmid.

Proposition 2.1.7 *For every* $\varepsilon \in (0, \kappa)$,

$$\lim_{n\to\infty} \sup\left\{2^{-n}E(|K_N|) \colon N \leqslant \exp((\kappa - \varepsilon)n)\right\} = 0.$$

Proof We choose $t = \kappa - \varepsilon/2$. From Lemma 2.1.6 we have that $\lim_{n\to\infty} 2^{-n} |F_t| = 0$. On the other hand, if $N \leqslant \exp((\kappa - \varepsilon)n)$, then Lemma 2.1.5 gives

$$N\varphi_+(F_t) \leqslant \exp(-\varepsilon n/2).$$

Applying Lemma 2.1.2 with $A = F_t$ we get

$$2^{-n}\mathbb{E}\left(|K_N|\right) \leqslant 2^{-n} |F_t| + \exp(-\varepsilon n/2),$$

and Lemma 2.1.6 shows that the right hand side tends to 0 as $n \to \infty$. □

For the proof of (2.3) we need to estimate $\varphi(x)$ from below in order to use Lemma 2.1.3. The basic technical step is the next proposition, which will be discussed, in a more general context, in Sect. 2.3.

Proposition 2.1.8 *For every* $\varepsilon > 0$, *there exists* $n(\varepsilon) \in \mathbb{N}$, *depending only on* ε, *such that for every* $0 < t < \ln 2$ *and every* $n \geqslant n(\varepsilon)$ *we have*

$$\varphi_-(F_t) \geqslant \exp(-t(1+\varepsilon)n).$$

Then, the proof of (2.3) is simple.

Proposition 2.1.9 *For every* $\varepsilon > 0$,

$$\lim_{n\to\infty} \inf\left\{2^{-n}\mathbb{E}\left(|K_N|\right) \colon N \geqslant \exp((\kappa + \varepsilon)n)\right\} = 1.$$

Proof Fix $\varepsilon > 0$ and choose $t = \kappa + \varepsilon/3$. Combining Lemma 2.1.3 with Proposition 2.1.8 we see that if $n \geqslant n(\varepsilon)$ and $N \geqslant \exp((\kappa + \varepsilon)n) \geqslant \exp((t + 2\varepsilon/3)n)$, then

$$\mathbb{E}\left(|K_N|\right) \geqslant |F_t| \;\mathrm{Prob}\big(K_N \supseteq F_t\big) \geqslant |F_t| \left(1 - 2^{-n+1}\right).$$

Since $t > \kappa$, Lemma 2.1.6 shows that $\lim_{n\to\infty} 2^{-n} |F_t| = 1$, and the result follows. □

Special cases of the threshold problem have been studied in various works. In addition to the case of the discrete cube, Dyer, Füredi and McDiarmid established in [DFM92] a sharp threshold for the expected volume of random polytopes with vertices uniformly distributed in the solid cube $Q_n = [-1, 1]^n$. If $\kappa = \ln(2\pi) - \gamma - \frac{1}{2}$, where γ is the Euler constant, then for every $\varepsilon \in (0, \kappa)$ one has

$$\lim_{n\to\infty} \sup \left\{2^{-n}\mathbb{E}|K_N| : N \leqslant \exp((\kappa - \varepsilon)n)\right\} = 0$$

and

$$\lim_{n\to\infty} \inf \left\{2^{-n}\mathbb{E}|K_N| : N \geqslant \exp((\kappa + \varepsilon)n)\right\} = 1.$$

These results were generalized in [GG09] to the setting where the vertices of K_N have independent coordinates whose distribution is a fixed even measure with compact support in $\mathbb{R}$ that satisfies some mild condition (see Sect. 2.8).

The articles [Piv07, BCG+19, BKT21] address the same question for a number of cases where X_i have a rotationally invariant density supported on the Euclidean unit ball B_2^n. More precisely, Pivovarov proved in [Piv07] that if the vertices of K_N are uniformly distributed on the unit ball B_2^n then, for any $\varepsilon \in (0, 1)$, if $N \leqslant \exp\left((1-\varepsilon)\frac{n}{2}\ln n\right)$ then $\mathbb{E}\,|K_N|/|B_2^n|$ tends to 0 and if $N \geqslant \exp\left((1+\varepsilon)\frac{n}{2}\ln n\right)$ then $\mathbb{E}\,|K_N|/|B_2^n|$ tends to 1 as $n \to \infty$. In the same work, he studied the case where the vertices of K_N are distributed according to the uniform measure on the unit sphere S^{n-1} or according to the standard Gaussian measure γ_n on $\mathbb{R}^n$; in the latter case, for a large range of $r = r_n > 0$ he established a sharp threshold for the ratio $\mathbb{E}\,|K \cap rB_2^n|/|rB_2^n|$ as $n \to \infty$. The work [BCG+19] of Bonnet, Chasapis, Grote, Temesvari and Turchi deals with the case where the vertices of K_N are distributed according to the measure with density $(1 - |x|^2)^\beta$ or $(1 - |x|^2/\sigma^2)^{-\beta}$ on B_2^n, where $\beta > -1$ in the first case and $\beta > \frac{n}{2}$ and $\sigma > 0$ in the second case. Sharper estimates for these models, describing the phase transition as well as its shape, were obtained in [BKT21].

Exponential in the dimension upper and lower thresholds are obtained in [FPT20] for the case where X_i are uniformly distributed in a simplex (then, the result can be extended to simplicial polytopes in $\mathbb{R}^n$). Let $\Omega_n = \{x = (x_1, \ldots, x_n) : x_1 + \cdots + x_n = 1,\ x_i \geqslant 0\}$ be the standard embedding of the $(n-1)$-dimensional simplex in n-dimensional space. If $N > C_0^n$, where $C_0 > 0$ is an absolute constant, then

$$\mathbb{E}\,|\mathrm{conv}\{x_1, \ldots, x_N\}| \geqslant (1 - e^{-c_0\sqrt{n}})|\Omega_n|.$$

A second main result of the same paper provides an upper threshold. For every $\varepsilon > 0$, if $N < e^{(\gamma-\varepsilon)n}$, where γ is the Euler constant, then the convex hull of N random points $x_1, \ldots, x_N$ uniformly distributed in Ω_n satisfies $\mathbb{E}\,|\mathrm{conv}\{x_1, \ldots, x_N\}|/|\Omega_n| \to 0$ as $n \to \infty$. To this end, the authors compute the Legendre transform of the log-moment generating function of a random vector X uniformly distributed in the simplex.

2.2 Log-Concave Probability Measures

We would like to formulate and study the question how to obtain a threshold for the expected measure of a random polytope defined as the convex hull of independent random points with a log-concave distribution. Consider a log-concave probability measure μ on $\mathbb{R}^n$ and let $X_1, X_2, \dots$ be a sequence of independent random points in $\mathbb{R}^n$ distributed according to μ. Then, for any $N > n$ we may define the random polytope

$$K_N = \operatorname{conv}\{X_1, \dots, X_N\}.$$

We are interested in the expectation $\mathbb{E}_{\mu^N}[\mu(K_N)]$ of the μ-measure of K_N with respect to the product measure $\mu^N = \mu \otimes \cdots \otimes \mu$ (N times). This is an affinely invariant quantity, and hence we may assume that μ is centered, i.e. the barycenter of μ is at the origin.

Given $\delta \in (0, 1)$ we say that μ satisfies a "δ-upper threshold" with constant ϱ_1 if

$$\sup\{\mathbb{E}_{\mu^N}[\mu(K_N)] : N \leqslant \exp(\varrho_1 n)\} \leqslant \delta \tag{2.7}$$

and that μ satisfies a "δ-lower threshold" with constant ϱ_2 if

$$\inf\{\mathbb{E}_{\mu^N}[\mu(K_N)] : N \geqslant \exp(\varrho_2 n)\} \geqslant 1 - \delta. \tag{2.8}$$

Then, we define $\varrho_1(\mu, \delta) = \sup\{\varrho_1 : \text{(2.7) holds true}\}$ and $\varrho_2(\mu, \delta) = \inf\{\varrho_2 : \text{(2.8) holds true}\}$. Our main goal is to obtain upper bounds for the difference

$$\varrho(\mu, \delta) := \varrho_2(\mu, \delta) - \varrho_1(\mu, \delta)$$

for any fixed $\delta \in \left(0, \frac{1}{2}\right)$.

One may also consider a sequence $\{\mu_n\}_{n=1}^{\infty}$ of log-concave probability measures, where μ_n is on $\mathbb{R}^n$, and say that $\{\mu_n\}_{n=1}^{\infty}$ exhibits a "sharp threshold" if there exists a sequence $\{\delta_n\}_{n=1}^{\infty}$ of positive reals such that $\delta_n \to 0$ and $\varrho(\mu_n, \delta_n) \to 0$ as $n \to \infty$.

We shall describe a general approach to the problem, that was proposed in [BGP24b], working with an arbitrary log-concave probability measure μ on $\mathbb{R}^n$. We present the main ideas, the progress that has been achieved (especially in the case of the uniform measure on a convex body) and several remaining open questions. In the remaining part of this section we provide the necessary background information.

Throughout this article we write $\langle \cdot, \cdot \rangle$ for the standard inner product in $\mathbb{R}^n$ and denote the Euclidean norm by $|\cdot|$, the Euclidean unit ball by B_2^n and the unit sphere by S^{n-1}. Lebesgue measure in $\mathbb{R}^n$ is denoted by $|\cdot|$ and σ is the rotationally invariant probability measure on S^{n-1}. We use the letters c, c', c_j, c'_j etc. to denote absolute positive constants whose value may change from line to line.

A convex body in $\mathbb{R}^n$ is a compact convex set $K \subset \mathbb{R}^n$ with non-empty interior. We often consider bounded convex sets K in $\mathbb{R}^n$ with $0 \in \operatorname{int}(K)$; since the closure of such a set is a convex body, we shall call these sets convex bodies too. We say that K is centrally symmetric if $-K = K$ and that K is centered if the barycenter $\mathrm{bar}(K) = \frac{1}{|K|}\int_K x\,dx$ of K is at the origin. The radial function ϱ_K of a convex body K with $0 \in \mathrm{int}(K)$ is defined for all $x \neq 0$ by $\varrho_K(x) = \sup\{\lambda > 0 : \lambda x \in K\}$ and the support function of K is given by $h_K(x) = \sup\{\langle x, y\rangle : y \in K\}$ for all $x \in \mathbb{R}^n$. The polar body K° of a convex body K in $\mathbb{R}^n$ with $0 \in \mathrm{int}(K)$ is the convex body

$$K^\circ := \big\{y \in \mathbb{R}^n : \langle x, y\rangle \leqslant 1 \text{ for all } x \in K\big\}.$$

A Borel measure μ on $\mathbb{R}^n$ is called log-concave if $\mu(H) < 1$ for every hyperplane H in $\mathbb{R}^n$ and $\mu(\lambda A + (1-\lambda)B) \geqslant \mu(A)^\lambda \mu(B)^{1-\lambda}$ for any compact subsets A, B of $\mathbb{R}^n$ and any $\lambda \in (0,1)$. A theorem of Borell [Bor74] shows that under these assumptions, μ has a log-concave density f_μ. A function $f : \mathbb{R}^n \to [0,\infty)$ is called log-concave if its support $\{f > 0\}$ is a convex set in $\mathbb{R}^n$ and the restriction of $\ln f$ to it is concave. If f has finite positive integral then there exist constants $A, B > 0$ such that $f(x) \leqslant Ae^{-B|x|}$ for all $x \in \mathbb{R}^n$ (see [BGVV14, Lemma 2.2.1]). In particular, f has finite moments of all orders. We say that μ is even if $\mu(-B) = \mu(B)$ for every Borel subset B of $\mathbb{R}^n$ and that μ is centered if

$$\int_{\mathbb{R}^n} \langle x, \xi\rangle d\mu(x) = \int_{\mathbb{R}^n} \langle x, \xi\rangle f_\mu(x)dx = 0$$

for all $\xi \in S^{n-1}$. We shall use the fact that if μ is a centered log-concave probability measure on $\mathbb{R}^k$ then

$$\|f_\mu\|_\infty \leqslant e^k f_\mu(0). \tag{2.9}$$

This is a result of Fradelizi from [Fra97]. Note that if K is a convex body in $\mathbb{R}^n$ then the Brunn-Minkowski inequality implies that the indicator function $\mathbb{1}_K$ of K is the density of a log-concave measure, the Lebesgue measure on K.

If μ is a log-concave measure on $\mathbb{R}^n$ with density f_μ, we define the isotropic constant of μ by

$$L_\mu := \left(\frac{\sup_{x\in\mathbb{R}^n} f_\mu(x)}{\int_{\mathbb{R}^n} f_\mu(x)dx}\right)^{\frac{1}{n}} [\det \mathrm{Cov}(\mu)]^{\frac{1}{2n}},$$

where $\mathrm{Cov}(\mu)$ is the covariance matrix of μ with entries

$$\mathrm{Cov}(\mu)_{ij} := \frac{\int_{\mathbb{R}^n} x_i x_j f_\mu(x)\,dx}{\int_{\mathbb{R}^n} f_\mu(x)\,dx} - \frac{\int_{\mathbb{R}^n} x_i f_\mu(x)\,dx}{\int_{\mathbb{R}^n} f_\mu(x)\,dx}\,\frac{\int_{\mathbb{R}^n} x_j f_\mu(x)\,dx}{\int_{\mathbb{R}^n} f_\mu(x)\,dx}.$$

We say that a log-concave probability measure μ on $\mathbb{R}^n$ is isotropic if it is centered and $\mathrm{Cov}(\mu) = I_n$, where I_n is the identity $n \times n$ matrix. In this case, $L_\mu = \|f_\mu\|_\infty^{1/n}$. For every μ there exists an invertible affine transformation T such that the push forward $T_*\mu$ of μ defined by $T_*\mu(B) = \mu(T^{-1}(B))$ for every Borel subset B of

$\mathbb{R}^n$ is isotropic. The hyperplane conjecture asks if there exists an absolute constant $C > 0$ such that

$$L_n := \max\{L_\mu : \mu \text{ is an isotropic log-concave probability measure on } \mathbb{R}^n\} \leqslant C$$

for all $n \geqslant 1$. Bourgain [Bou91] established the upper bound $L_n \leqslant c\sqrt[4]{n}\ln n$; later, Klartag, in [Kla06], improved this estimate to $L_n \leqslant c\sqrt[4]{n}$. In a breakthrough work, Chen [Che21] proved that for any $\varepsilon > 0$ there exists $n_0(\varepsilon) \in \mathbb{N}$ such that $L_n \leqslant n^\varepsilon$ for every $n \geqslant n_0(\varepsilon)$. Subsequently, Klartag and Lehec [KL22] showed that $L_n \leqslant c(\ln n)^4$, and very recently Klartag [Kla23] achieved the best known bound $L_n \leqslant c\sqrt{\ln n}$.

Let μ be a centered log-concave probability measure on $\mathbb{R}^n$. The logarithmic Laplace transform of μ is the function

$$\Lambda_\mu(\xi) = \ln\left(\int_{\mathbb{R}^n} e^{\langle \xi, z\rangle} f_\mu(z)dz\right).$$

It is easy to check that Λ_μ is convex and $\Lambda_\mu(0) = 0$. Since $\mathrm{bar}(\mu) = 0$, Jensen's inequality shows that $\Lambda_\mu(\xi) \geqslant 0$ for all ξ. One can also check that the set $A(\mu) = \{\Lambda_\mu < \infty\}$ is open and Λ_μ is C^∞ and strictly convex on $A(\mu)$. The Legendre transform of Λ_μ defined by

$$\Lambda_\mu^*(x) = \sup_{\xi\in\mathbb{R}^n}\left\{\langle x, \xi\rangle - \Lambda_\mu(\xi)\right\}$$

is called the Cramér transform of μ and plays a crucial role in the theory of large deviations (see [DZ98]). For every $t > 0$ we also define the convex set

$$B_t(\mu) := \{x \in \mathbb{R}^n : \Lambda_\mu^*(x) \leqslant t\}.$$

A second important family of convex bodies associated to any log-concave probability measure μ on $\mathbb{R}^n$ is the family of L_t-centroid bodies $Z_t(\mu)$. For every $t \geqslant 1$ the body $Z_t(\mu)$ is the convex body with support function

$$h_{Z_t(\mu)}(y) := \left(\int_{\mathbb{R}^n} |\langle x, y\rangle|^t \, d\mu(x)\right)^{1/t}.$$

Note that $Z_t(\mu)$ is always centrally symmetric, and $Z_t(T_*\mu) = T(Z_t(\mu))$ for every $T \in GL(n)$ and $t \geqslant 1$. A centered log-concave probability measure μ is isotropic if and only if $Z_2(\mu) = B_2^n$. Paouris (see [BGVV14, Theorem 5.1.17]) established upper bounds for the volume of the L_t-centroid bodies of isotropic log-concave probability measures.

Theorem 2.2.1 *If μ is a centered log-concave probability measure on $\mathbb{R}^n$, then for every $2 \leqslant t \leqslant n$ we have that*

$$|Z_t(\mu)|^{1/n} \leqslant c\sqrt{t/n}[\det \mathrm{Cov}(\mu)]^{\frac{1}{2n}},$$

where $c > 0$ is an absolute constant. In particular, if μ is isotropic then $|Z_t(\mu)|^{1/n} \leqslant c\sqrt{t/n}$ for all $2 \leqslant t \leqslant n$.

A variant of the L_t-centroid bodies of μ is defined as follows. For every $t \geqslant 1$ we consider the convex body $Z_t^+(\mu)$ with support function

$$h_{Z_t^+(\mu)}(y) = \left(2\int_{\mathbb{R}^n} \langle x, y\rangle_+^t f_\mu(x)dx\right)^{1/t},$$

where $a_+ = \max\{a, 0\}$. When f_μ is even, it is clear that $Z_t^+(\mu) = Z_t(\mu)$. In any case, we easily verify that $Z_t^+(\mu) \subseteq 2^{1/t} Z_t(\mu)$. Moreover, if μ is isotropic then $Z_2^+(\mu) \supseteq cB_2^n$ for an absolute constant $c > 0$. One can also check that if $1 \leqslant t < s$ then

$$\left(\frac{2}{e}\right)^{\frac{1}{t}-\frac{1}{s}} Z_t^+(\mu) \subseteq Z_s^+(\mu) \subseteq c_1 \left(\frac{2e-2}{e}\right)^{\frac{1}{t}-\frac{1}{s}} \frac{s}{t} Z_t^+(\mu).$$

The right-hand side inequality gives

$$\mathbb{E}_\mu(2\langle z, \xi\rangle_+^{2t}) = [h_{Z_{2t}^+(\mu)}(\xi)]^{2t} \leqslant C^{2t}[h_{Z_t^+(\mu)}(\xi)]^{2t} = C^{2t}[\mathbb{E}_\mu(2\langle z, \xi\rangle_+^t)]^2, \quad (2.10)$$

for all $\xi \in S^{n-1}$, where $C > 1$ is an absolute constant. For a proof of all these claims see [GM11].

The next proposition compares $B_t(\mu)$ with $Z_t(\mu)$ (for a proof see [BGVV14, Proposition 15.3.7]).

Proposition 2.2.2 *Let μ be a centered log-concave probability measure on $\mathbb{R}^n$. Then, for any $t \geqslant 2$ we have*

$$B_t(\mu) \subseteq cZ_t(\mu)$$

where $c > 0$ is an absolute constant.

Finally, if μ is a log-concave probability measure on $\mathbb{R}^n$ such that $f_\mu(0) > 0$ then, for every $t > 0$, we define

$$K_t(\mu) := K_t(f_\mu) = \left\{x \in \mathbb{R}^n : \int_0^\infty r^{t-1} f_\mu(rx)\, dr \geqslant \frac{f_\mu(0)}{t}\right\}.$$

From the definition it follows that the radial function of $K_t(\mu)$ is given by

$$\varrho_{K_t(\mu)}(x) = \left(\frac{1}{f_\mu(0)}\int_0^\infty t r^{t-1} f_\mu(rx)\, dr\right)^{1/t} \quad (2.11)$$

for $x \neq 0$. The bodies $K_t(\mu)$ were introduced by Ball [Bal88] who also established their convexity. If μ is additionally assumed centered then, for every $0 < t \leqslant s$,

$$\frac{\Gamma(t+1)^{\frac{1}{t}}}{\Gamma(s+1)^{\frac{1}{s}}} K_s(\mu) \subseteq K_t(\mu) \subseteq e^{\frac{n}{t}-\frac{n}{s}} K_s(\mu). \tag{2.12}$$

A proof is given in [BGVV14, Proposition 2.5.7]. It is easily checked that

$$|K_n(f)|\, f_\mu(0) = \int_{\mathbb{R}^n} f_\mu(x)dx = 1 \tag{2.13}$$

(see e.g. [BGVV14, Lemma 2.5.6]) and then we can use the inclusions (2.12) in order to estimate the volume of $K_t(\mu)$. For every $t > 0$ we have

$$e^{-1} \leqslant f_\mu(0)^{\frac{1}{n}+\frac{1}{t}} |K_{n+t}(\mu)|^{\frac{1}{n}+\frac{1}{t}} \leqslant e\frac{n+t}{n}. \tag{2.14}$$

We are mainly interested in the convex body $K_{n+1}(\mu)$. We shall use the fact that $K_{n+1}(\mu)$ is centered (see [BGVV14, Proposition 2.5.3 (v)]) and that

$$f_\mu(0)|K_{n+1}(\mu)| \approx 1. \tag{2.15}$$

The last estimate follows immediately from (2.13) and (2.14).

We conclude this section with some basic facts about κ-concave measures. Given $\kappa \in [-\infty, 1/n]$ we say that a measure μ on $\mathbb{R}^n$ is κ-concave if

$$\mu((1-\lambda)A + \lambda B) \geqslant ((1-\lambda)\mu^\kappa(A) + \lambda\mu^\kappa(B))^{1/\kappa} \tag{2.16}$$

for all compact subsets A, B of $\mathbb{R}^n$ with $\mu(A)\mu(B) > 0$ and all $\lambda \in (0, 1)$. The limiting cases are defined appropriately. For $\kappa = 0$ the right hand side in (2.16) becomes $\mu(A)^{1-\lambda}\mu(B)^\lambda$ (therefore, 0-concave measures are the log-concave measures). In the case $\kappa = -\infty$ the right hand side in (2.16) becomes $\min\{\mu(A), \mu(B)\}$. Note that if μ is κ-concave and $\kappa_1 \leqslant \kappa$ then μ is κ_1-concave.

Next, let $\gamma \in [-\infty, \infty]$. A function $f : \mathbb{R}^n \to [0, \infty)$ is called γ-concave if

$$f((1-\lambda)x + \lambda y) \geqslant ((1-\lambda)f^\gamma(x) + \lambda f^\gamma(y))^{1/\gamma}$$

for all $x, y \in \mathbb{R}^n$ with $f(x)f(y) > 0$ and all $\lambda \in (0, 1)$. Again, we define the cases $\gamma = 0, +\infty$ appropriately. Borell [Bor75] studied the relation between κ-concave probability measures and γ-concave functions and showed that if μ is a measure on $\mathbb{R}^n$ and the affine subspace F spanned by the support $\mathrm{supp}(\mu)$ of μ has dimension $\dim(F) = n$ then for every $-\infty \leqslant \kappa < 1/n$ we have that μ is κ-concave if and only if it has a non-negative density $\psi \in L^1_{\mathrm{loc}}(\mathbb{R}^n, dx)$ and ψ is γ-concave, where $\gamma = \frac{\kappa}{1-\kappa n} \in [-1/n, +\infty)$.

We refer to Schneider's book [Sch14] for basic facts from the Brunn-Minkowski theory and to the book [AAGM15] for basic facts from asymptotic convex geometry. We also refer to [BGVV14] for more information on isotropic convex bodies and log-concave probability measures.

2.3 Tukey's Half-Space Depth

Let μ be a probability measure on $\mathbb{R}^n$. For any $x \in \mathbb{R}^n$ we denote by $\mathcal{H}(x)$ the set of all half-spaces H of $\mathbb{R}^n$ containing x. The function

$$\varphi_\mu(x) = \inf\{\mu(H) : H \in \mathcal{H}(x)\}$$

is called Tukey's half-space depth. Tukey introduced the half-space depth for data sets in [Tuk75] as a measure of centrality for multivariate data that enables efficient visualization of random samples (some form of this notion had appeared in [H55]). The term "depth" also comes from Tukey's article. The survey article of Nagy, Schütt and Werner [NSW19] provides an overview of this topic, with an emphasis on its connections with convex geometry, and many references.

Tukey's half-space depth plays a key role in the study of the problem that we address in these notes. In this section we prove the basic results that we need, starting with estimates for the expectation

$$\mathbb{E}_\mu(\varphi_\mu) := \int_{\mathbb{R}^n} \varphi_\mu(x)\, d\mu(x)$$

of φ_μ with respect to μ. The question to provide an upper bound for this quantity was asked in [Mat] in connection with stochastic separability and applications to machine learning and error-correction mechanisms in artificial intelligence systems; motivation is given in [GGT23] and in the references therein. More precisely, it was asked if there exists an absolute constant $c \in (0, 1)$ such that $\mathbb{E}_\mu(\varphi_\mu) \leqslant c^n$ for all $n \geqslant 1$ and all log-concave probability measures μ on $\mathbb{R}^n$. The next theorem from [BGP24a] provides an affirmative answer (up to a $\ln n$-term).

Theorem 2.3.1 *Let μ be a log-concave probability measure on $\mathbb{R}^n$, $n \geqslant n_0$. Then, $\mathbb{E}_\mu(\varphi_\mu) \leqslant \exp\left(-cn/L_\mu^2\right)$ where L_μ is the isotropic constant of μ and $c > 0$, $n_0 \in \mathbb{N}$ are absolute constants.*

We shall use the next basic (and simple) lemma that generalizes Lemma 2.1.5.

Lemma 2.3.2 *Let μ be a Borel probability measure on $\mathbb{R}^n$. For every $x \in \mathbb{R}^n$ we have $\varphi_\mu(x) \leqslant \exp(-\Lambda_\mu^*(x))$. In particular, for any $t > 0$ and for all $x \notin B_t(\mu)$ we have that $\varphi_\mu(x) \leqslant \exp(-t)$.*

Proof Let $x \in \mathbb{R}^n$. For any $\xi \in \mathbb{R}^n$ the half-space $\{z : \langle z - x, \xi \rangle \geqslant 0\}$ is in $\mathcal{H}(x)$, therefore

$$\varphi_\mu(x) \leqslant \mu(\{z : \langle z, \xi \rangle \geqslant \langle x, \xi \rangle\}) \leqslant e^{-\langle x,\xi\rangle}\mathbb{E}_\mu\big(e^{\langle z,\xi\rangle}\big) = \exp\big(-[\langle x, \xi\rangle - \Lambda_\mu(\xi)]\big),$$

and taking the infimum over all $\xi \in \mathbb{R}^n$ we see that $\varphi_\mu(x) \leqslant \exp(-\Lambda_\mu^*(x))$, as claimed. □

Proof of Theorem 2.3.1 The quantity $\mathbb{E}_\mu(\varphi_\mu)$ is affinely invariant and hence we may assume that μ is isotropic. Using Lemma 2.3.2 we write

$$\begin{aligned}\int_{\mathbb{R}^n} \varphi_\mu(x)\, d\mu(x) &\leqslant \int_{\mathbb{R}^n} e^{-\Lambda_\mu^*(x)} f_\mu(x)\, dx = \int_{\mathbb{R}^n} \left(\int_{\Lambda_\mu^*(x)}^{\infty} e^{-t} dt \right) f_\mu(x) dx \\ &= \int_0^\infty e^{-t} \int_{\mathbb{R}^n} \mathbb{1}_{B_t(\mu)}(x) f_\mu(x) dx\, dt = \int_0^\infty e^{-t} \mu(B_t(\mu))\, dt.\end{aligned}$$

Fix $b \in (2/n, 1/2]$ that will be specified in the end. Since $\mu(B_t(\mu)) \leqslant 1$ and also $\mu(B_t(\mu)) \leqslant \|f_\mu\|_\infty |B_t(\mu)|$ for all $t > 0$, we may write

$$\begin{aligned}\int_{\mathbb{R}^n} \varphi_\mu(x)\, d\mu(x) &\leqslant \int_{bn}^\infty e^{-t} \mu(B_t(\mu)) dt + \|f_\mu\|_\infty \int_0^{bn} e^{-t} |B_t(\mu)|\, dt \\ &\leqslant \int_{bn}^\infty e^{-t}\, dt + L_\mu^n \int_0^2 e^{-t} |B_t(\mu)|\, dt + L_\mu^n \int_2^{bn} e^{-t} |B_t(\mu)|\, dt \\ &\leqslant e^{-bn} + L_\mu^n |B_2(\mu)| + L_\mu^n \int_2^{bn} e^{-t} |B_t(\mu)|\, dt.\end{aligned}$$

Applying Proposition 2.2.2 and Theorem 2.2.1 we get

$$|B_t(\mu)|^{1/n} \leqslant c_1 |Z_t(\mu)|^{1/n} \leqslant c_2 \sqrt{t/n}$$

for all $2 \leqslant t \leqslant n$, where $c_1, c_2 > 0$ are absolute constants. It is also known that $L_\mu \geqslant c_3$ where $c_3 > 0$ is an absolute constant (see [BGVV14, Proposition 2.3.12] for a proof). So, we may assume that $c_2 L_\mu \geqslant \sqrt{2}$. Choosing $b_0 := 1/(c_2 L_\mu)^2 \leqslant 1/2$ we write

$$L_\mu^n \int_2^{b_0 n} e^{-t} |B_t(\mu)| dt \leqslant c_2^n L_\mu^n \int_2^{b_0 n} (t/n)^{n/2} e^{-t} dt = (c_2 L_\mu)^n \int_2^{b_0 n} (t/n)^{n/2} e^{-t} dt,$$

and since $b_0 n \leqslant n/2$ and the function $t \mapsto t^{n/2} e^{-t}$ is increasing on $[0, n/2]$, we get

$$(c_2 L_\mu)^n \int_2^{b_0 n} e^{-t} |B_t(\mu)|\, dt \leqslant (b_0 n - 2)\, (c_2 L_\mu)^n b_0^{n/2} e^{-b_0 n} = (b_0 n - 2) e^{-b_0 n}.$$

Moreover, $|B_2(\mu)|^{1/n} \leqslant c_2\sqrt{2/n}$, therefore

$$L_\mu^n |B_2(\mu)| \leqslant (c_3 L_\mu^2/n)^{n/2} \leqslant e^{-b_0 n},$$

because $c_3 L_\mu^2/n \leqslant e^{-2}$ if $n \geqslant n_0$. Combining the above we see that

$$\int_{\mathbb{R}^n} \varphi_\mu(x)\, d\mu(x) \leqslant e^{-b_0 n} + e^{-b_0 n} + (b_0 n - 2)e^{-b_0 n} = b_0 n e^{-b_0 n},$$

and hence

$$\int_{\mathbb{R}^n} \varphi_\mu(x)\, d\mu(x) \leqslant n \exp\left(-n/(c_2 L_\mu)^2\right)$$

which implies the result. □

The next theorem shows that, modulo the isotropic constant L_μ, the exponential estimate of Theorem 2.3.1 is sharp.

Theorem 2.3.3 *Let μ be a log-concave probability measure on $\mathbb{R}^n$. Then,*

$$\int_{\mathbb{R}^n} \varphi_\mu(x) d\mu(x) \geqslant e^{-cn},$$

where $c > 0$ is an absolute constant.

The proof is based on a number of observations. First, by the affine invariance of $\mathbb{E}_\mu(\varphi_\mu)$, we may assume that μ is centered. As an application of the Paley-Zygmund inequality we obtain the next lemma.

Lemma 2.3.4 *Let $t \geqslant 1$ and $\delta \in (0,1)$. For every $x \in \delta Z_t^+(\mu)$ we have that*

$$\varphi_\mu(x) \geqslant (1-\delta^t)^2/C_1^t,$$

where $C_1 > 1$ is an absolute constant.

Proof Let $x \in \delta Z_t^+(\mu)$. It is enough to show that

$$\inf \mu(\{z \in \mathbb{R}^n : \langle z, \xi\rangle \geqslant \langle x, \xi\rangle\}) \geqslant (1-\delta^t)^2/C_1^t \tag{2.17}$$

where the infimum is over all $\xi \in S^{n-1}$ with $\langle x, \xi\rangle \geqslant 0$, because if $\langle x, \xi\rangle < 0$ then Grünbaum's lemma (see [BGVV14, Lemma 2.2.6]) implies that $\mu_\xi(\{z : \langle z - x, \xi\rangle \geqslant 0\}) \geqslant 1/e$.

Since $x \in \delta Z_t^+(\mu)$, we have $\langle x, \xi\rangle \leqslant \delta h_{Z_t^+(\mu)}(\xi)$ for any such $\xi \in S^{n-1}$, so it is enough to show that

$$\mu(\{z \in \mathbb{R}^n : \langle z, \xi\rangle \geqslant \delta h_{Z_t^+(\mu)}(\xi)\}) \geqslant (1-\delta^t)^2/C_1^t. \tag{2.18}$$

We apply the Paley-Zygmund inequality

$$\mu(\{z : g(z) \geqslant \delta^t \mathbb{E}_\mu(g)\}) \geqslant (1-\delta^t)^2 \frac{[\mathbb{E}_\mu(g)]^2}{\mathbb{E}_\mu(g^2)}$$

for the function $g(z) = \langle z, \xi\rangle_+^t$. From (2.10) we see that

$$\mathbb{E}_\mu(g^2) \leqslant C_1^t \, [\mathbb{E}_\mu(g)]^2$$

for some absolute constant $C_1 > 0$, and the lemma follows. □

For every $t \geqslant 1$ we consider the convex set

$$R_t(\mu) = \{x \in \mathbb{R}^n : f_\mu(x) \geqslant e^{-t} f_\mu(0)\}.$$

Since f_μ is log-concave, we easily check that $R_t(\mu)$ is convex. Note also that $R_t(\mu)$ is bounded and $0 \in \mathrm{int}(R_t(\mu))$.

Lemma 2.3.5 *For every $t \geqslant 5n$ we have $R_t(\mu) \supseteq c_0 K_{n+1}(\mu)$, where $c_0 > 0$ is an absolute constant.*

Proof Let $t \geqslant 5n$. For any $\xi \in S^{n-1}$ consider the log-concave function $h : [0,\infty) \to [0,\infty)$ with $h(t) = f_\mu(t\xi)$. Klartag has proved in [Kla07, Lemma 5.2] that

$$\int_0^{\varrho_{R_t(\mu)}(\xi)} r^{n-1} h(r)dr \geqslant (1-e^{-t/8}) \int_0^\infty r^{n-1} h(r)dr.$$

The definition of $K_n(\mu)$ gives

$$\int_0^\infty r^{n-1} h(r)dr = \frac{f_\mu(0)}{n} [\varrho_{K_n(\mu)}(\xi)]^n$$

and

$$\int_0^{\varrho_{R_t(\mu)}(\xi)} r^{n-1} h(r)dr \leqslant \|f\|_\infty \int_0^{\varrho_{R_t(\mu)}(\xi)} r^{n-1} dr = \frac{\|f\|_\infty}{n} [\varrho_{R_t(\mu)}(\xi)]^n.$$

Combining the above with the inequality $\|f\|_\infty \leqslant e^n f_\mu(0)$ from (2.9) we get

$$e^n [\varrho_{R_t(\mu)}(\xi)]^n \geqslant (1-e^{-t/8}) [\varrho_{K_n(\mu)}(\xi)]^n.$$

This shows that $R_t(\mu) \supseteq c_0 K_n(\mu)$, where $c_0 > 0$ is an absolute constant. From (2.12) we know that $K_n(\mu) \approx K_{n+1}(\mu)$, and the lemma follows. □

We can also compare $Z_t^+(\mu)$ with $K_{n+1}(\mu)$ when $t \geqslant 5n$.

Lemma 2.3.6 *For every $t \geqslant 5n$ we have that $Z_t^+(\mu) \supseteq c_0' K_{n+1}(\mu)$, where $c_0' > 0$ is an absolute constant.*

Proof From Lemma 2.3.5 we know that $c_0 K_{n+1}(\mu) \subseteq R_t(\mu)$ for all $t \geqslant 5n$, where $c_0 > 0$ is an absolute constant. Let $\xi \in S^{n-1}$ and set $m_\xi := h_{c_0 K_{n+1}(\mu)}(\xi) = c_0 h_{K_{n+1}(\mu)}(\xi)$. Define

$$A_\xi = c_0 K_{n+1}(\mu) \cap \{x : \langle x, \xi\rangle \geqslant m_\xi/2\}.$$

Since $K_{n+1}(\mu)$ is centered, one can check (see e.g. [Kla04, Lemma 2.2] or [BGP24a, Proposition 4.1]) that

$$|A_\xi| \geqslant |c_0 K_{n+1}(\mu)|/C^n$$

for some absolute constant $C > c_0$. Moreover, if $x \in A_\xi$ then $x \in R_t(\mu)$ and hence $f_\mu(x) \geqslant e^{-t} f_\mu(0)$. We write

$$\begin{aligned}\int_{\mathbb{R}^n} \langle x, \xi\rangle_+^t d\mu(x) &\geqslant \int_{A_\xi} \langle x, \xi\rangle_+^t d\mu(x)\\ &\geqslant \left(\frac{m_\xi}{2}\right)^t e^{-t} f_\mu(0)|A_\xi| \geqslant \left(\frac{m_\xi}{2e}\right)^t \left(\frac{c_0}{C}\right)^n f_\mu(0)|K_{n+1}(\mu)|.\end{aligned}$$

Using also the fact that $(c_0/C)^n \geqslant (c_0/C)^t$ because $t \geqslant 5n$, we get

$$\int_{\mathbb{R}^n} \langle x, \xi\rangle_+^t d\mu(x) \geqslant (c_1 m_\xi)^t f_\mu(0)|K_{n+1}(\mu)|,$$

where $c_1 > 0$ is an absolute constant. Finally, $f_\mu(0)|K_{n+1}(\mu)| \approx 1$ by (2.15), which implies that

$$h_{Z_t^+(\mu)}(\xi) \geqslant c_2 m_\xi = c_0' h_{K_{n+1}(\mu)}(\xi),$$

where $c_0' = c_2 c_0$. □

Proof of Theorem 2.3.3 Combining Lemmas 2.3.5 and 2.3.6 we see that

$$R_{5n}(\mu) \cap Z_{5n}^+(\mu) \supseteq c_1 K_{n+1}(\mu)$$

for some absolute constant $c_1 > 0$. We apply Lemma 2.3.4 with $t = 5n$ and $\delta = \frac{1}{2}$. For every $x \in \frac{1}{2} Z_{5n}^+(\mu)$ we have

$$\varphi_\mu(x) \geqslant C_1^{-n}$$

for some absolute constant $C_1 > 1$. It follows that

$$\int_{\mathbb{R}^n} \varphi_\mu(x)\, d\mu(x) \geqslant C_1^{-n} \mu\left(\tfrac{1}{2} Z_{5n}^+(\mu)\right).$$

Then, by Lemma 2.3.6 we have $\frac{1}{2}Z_{5n}^{+}(\mu) \supseteq \frac{c_1}{2}K_{n+1}(\mu)$. Since $\frac{c_1}{2}K_{n+1}(\mu) \subseteq R_{5n}(\mu)$, we know that $f_\mu(x) \geqslant e^{-5n}f_\mu(0)$ for all $x \in \frac{c_1}{2}K_{n+1}(\mu)$. Using also (2.15), we get

$$\mu\left(\tfrac{1}{2}Z_{5n}^{+}(\mu)\right) \geqslant \mu\left(\frac{c_1}{2}K_{n+1}(\mu)\right) = \int_{\frac{c_1}{2}K_{n+1}(\mu)} f_\mu(x)\,dx \geqslant e^{-5n}f_\mu(0)\left|\frac{c_1}{2}K_{n+1}(\mu)\right|$$
$$= e^{-5n}(c_1/2)^n f_\mu(0)|K_{n+1}(\mu)| \geqslant e^{-5n}c_2^n.$$

Combining the above we conclude that

$$\int_{\mathbb{R}^n} \varphi_\mu(x)\,d\mu(x) \geqslant C_1^{-n}e^{-5n}c_2^n \geqslant e^{-cn},$$

for some absolute constant $c > 0$. □

The half-space depth plays a key role in the study of the threshold problem. Let μ be a log-concave probability measure on $\mathbb{R}^n$. Let $X_1, X_2, \ldots$ be independent random points in $\mathbb{R}^n$ distributed according to μ and for any $N > n$ consider the random polytope $K_N = \mathrm{conv}\{X_1, \ldots, X_N\}$. We shall generalize Lemmas 2.1.2 and 2.1.3 in this setting. To this end, for every convex body A in $\mathbb{R}^n$ with $0 \in \mathrm{int}(A)$ we define

$$\varphi_+(A) = \sup_{x \notin A} \varphi_\mu(x) \quad \text{and} \quad \varphi_-(A) = \inf_{x \in A} \varphi_\mu(x).$$

Recall that $B_t(\mu) = \{v \in \mathbb{R}^n : \Lambda_\mu^*(v) \leqslant t\}$, where Λ_μ^* is the Cramér transform of μ.

Lemma 2.3.7 *Let μ be a log-concave probability measure on $\mathbb{R}^n$. For every convex body A in $\mathbb{R}^n$ and every $N > n$ we have that*

$$\mathbb{E}_{\mu^N}(\mu(K_N)) \leqslant \mu(A) + N\varphi_+(A).$$

In particular, for every $t > 0$,

$$\mathbb{E}_{\mu^N}(\mu(K_N)) \leqslant \mu(B_t(\mu)) + N\exp(-t).$$

Proof We write

$$\mathbb{E}_{\mu^N}(\mu(K_N)) = \mathbb{E}_{\mu^N}(\mu(K_N \cap A)) + \mathbb{E}_{\mu^N}(\mu(K_N \setminus A)) \leqslant \mu(A) + \mathbb{E}_{\mu^N}(\mu(K_N \setminus A)).$$

Arguing as in the proof of Lemma 2.1.2 we see that $\mu^N\big(x \in K_N\big) \leqslant N\varphi_\mu(x)$ for any $x \in \mathbb{R}^n$. Then, Fubini's theorem shows that

$$\mathbb{E}_{\mu^N}(\mu(K_N \setminus A)) = \int_{\mathbb{R}^n \setminus A} \mu^N(x \in K_N)\,d\mu(x) \leqslant \int_{\mathbb{R}^n \setminus A} N\varphi_\mu(x)\,d\mu(x) \leqslant N\varphi_+(A).$$

The last claim follows if we set $A = B_t(\mu)$ because, by Lemma 2.3.2, $\varphi_\mu(x) \leqslant \exp(-\Lambda_\mu^*(x)) \leqslant e^{-t}$ for all $x \notin B_t(\mu)$. □

For the lower threshold we shall use the next lemma which is in the spirit of Lemma 2.1.3.

Lemma 2.3.8 *Let μ be a log-concave probability measure on $\mathbb{R}^n$. For every convex body A in $\mathbb{R}^n$ and every $N > n$ we have that*

$$1 - \mu^N(K_N \supseteq A) \leqslant 2\binom{N}{n}(1 - \varphi_-(A))^{N-n}.$$

Therefore,

$$\mathbb{E}_{\mu^N}(\mu(K_N)) \geqslant \mu(A)\left(1 - 2\binom{N}{n}(1 - \varphi_-(A))^{N-n}\right).$$

Proof Note that, with probability equal to 1 the random polytope K_N has non-empty interior. For every subset $J = \{j_1, \ldots, j_n\}$ of $\{1, \ldots, N\}$, of cardinality n, note that $X_{j_1}, \ldots, X_{j_n}$ are affinely independent with probability 1, and define the event L_J as follows: for one of the two closed half-spaces H_1, H_2 they determine, say H_i, we have simultaneously $K_N \subset H_i$ and $\mu(\mathbb{R}^n \setminus H_i) \geqslant \varphi_-(A)$.

If $A \nsubseteq K_N$, then there exists $x \in \partial(A) \setminus K_N$. Since $x \notin K_N$, there exists a facet F of K_N with the following property: one of the two closed half-spaces H_1 and H_2 determined by F contains K_N but does not contain x. Thus, if H_i is this half-space, we have simultaneously $K_N \subset H_i$ and $\mu(\mathbb{R}^n \setminus H_i) \geqslant \varphi_\mu(x) \geqslant \varphi_-(A)$. Since the hyperplane bounding H_i is determined by some affinely independent vertices $X_{j_1}, \ldots, X_{j_n}$ of K_N which lie in F, this shows that

$$\{A \nsubseteq K_N\} \subseteq \bigcup_J L_J.$$

It follows that

$$\mathrm{Prob}(A \nsubseteq K_N) \leqslant \sum_J \mathrm{Prob}(L_J) = \binom{N}{n}\mathrm{Prob}(L'),$$

where $L' := L_{\{1,\ldots,n\}}$. It is not hard to see that

$$\mathrm{Prob}(L') \leqslant 2(1 - \varphi_-(A))^{N-n}.$$

Indeed, $X_1, \ldots, X_n$ determine two closed half-spaces $H_i = H_i(X_1, \ldots, X_n)$, $i = 1, 2$. Let L^i be the event that $\mu(\mathbb{R}^n \setminus H_i) \geqslant \varphi_-(A)$. Then, with Exp denoting expectation with respect to the measure Prob,

$$\begin{aligned}\mathrm{Prob}(L') &\leqslant \sum_{i=1}^{2} \mathrm{Prob}\big(\{X_{n+1},\dots,X_N \in H_i\} \cap L^i\big)\\ &= \sum_{i=1}^{2} \mathrm{Exp}\big(\mathrm{Prob}\big(\{X_{n+1},\dots,X_N \in H_i\} \mid X_1,\dots,X_n\big)\mathbb{1}_{L^i}\big)\\ &\leqslant (1-\varphi_-(A))^{N-n} \sum_{i=1}^{2} \mathrm{Prob}(L^i).\end{aligned}$$

The second claim of the lemma follows from Markov's inequality. □

2.4 Rough Upper and Lower Thresholds

Rough upper and lower thresholds were obtained by Chakraborti et al. in [CTV21] for some general families of distributions. If μ is an even log-concave probability measure supported on a convex body K in $\mathbb{R}^n$ and if $X_1, X_2, \dots$ are independent random points distributed according to μ, then for any $n < N \leqslant \exp(c_1 n/L_\mu^2)$ we have that

$$\frac{\mathbb{E}_{\mu^N}(|K_N|)}{|K|} \leqslant \exp\left(-c_2 n/L_\mu^2\right),$$

where $c_1, c_2 > 0$ are absolute constants.

We shall describe a variant of this result for log-concave probability measures. We consider independent random points $X_1, X_2, \dots$ in $\mathbb{R}^n$ distributed according to a log-concave probability measure μ and the expectation $\mathbb{E}_{\mu^N}[\mu(K_N)]$ of the μ-measure of K_N. Recall that if $T : \mathbb{R}^n \to \mathbb{R}^n$ is an invertible affine transformation and $T_*\mu$ is the push-forward of μ then

$$\mathbb{E}_{(T_*\mu)^N}[(T_*\mu)(K_N)] = \mathbb{E}_{\mu^N}[\mu(K_N)].$$

So, we may assume that μ is isotropic.

Theorem 2.4.1 *Let μ be an isotropic log-concave probability measure on $\mathbb{R}^n$, $n \geqslant n_0$. For any $N \leqslant \exp(c_1 n/L_\mu^2)$ we have that*

$$\mathbb{E}_{\mu^N}(\mu(K_N)) \leqslant 2\exp\left(-c_2 n/L_\mu^2\right),$$

where $c_1, c_2 > 0$ and $n_0 \in \mathbb{N}$ are absolute constants.

Proof Using the estimate $\mu(B_t(\mu)) \leqslant \|f_\mu\|_\infty |B_t(\mu)|$, Proposition 2.2.2 and Theorem 2.2.1, from Lemma 2.3.7 we get

$$\mathbb{E}_{\mu^N}(\mu(K_N)) \leqslant \left(c_1 \|f_\mu\|_\infty^{1/n} \sqrt{t/n}\right)^n + N\exp(-t)$$

for every $N > n$ and $2 \leqslant t \leqslant n$. Recall that μ is isotropic, therefore $\|f_\mu\|_\infty^{2/n} = L_\mu^2 = O(\ln n)$. Then, if $n \geqslant n_0$ where $n_0 \in \mathbb{N}$ is an absolute constant, we see that $t := (c_1 e)^{-2} n/\|f_\mu\|_\infty^{2/n}$ satisfies $2 \leqslant t \leqslant n$ and

$$\left(c_1 \|f_\mu\|_\infty^{1/n} \sqrt{t/n}\right)^n \leqslant e^{-n}.$$

It follows that

$$\mathbb{E}_{\mu^N}(\mu(K_N)) \leqslant e^{-n} + N \exp(-c_2 n/\|f_\mu\|_\infty^{2/n}),$$

where $c_2 = (c_1 e)^{-2}$. Then, if $N \leqslant \exp(c_3 n/\|f_\mu\|_\infty^{2/n})$ where $c_3 = c_2/2$, we see that

$$\mathbb{E}_{\mu^N}(\mu(K_N)) \leqslant e^{-n} + \exp(-c_3 n/\|f_\mu\|_\infty^{2/n})$$

and the result follows from the fact that $\|f_\mu\|_\infty^{2/n} = L_\mu^2 \geqslant c$. □

We pass now to the lower threshold. It was proved in [CTV21] that if μ is an even κ-concave measure on $\mathbb{R}^n$ with $0 < \kappa < 1/n$, supported on a convex body K in $\mathbb{R}^n$, if $X_1, X_2, \dots$ are independent random points in $\mathbb{R}^n$ distributed according to μ and $K_N = \mathrm{conv}\{X_1, \dots, X_N\}$ as before, then for any $M \geqslant C$ and any $N \geqslant \exp\left(\frac{1}{\kappa}(\ln n + 2\ln M)\right)$ we have that

$$\frac{\mathbb{E}_{\mu^N}(|K_N|)}{|K|} \geqslant 1 - \frac{1}{M}, \tag{2.19}$$

where $C > 0$ is an absolute constant.

Since the family of log-concave probability measures corresponds to the case $\kappa = 0$, it is natural to ask for analogues of this result for 0-concave, i.e. log-concave, probability measures. To get an understanding of what happens, we should note that in the case where $X_1, X_2, \dots$ are uniformly distributed in the Euclidean unit ball the sharp threshold for the problem (see [Piv07, BCG+19]) is

$$\exp\left((1 \pm \varepsilon)\tfrac{1}{2} n \ln n\right), \qquad \varepsilon > 0.$$

We shall establish a weak lower threshold of this order.

Theorem 2.4.2 *Let $\delta \in (0, 1)$. Then,*

$$\inf_\mu \left(\inf \left\{\mathbb{E}_{\mu^N}\left[\mu((1+\delta)K_N)\right] : N \geqslant \exp\left(C\delta^{-1} \ln(2/\delta)\, n \ln n\right)\right\}\right) \longrightarrow 1$$

as $n \to \infty$, where the first infimum is over all centered log-concave probability measures μ on $\mathbb{R}^n$ and $C > 0$ is an absolute constant.

This is a weak threshold in the sense that we consider the expected measure of $(1+\delta)K_N$ instead of K_N, where $\delta > 0$ is arbitrarily small. The reason for this is the dependence on δ in the next technical proposition (we omit the proof; see [BGP24a, Proposition 5.6] for the details).

Proposition 2.4.3 *Let μ be an isotropic log-concave probability measure on $\mathbb{R}^n$. For any $\delta \in (0,1)$ and any $t \geqslant C_\delta n \ln n$ we have that*

$$\mu((1+\delta)Z_t^+(\mu)) \geqslant 1 - e^{-c_\delta t}$$

where $C_\delta = C\delta^{-1}\ln(2/\delta)$ and $c_\delta = c\delta$ are positive constants depending only on δ.

Proof of Theorem 2.4.2 Let $0 < \delta < 1$ and set $\varepsilon = \delta/3$. Let μ be a centered log-concave probability measure on $\mathbb{R}^n$. Since the expectation $\mathbb{E}_{\mu^N}\big[\mu((1+\delta)K_N)\big]$ is a linearly invariant quantity, we may assume that μ is isotropic. From Lemma 2.3.4 we know that for every $x \in (1-\varepsilon)Z_t^+(\mu)$ we have

$$\varphi_\mu(x) \geqslant \frac{(1-(1-\varepsilon)^t)^2}{C_1^t},$$

where $C_1 > 1$ is an absolute constant. Then, taking into account the fact that $1-\varepsilon > 2/3$, we get

$$\mu^N\Big(K_N \supseteq (1-\varepsilon)Z_t^+(\mu)\Big) \geqslant 1 - 2\binom{N}{n}\left[1 - \frac{(1-(1-\varepsilon)^t)^2}{C_1^t}\right]^{N-n}.$$

By the mean value theorem we have $1-(1-\varepsilon)^t = t\varepsilon z^{t-1}$ for some $z \in (1-\varepsilon, 1)$, and hence $1-(1-\varepsilon)^t \geqslant t\varepsilon(1-\varepsilon)^{t-1}$. Taking also into account the fact that $1-\varepsilon > 2/3$, we get

$$\begin{aligned}\mu^N\Big(K_N \supseteq (1-\varepsilon)Z_t^+(\mu)\Big) &\geqslant 1 - 2\binom{N}{n}\left[1 - \frac{(t\varepsilon(1-\varepsilon)^{t-1})^2}{C_1^t}\right]^{N-n}\\ &\geqslant 1 - \left(\frac{2eN}{n}\right)^n \exp\left(-(N-n)\frac{(t\varepsilon)^2}{(3C_1)^t}\right).\end{aligned}$$

This last quantity tends to 1 as $n \to \infty$ if

$$(3C_1)^t n \ln(4eN/n) < (N-n)(t\varepsilon)^2, \tag{2.20}$$

and assuming that $\delta \in (1/n^2, 1)$ and $t \geqslant C_\varepsilon n \ln n$ where C_ε is the constant from Proposition 2.4.3, we check that (2.20) holds true if $N \geqslant \exp(C_2 t)$ for a large enough absolute constant $C_2 > 0$.

Note that $\varepsilon = \delta/3$ implies that $1+\delta > \frac{1+\varepsilon}{1-\varepsilon}$. Then, if $N \geqslant \exp(C_2 C_\varepsilon n \ln n)$ we see that

$$\begin{aligned}\mathbb{E}_{\mu^N}\left[\mu\left((1+\delta)K_N\right)\right] &\geqslant \mathbb{E}_{\mu^N}\left[\mu\left(\frac{1+\varepsilon}{1-\varepsilon}K_N\right)\right]\\ &\geqslant \mu((1+\varepsilon)Z_t^+(\mu)) \times \mu^N\left(K_N \supseteq (1-\varepsilon)Z_t^+(\mu)\right)\\ &\geqslant \left(1-e^{-c\varepsilon t}\right)\left[1-\left(\frac{2eN}{n}\right)^n \exp\left(-(N-n)\frac{(t\varepsilon)^2}{(3C_1)^t}\right)\right] \longrightarrow 1\end{aligned}$$

as $n \to \infty$. □

The next theorem provides an estimate where "δ is removed", however the dependence on n becomes worse.

Theorem 2.4.4 *There exists an absolute constant $C > 0$ such that*

$$\inf_\mu \left(\inf\left\{\mathbb{E}_{\mu^N}\big[\mu(K_N)\big] : N \geqslant \exp(C(n\ln n)^2 u(n))\right\}\right) \longrightarrow 1$$

as $n \to \infty$, where the first infimum is over all log-concave probability measures μ on $\mathbb{R}^n$ and $u(n)$ is any function with $u(n) \to \infty$ as $n \to \infty$.

Proof Let μ be a log-concave probability measure on $\mathbb{R}^n$. Since the expectation $\mathbb{E}_{\mu^N}\big[\mu(K_N)\big]$ is an affinely invariant quantity, we may assume that μ is centered. Note that if $A \subset \mathbb{R}^n$ is a Borel set, then

$$\mu((1+\delta)A) = \int_{(1+\delta)A} f_\mu(x)\,dx = (1+\delta)^n \int_A f_\mu((1+\delta)x)\,dx.$$

Since f_μ is log-concave, we see that

$$f_\mu((1+\delta)x) \leqslant f_\mu(x)\left(\frac{f_\mu(x)}{f_\mu(0)}\right)^\delta \leqslant e^{n\delta} f_\mu(x)$$

for every $x \in \mathbb{R}^n$, because $f_\mu(x) \leqslant e^n f_\mu(0)$ by (2.9). It follows that

$$\mu((1+\delta)A) \leqslant (1+\delta)^n e^{n\delta}\mu(A) \leqslant e^{2n\delta}\mu(A). \tag{2.21}$$

Given a function $u(n)$ with $u(n) \to \infty$ as $n \to \infty$, choose $\delta_n = (nu(n))^{-1}$. From (2.21) we see that

$$\mathbb{E}_{\mu^N}\big[\mu(K_N)\big] \geqslant e^{-2n\delta_n}\mathbb{E}_{\mu^N}\big[\mu((1+\delta_n)K_N)\big].$$

Therefore, we see that

$$\inf_{\mu}\left(\inf\left\{\mathbb{E}_{\mu^N}\left[\mu(K_N)\right]:N\geqslant\exp\left(C\delta_n^{-1}\ln\left(2/\delta_n\right)n\ln n\right)\right\}\right)$$

$$\geqslant e^{-2n\delta_n}\inf_{\mu}\left(\inf\left\{\mathbb{E}_{\mu^N}\left[\mu((1+\delta_n)K_N)\right]:N\geqslant\exp\left(C\delta_n^{-1}\ln\left(2/\delta_n\right)n\ln n\right)\right\}\right)\longrightarrow 1$$

as $n\to\infty$, using Theorem 2.4.2 and the fact that $e^{-2n\delta_n}=e^{-2/u(n)}\to 1$. We may clearly assume that $u(n)=O(n)$. Then,

$$\delta_n^{-1}\ln\left(2/\delta_n\right)n\ln n=n^2\ln n\ln(2nu(n))u(n)\approx(n\ln n)^2u(n),$$

and the result follows. □

2.5 Comparing Half-Space Depth with the Cramér Transform

Let μ be a centered log-concave probability measure on $\mathbb{R}^n$ with density $f:=f_{\mu}$. Recall that for every $t>0$ we consider the convex set $B_t(\mu):=\{x\in\mathbb{R}^n:\Lambda_{\mu}^*(x)\leqslant t\}$, and for any $x\in\mathbb{R}^n$ we denote by $\mathcal{H}(x)$ the set of all half-spaces H of $\mathbb{R}^n$ containing x and consider Tukey's half-space depth $\varphi_{\mu}(x)=\inf\{\mu(H):H\in\mathcal{H}(x)\}$. In Lemma 2.3.2 we showed that for every $x\in\mathbb{R}^n$ we have $\varphi_{\mu}(x)\leqslant\exp(-\Lambda_{\mu}^*(x))$, which implies that

$$\varphi_+(B_t(\mu))\leqslant e^{-t}$$

for any $t>0$. Our aim in this section is to obtain a lower bound for $\varphi_-(B_t(\mu))$, or equivalently for $\varphi_{\mu}(x)$ when $x\in B_t(\mu)$. First, we consider the case where $\mu=\mu_K$ is the uniform measure on a centered convex body K of volume 1 in $\mathbb{R}^n$, and prove the following.

Theorem 2.5.1 *Let K be a centered convex body of volume 1 in $\mathbb{R}^n$. Then, for every $t>0$ we have that*

$$\inf\{\varphi_{\mu_K}(x):x\in B_t(\mu_K)\}\geqslant\frac{1}{10}\exp(-t-2\sqrt{n}).$$

The first part of the argument works for any centered log-concave probability measure μ with density f on $\mathbb{R}^n$. For every $\xi\in\mathbb{R}^n$ we define the probability measure μ_{ξ} with density

$$f_{\xi}(z)=e^{-\Lambda_{\mu}(\xi)+\langle\xi,z\rangle}f(z).$$

One can compute that the barycenter of μ_ξ is $x = \nabla\Lambda_\mu(\xi)$ and $\mathrm{Cov}(\mu_\xi) = \mathrm{Hess}\,(\Lambda_\mu)(\xi)$ (a proof is given in [BGVV14, Proposition 7.2.1]). Next, we set

$$\sigma_\xi^2 = \int_{\mathbb{R}^n} \langle z - x, \xi\rangle^2 d\mu_\xi(z) = \mathrm{Var}_{\mu_\xi}(\langle \xi, z\rangle).$$

Let $t > 0$. Since $B_t(\mu)$ is convex, in order to give a lower bound for $\inf\{\varphi_\mu(x) : x \in B_t(\mu)\}$ it suffices to give a lower bound for $\mu(H)$, where H is any closed half-space whose bounding hyperplane supports $B_t(\mu)$. In that case,

$$\mu(H) = \mu(\{z : \langle z - x, \xi\rangle \geqslant 0\}) \tag{2.22}$$

for some $x \in \partial(B_t(\mu))$, with $\xi = \nabla\Lambda_\mu^*(x)$, or equivalently $x = \nabla\Lambda_\mu(\xi)$ (see e.g. Theorem 23.5 and Corollary 23.5.1 in [Roc70]). Note that

$$\begin{aligned}\mu(\{z : \langle z - x, \xi\rangle \geqslant 0\}) &= \int_{\mathbb{R}^n} \mathbb{1}_{[0,\infty)}(\langle z - x, \xi\rangle) f(z)\,dz \\ &= e^{\Lambda_\mu(\xi)} \int_{\mathbb{R}^n} \mathbb{1}_{[0,\infty)}(\langle z - x, \xi\rangle) e^{-\langle z,\xi\rangle}\,d\mu_\xi(z) \\ &= e^{\Lambda_\mu(\xi)} e^{-\langle x,\xi\rangle} \int_{\mathbb{R}^n} \mathbb{1}_{[0,\infty)}(\langle z - x, \xi\rangle) e^{-\langle z-x,\xi\rangle}\,d\mu_\xi(z) \\ &\geqslant e^{-\Lambda_\mu^*(x)} \int_0^\infty \sigma_\xi e^{-\sigma_\xi t} \mu_\xi(\{z : 0 \leqslant \langle z - x, \xi\rangle \leqslant \sigma_\xi t\})\,dt.\end{aligned} \tag{2.23}$$

Using Markov's inequality we check that $\mu_\xi(\{z : \langle z - x, \xi\rangle \geqslant 2\sigma_\xi\}) \leqslant \frac{1}{4}$, and since x is the barycenter of μ_ξ, from Grünbaum's lemma (see [BGVV14, Lemma 2.2.6]) we get that $\mu_\xi(\{z : \langle z - x, \xi\rangle \geqslant 0\}) \geqslant \frac{1}{e}$. Therefore,

$$\begin{aligned}&\int_0^\infty \sigma_\xi e^{-\sigma_\xi t} \mu_\xi(\{z : 0 \leqslant \langle z - x, \xi\rangle \leqslant \sigma_\xi t\})\,dt \\ &\geqslant \int_2^\infty \sigma_\xi e^{-\sigma_\xi t}\left(\frac{1}{e} - \frac{1}{4}\right) dt \geqslant \frac{4-e}{4e} e^{-2\sigma_\xi}.\end{aligned} \tag{2.24}$$

We would like to have an upper bound for $\sup_\xi \sigma_\xi$. This is the point where we need to restrict ourselves to the case where $\mu = \mu_K$ is the uniform measure on a centered convex body K of volume 1 on $\mathbb{R}^n$: then, we can exploit a theorem of Nguyen [Ngu14] which was proved independently by Wang [Wan14] (the sketch of its proof below follows [FMW16]).

Theorem 2.5.2 *Let ν be a log-concave probability measure on $\mathbb{R}^n$ with density $g = \exp(-p)$, where $p : \mathbb{R}^n \to (-\infty, \infty]$ is a convex function. Then,*

$$\mathrm{Var}_\nu(p) \leqslant n.$$

Sketch of the proof Note that

$$\mathrm{Var}_{\nu}(p) = V(g) := \int_{\mathbb{R}^n} g(\ln g)^2 - \left(\int_{\mathbb{R}^n} g \ln g\right)^2.$$

Define $F : (0, \infty) \to \mathbb{R}$ with $F(s) = \ln\left(\int_{\mathbb{R}^n} g^s(x)\,dx\right)$. A careful computation of $F''(s)$ and $V(g_s)$ shows that $F''(s) = V(g_s)/s^2$, where g_s is the log-concave density

$$g_s = \frac{g^s}{\int_{\mathbb{R}^n} g^s}.$$

Next, observe that the function $w : \mathbb{R}^n \times (0, \infty) \to (-\infty, \infty]$ with $w(z, s) = s\psi(z/s)$ is convex. It follows that the function $G : (0, \infty) \to \mathbb{R}$ defined by

$$G(s) = s^n \int_{\mathbb{R}^n} g^s(x)\,dx$$

is log-concave. In order to check this, one can make the change of variables $x = z/s$ and use the convexity of w as well as the fact that marginals of a log-concave measure are log-concave, therefore they have a log-concave density. This implies that $V(g_s) \leqslant n$ for every $s > 0$. To see this, note that $\ln G(s) = n \ln s + F(s)$ and differentiate twice. In particular, for $s = 1$, we get $V(g) \leqslant n$.

In the case where $g = e^{-p}$ as above, and p is positively homogeneous of degree 1, one can check that $G(s) = 1$ for all $s > 0$, and hence $V(g_s) = n$. In particular, we have that $V(g) = n$, which shows that the inequality of the theorem is sharp. □

Proof of Theorem 2.5.1 Set $\mu := \mu_K$. Since $f(z) = \mathbb{1}_K(z)$, the density f_ξ of μ_ξ is proportional to $e^{\langle \xi, z\rangle}\mathbb{1}_K(z)$. From Theorem 2.5.2 we get

$$\sigma_\xi^2 = \mathbb{E}_{\mu_\xi}(\langle z - x, \xi\rangle)^2 = \mathrm{Var}_{\mu_\xi}(\langle \xi, z\rangle) = \mathrm{Var}_{\mu_\xi}(-\ln f_\xi) \leqslant n.$$

Then, combining (2.22), (2.23) and (2.24), for any bounding hyperplane H of $B_t(\mu)$ we have

$$\begin{aligned}\mu(H) &\geqslant e^{-\Lambda_\mu^*(x)} \int_0^\infty \sigma_\xi e^{-\sigma_\xi t} \mu_\xi(0 \leqslant \langle z - x, \xi\rangle \leqslant \sigma_\xi t)\,dt \\ &\geqslant \frac{4-e}{4e} e^{-\Lambda_\mu^*(x) - 2\sigma_\xi} \geqslant \frac{1}{10}\exp(-t - 2\sqrt{n}),\end{aligned}$$

as claimed. □

Theorem 2.5.1 shows that if K is a centered convex body of volume 1 in $\mathbb{R}^n$ then

$$10\varphi_{\mu_K}(x) \geqslant \exp(-\Lambda_{\mu_K}^*(x) - 2\sqrt{n})$$

for all $x \in \mathbb{R}^n$. Setting

$$\omega_{\mu_K}(x) = \ln\left(\frac{1}{\varphi_{\mu_K}(x)}\right) \tag{2.25}$$

and taking into account Lemma 2.3.2 we have the next two-sided estimate.

Corollary 2.5.3 *Let K be a centered convex body of volume 1 in $\mathbb{R}^n$. Then, for every $x \in \mathrm{int}(K)$ we have that*

$$\omega_{\mu_K}(x) - 5\sqrt{n} \leqslant \Lambda^*_{\mu_K}(x) \leqslant \omega_{\mu_K}(x). \tag{2.26}$$

A basic question that arises is whether an analogue of (2.26) holds true for any centered log-concave probability measure μ on $\mathbb{R}^n$. This would allow us to apply the next steps of our procedure to all log-concave probability measures. Brazitikos and Chasapis [BC24] have recently obtained such a variant of Theorem 2.5.1 which is valid in this more general setting.

Theorem 2.5.4 *Let μ be a log-concave probability measure on $\mathbb{R}^n$. For every $x \in \mathrm{supp}(\mu)$ and any $\varepsilon \in (0, 1)$ we have that*

$$\Lambda^*_\mu(x) \geqslant (1-\varepsilon)\ln\left(\frac{1}{\varphi_\mu(x)}\right) + \ln\left(\frac{\varepsilon}{2^{1-\varepsilon}}\right).$$

Sketch of the proof We start with the case $n = 1$. Note that

$$\varphi_\mu(y) = \min\{\mu((-\infty, y]), \mu([y, \infty))\}$$

is a log-concave function. Then, the function $g(y) = -(1-\varepsilon)\ln\varphi_\mu(y)$ is convex, and hence we may find $t, b \in \mathbb{R}$ such that the $\ell(y) = ty + b$ satisfies $\ell(x) = g(x)$ and $\ell(y) \leqslant g(y)$ for all y. Then, $\mathbb{E}_\mu(e^{\ell(y)}) \leqslant \mathbb{E}_\mu(e^{g(y)})$, or equivalently $-\ln\left(\mathbb{E}_\mu(e^{g(y)})\right) \leqslant -\ln\left(\mathbb{E}_\mu(e^{\ell(y)})\right) = -b - \Lambda_\mu(t)$. Since $\ell(x) = b(x)$, we see that

$$g(x) - \ln\left(\mathbb{E}_\mu(e^{g(y)})\right) \leqslant \ell(x) - b - \Lambda_\mu(t) = xt - \Lambda_\mu(t) \leqslant \Lambda^*_\mu(x).$$

Finally, one can check that $\mathbb{E}(\varphi_\mu(y)^{\varepsilon-1}) \leqslant (\varepsilon 2^{\varepsilon-1})^{-1}$, which implies that

$$-\ln\left(\mathbb{E}_\mu(e^{g(y)})\right) = -\ln\left(\mathbb{E}_\mu(\phi_\mu(y)^{\varepsilon-1})\right) \geqslant \ln\left(\frac{\varepsilon}{2^{1-\varepsilon}}\right)$$

and the claim follows.

Next, assume that $n > 1$. If X is a random vector in $\mathbb{R}^n$ which is distributed according to μ, for any $\xi \in S^{n-1}$ consider the random variable $\xi_X = \langle X, \xi\rangle$. Then, the distribution μ_ξ of ξ_X is log-concave and from the one-dimensional result we see that

$$\Lambda^*_{\mu_\xi}(\langle x, \xi\rangle) \geqslant (1-\varepsilon)\ln\left(\frac{1}{\varphi_{\mu_\xi}(\langle x, \xi\rangle)}\right) + \ln\left(\frac{\varepsilon}{2^{1-\varepsilon}}\right)$$

for all $x \in \mathbb{R}^n$. We observe that, for any $x \in \mathbb{R}^n$,

$$\begin{aligned}\Lambda_\mu^*(x) &= \sup_{y\in\mathbb{R}^n} \big(\langle x, y\rangle - \Lambda_\mu(y)\big) = \sup_{(\xi,t)\in S^{n-1}\times\mathbb{R}} \big(t\langle x,\xi\rangle - \Lambda_{\mu_\xi}(t)\big)\\ &= \sup_{\xi\in S^{n-1}} \Lambda_{\mu_\xi}^*(\langle x,\xi\rangle).\end{aligned}$$

Since

$$\sup_{\xi\in S^{n-1}} \ln\left(\frac{1}{\varphi_{\mu_\xi}(\langle x,\xi\rangle)}\right) = \ln\left(\frac{1}{\inf_{\xi\in S^{n-1}} \mu(\{z : \langle z,\xi\rangle \geqslant \langle x,\xi\rangle\})}\right) = \ln\left(\frac{1}{\varphi_\mu(x)}\right),$$

combining the above we obtain the result. □

Note Theorem 2.5.4 can be rewritten as $\varphi_\mu(x)^{1-\varepsilon} \geqslant (\varepsilon/2^{1-\varepsilon})e^{-\Lambda_\mu^*(x)}$, which gives

$$\big[\varphi_-(B_t(\mu))\big]^{1-\varepsilon} \geqslant \frac{\varepsilon}{2^{1-\varepsilon}}e^{-t} \tag{2.27}$$

for all $\varepsilon \in (0,1)$. This last inequality, which is valid for all log-concave probability measures, may be viewed as a substitute of the estimate in Theorem 2.5.1.

2.6 Moments of the Cramér Transform

Our approach to the threshold problem for a given centered log-concave probability measure μ on $\mathbb{R}^n$ requires to know that the Cramér transform Λ_μ^* has finite variance. We can give an affirmative answer to this question in the case where $\mu = \mu_K$ is the uniform measure on a centered convex body K of volume 1 in $\mathbb{R}^n$. Actually, one can show that, for a more general class of measures, it is still true that Λ_μ^* has finite moments of all orders.

Theorem 2.6.1 *Let K be a centered convex body of volume* 1 *in* $\mathbb{R}^n$. *Let $\kappa \in (0, 1/n]$ and let μ be a centered κ-concave probability measure with* $\mathrm{supp}(\mu) = K$. *Then,*

$$\int_{\mathbb{R}^n} \exp(\kappa\Lambda_\mu^*(x)/2)\, d\mu(x) < \infty.$$

In particular, for all $p \geqslant 1$ we have that $\mathbb{E}_\mu\big((\Lambda_\mu^*(x))^p\big) < \infty$.

For the proof we need a lemma, which is proved in [CTV21, Lemma 7] in the symmetric case.

Lemma 2.6.2 *Let K be a centered convex body of volume 1 in $\mathbb{R}^n$. Let $\kappa \in (0, 1/n]$ and let μ be a centered κ-concave probability measure with* $\mathrm{supp}(\mu) = K$*. Then,*

$$\varphi_\mu(x) \geqslant e^{-2}\kappa(1 - \|x\|_K)^{1/\kappa} \tag{2.28}$$

for every $x \in K$, where $\|x\|_K$ is the Minkowski functional of K.

Sketch of the proof Let X be a random vector distributed according to μ. Given $\theta \in S^{n-1}$ let $b = h_K(\theta)$ and $a = h_K(-\theta)$. If g_θ is the density of $\langle X, \theta\rangle$ then $g_\theta^{\frac{\kappa}{1-\kappa}}$ is concave on $[-a, b]$, therefore

$$g_\theta(t) \geqslant g_\theta(0)\left(1 - \frac{t}{b}\right)^{\frac{1-\kappa}{\kappa}}$$

for all $t \in [0, b]$. It follows that, for every $0 < s < b$,

$$\mathbb{P}(\langle X, \theta\rangle \geqslant s) = \int_s^b g_\theta(t)\,dt \geqslant g_\theta(0)\int_s^b \left(1 - \frac{t}{b}\right)^{\frac{1-\kappa}{\kappa}} dt = \kappa g_\theta(0) b\left(1 - \frac{s}{b}\right)^{\frac{1}{\kappa}}.$$

Note that g_θ is a centered log-concave density. Therefore, $g_\theta(0) \geqslant e^{-1}\|g_\theta\|_\infty$ by (2.9) and $\|g_\theta\|_\infty b \geqslant \mathbb{P}(\langle X, \theta\rangle \geqslant 0) \geqslant e^{-1}$ by Grünbaum's lemma [BGVV14, Lemma 2.2.6], which implies that $g_\theta(0)b \geqslant e^{-2}$. It follows that

$$\mathbb{P}(\langle X, \theta\rangle \geqslant s) = \int_s^b g_\theta(t)\,dt \geqslant e^{-2}\kappa\left(1 - \frac{s}{b}\right)^{\frac{1}{\kappa}}.$$

Now, let $x \in K$. Then $\langle x, \theta\rangle \leqslant \|x\|_K h_K(\theta) = \|x\|_K b$, therefore

$$\mathbb{P}(\langle X, \theta\rangle \geqslant \langle x, \theta\rangle) \geqslant \mathbb{P}(\langle X, \theta\rangle \geqslant \|x\|_K b) \geqslant e^{-2}\kappa\,(1 - \|x\|_K)^{\frac{1}{\kappa}}$$

as claimed. □

Proof of Theorem 2.6.1 From Lemma 2.3.2 we know that $\varphi_\mu(x) \leqslant \exp(-\Lambda_\mu^*(x))$, or equivalently,

$$\exp(\kappa\Lambda_\mu^*(x)/2) \leqslant \varphi_\mu(x)^{-\kappa/2}$$

for all $x \in K$. Lemma 2.6.2 shows that

$$\varphi_\mu(x) \geqslant e^{-2}\kappa(1 - \|x\|_K)^{1/\kappa}$$

for every $x \in K$, and hence

$$\int_K \exp(\kappa\Lambda_\mu^*(x)/2)\,d\mu(x) \leqslant (e^2/\kappa)^{\kappa/2}\int_K \frac{1}{(1 - \|x\|_K)^{1/2}}d\mu(x).$$

Recall that the cone probability measure ν_K on the boundary $\partial(K)$ of a convex body K with $0 \in \mathrm{int}(K)$ is defined by

$$\nu_K(B) = |\{rx : x \in B, 0 \leqslant r \leqslant 1\}|/|K|$$

for all Borel subsets B of $\partial(K)$. We shall use the identity

$$\int_{\mathbb{R}^n} g(x)\,dx = n|K| \int_0^\infty r^{n-1} \int_{\partial(K)} g(rx)\,d\nu_K(x)\,dr$$

which holds for every integrable function $g : \mathbb{R}^n \to \mathbb{R}$ (see [NR03, Proposition 1]). Let f denote the density of μ on K. We write

$$\begin{aligned}
\int_K \frac{1}{(1-\|x\|_K)^{1/2}}\,d\mu(x) &= \int_{\mathbb{R}^n} \frac{f(x)}{(1-\|x\|_K)^{1/2}} \mathbb{1}_K(x)\,dx \\
&= n|K| \int_0^\infty r^{n-1} \int_{\partial(K)} \frac{f(ry)}{(1-\|ry\|_K)^{1/2}} \mathbb{1}_K(ry)\,d\nu_K(y)\,dr \\
&= n|K| \int_0^1 \frac{r^{n-1}}{\sqrt{1-r}} \int_{\partial(K)} f(ry)\,d\nu_K(y)\,dr \\
&\leqslant n|K|\|f\|_\infty \int_0^1 \frac{r^{n-1}}{\sqrt{1-r}}\,dr = n|K|B(n,1/2)\|f\|_\infty \\
&\leqslant c\sqrt{n}\|f\|_\infty < +\infty,
\end{aligned}$$

and the proof is complete. □

In the case of the uniform measure $\mu = \mu_K$ on a centered convex body K of volume 1 in $\mathbb{R}^n$ we see that

$$\int_K (\Lambda^*_{\mu_K}(x)/2n)^p\,dx \leqslant (c_1 p)^p \int_K \exp(\Lambda^*_{\mu_K}(x)/2n)\,dx \leqslant (c_2 p)^p \sqrt{n},$$

where $c_1, c_2 > 0$ are absolute constants. This shows that

$$\|\Lambda^*_{\mu_K}\|_{L^p(\mu_K)} \leqslant cpn^{1+\frac{1}{2p}}$$

for all $p \geqslant 1$. However, the argument that we used for Theorem 2.6.1 leads to sharp estimates in the cases $p = 1, 2$. We shall use the fact that

$$\int_0^1 r^{n-1} \ln(1-r)\,dr = -\frac{1}{n} H_n \quad \text{and} \quad \int_0^1 r^{n-1} \ln^2(1-r)\,dr = \frac{1}{n} H_n^2 + \frac{1}{n} \sum_{k=1}^n \frac{1}{k^2}, \tag{2.29}$$

where $H_n = 1 + \frac{1}{2} + \cdots + \frac{1}{n}$.

Theorem 2.6.3 *Let K be a centered convex body of volume 1 in $\mathbb{R}^n$, $n \geqslant 2$. Let $\kappa \in (0, 1/n]$ and let μ be a centered κ-concave probability measure with* $\mathrm{supp}(\mu) = K$. *Then,*

$$\mathbb{E}_\mu(\Lambda_\mu^*) \leqslant \left(\mathbb{E}_\mu[(\Lambda_\mu^*)^2]\right)^{1/2} \leqslant \frac{c \ln n}{\kappa} \|f\|_\infty^{1/2},$$

where $c > 0$ is an absolute constant and f is the density of μ.

Proof As in the proof of Theorem 2.6.1 we write

$$\int_K \left(\Lambda_\mu^*(x)\right)^2 d\mu(x) \leqslant \int_K \ln^2\left(\frac{e^2}{\kappa}\frac{1}{(1-\|x\|_K)^{1/\kappa}}\right) d\mu(x).$$

If f is the density of μ on K and ν_K is the cone measure of K, using the inequality $\ln^2(ab) \leqslant 2(\ln^2 a + \ln^2 b)$ where $a, b > 0$, we may write

$$\begin{aligned}
&\frac{1}{2}\int_K \ln^2\left(\frac{e^2}{\kappa}\frac{1}{(1-\|x\|_K)^{1/\kappa}}\right) d\mu(x) - \ln^2\left(\frac{e^2}{\kappa}\right)\\
&\leqslant \int_{\mathbb{R}^n} f(x)\ln^2\left(\frac{1}{(1-\|x\|_K)^{1/\kappa}}\right) \mathbb{1}_K(x)\,dx\\
&= n|K|\int_0^\infty r^{n-1}\int_{\partial(K)} f(ry)\ln^2\left(\frac{1}{(1-\|ry\|_K)^{1/\kappa}}\right)\mathbb{1}_K(ry)\,d\nu_K(y)\,dr\\
&= \frac{n}{\kappa^2}\int_0^1 r^{n-1}\ln^2(1-r)\int_{\partial(K)} f(ry)\,d\nu_K(y)\,dr\\
&\leqslant \frac{n}{\kappa^2}\|f\|_\infty \int_0^1 r^{n-1}\ln^2(1-r)\,dr.
\end{aligned}$$

Since $1 \leqslant \int_K f(x)\,dx \leqslant \|f\|_\infty$, using also (2.29) we get

$$\begin{aligned}
\int_K \left(\Lambda_\mu^*(x)\right)^2 d\mu(x) &\leqslant \frac{2n}{\kappa^2}\left(\frac{1}{n}H_n^2 + \frac{1}{n}\sum_{k=1}^n \frac{1}{k^2}\right)\|f\|_\infty + 2\ln^2\left(\frac{e^2}{\kappa}\right)\\
&\leqslant \left(\frac{4H_n^2}{\kappa^2} + 2\ln^2(e^2/\kappa)\right)\|f\|_\infty \leqslant \frac{c_1 \ln^2 n}{\kappa^2}\|f\|_\infty,
\end{aligned}$$

where $c_1 > 0$ is an absolute constant. □

In fact, for the uniform measure on a convex body, we can summarize in a sharp two sided estimate.

Theorem 2.6.4 *Let K be a centered convex body of volume 1 in $\mathbb{R}^n$, $n \geqslant 2$. Then,*

$$c_1 n/L_{\mu_K}^2 \leqslant \|\Lambda_{\mu_K}^*\|_{L^1(\mu_K)} \leqslant \|\Lambda_{\mu_K}^*\|_{L^2(\mu_K)} \leqslant c_2 n \ln n,$$

where L_{μ_K} is the isotropic constant of the uniform measure μ_K on K and $c_1, c_2 > 0$ are absolute constants.

Proof For the left-hand side inequality recall that in Theorem 2.3.1 we saw that

$$\int_{\mathbb{R}^n} \varphi_\mu(x)\, d\mu(x) \leqslant \exp\left(-cn/L_\mu^2\right),$$

where $c > 0$ is an absolute constant. In fact, the proof of this estimate starts with Lemma 2.3.2 and follows from the next stronger result: If $n \geqslant n_0$ then

$$\int_{\mathbb{R}^n} \exp(-\Lambda_\mu^*(x))\, d\mu(x) \leqslant \exp\left(-cn/L_\mu^2\right)$$

where L_μ is the isotropic constant of μ and $c > 0$, $n_0 \in \mathbb{N}$ are absolute constants. Then, Jensen's inequality implies that

$$e^{-\mathbb{E}_\mu(\Lambda_\mu^*)} \leqslant \int_{\mathbb{R}^n} \exp(-\Lambda_\mu^*(x))\, d\mu(x) \leqslant \exp\left(-cn/L_\mu^2\right)$$

and the result follows. □

Both the lower and the upper bound are of optimal order with respect to the dimension. This can be seen e.g. from the example of the uniform measure on the cube or the Euclidean ball, respectively.

An alternative approach to the question of moments of Λ_μ^* may be based on the notion of affine surface area. Let K be a convex body in $\mathbb{R}^n$. The affine surface area of K is the quantity

$$\mathrm{as}(K) = \int_{\partial(K)} \kappa(x)^{\frac{1}{n+1}}\, d\mu_{\partial(K)}(x),$$

where $\kappa(x)$ is the generalized Gauss-Kronecker curvature at x and $\mu_{\partial(K)}$ is the surface measure on $\partial(K)$ (see [NSW19] and the references therein). The affine surface area of a Euclidean ball of radius 1 is equal to its surface area and the affine surface area of all polytopes is equal to 0. The affine isoperimetric inequality states that

$$\left(\frac{\mathrm{as}(K)}{\mathrm{as}(B_2^n)}\right)^{n+1} \leqslant \left(\frac{|K|}{|B_2^n|}\right)^{n-1}$$

with equality if and only if K is an ellipsoid (see [Sch14, Sect. 10.5]). Since $\mathrm{as}(B_2^n) = n\omega_n$ where $\omega_n = |B_2^n|$, we see that if $|K| = |B_2^n|$ then $\mathrm{as}(K) \leqslant \mathrm{as}(B_2^n)$. Thus, if $|K| = 1$ then

$$\mathrm{as}(K) \leqslant c_1.$$

For every $\delta \in (0, 1/2)$ we define the floating body

$$K_\delta = \bigcap \{H^+ : H^+ \text{ is a closed half-space with } |K \cap H^-| = \delta\},$$

where H^- is the complementary half-space of H^+. We also define

$$T_\delta = \{x \in K : \varphi_{\mu_K}(x) \geqslant \delta\}.$$

Note that T_δ is convex: if $x, y \in T_\delta$ then for any $z \in [x, y]$ and any $H \in \mathcal{H}(z)$ we have that either x or y belongs to H, and hence $|K \cap H| \geqslant \delta$, therefore $\varphi_{\mu_K}(z) \geqslant \delta$, In fact, it is not hard to show that

$$T_\delta = K_\delta.$$

Schütt and Werner proved in [SW90] that for every convex body K in $\mathbb{R}^n$ one has that

$$\lim_{\delta\to 0} \frac{|K| - |K_\delta|}{\delta^{\frac{2}{n+1}}} = \frac{1}{2}\left(\frac{n+1}{\omega_{n-1}}\right)^{\frac{2}{n+1}} \mathrm{as}(K).$$

In particular, if $|K| = 1$ then there exists $\delta_0 > 0$ such that if $0 < \delta < \delta_0$ then

$$1 - |K_\delta| \leqslant c_2 n \delta^{\frac{2}{n+1}}.$$

We can exploit these results as follows. Recall that $\varphi_{\mu_K}(x) \leqslant \exp(-\Lambda^*_{\mu_K}(x))$ and hence, for any $s > 0$ we have that

$$\exp(s\Lambda^*_{\mu_K}(x)) \leqslant (\varphi_{\mu_K}(x))^{-s}.$$

Then,

$$\begin{aligned}\int_K \exp(s\Lambda^*_{\mu_K}(x))\,dx &\leqslant \int_K \frac{1}{(\varphi_{\mu_K}(x))^s}dx = \int_0^\infty \mu_K(\{1/\varphi^s_{\mu_K} \geqslant t\})\,dt \\ &= \int_0^\infty \mu_K(\{\varphi_{\mu_K} \leqslant 1/t^{1/s}\})\,dt = \int_0^\infty (1 - \mu_K(T_{t^{-1/s}}))\,dt \\ &= \int_0^\infty (1 - \mu_K(T_{t^{-1/s}}))\,dt.\end{aligned}$$

Let $t_0 = \delta_0^{-s}$. Then,

$$\int_{t_0}^{\infty} (1 - \mu_K(T_{t^{-1/s}}))\, dt \leqslant c_2 n \int_{t_0}^{\infty} t^{-\frac{2}{s(n+1)}}\, dt < +\infty$$

provided that $s < \frac{2}{n+1}$. This proves the following.

Theorem 2.6.5 *Let K be a convex body of volume 1 in $\mathbb{R}^n$. Then, for any $0 < s < \frac{2}{n+1}$,*

$$\int_{\mathbb{R}^n} \exp(s\Lambda^*_{\mu_K}(x))\, d\mu_K(x) < \infty.$$

In particular, for all $p \geqslant 1$ we have that $\mathbb{E}_{\mu_K}\left((\Lambda^*_{\mu_K}(x))^p\right) < \infty$.

One can also obtain a generalization of this fact, with a similar argument, using results of Besau, Ludwig and Werner on weighted floating bodies. More precisely, from [BLW18, Theorem 1.1] it follows that the conclusion of Theorem 2.6.5 still holds if we replace μ_K by any probability measure with continuous and strictly positive density $\psi : K \to (0, +\infty)$. These ideas might prove useful in the study of the question for an arbitrary log-concave probability measure.

The next result concerns the one-dimensional case. Let μ be a centered probability measure on $\mathbb{R}$ which is absolutely continuous with respect to Lebesgue measure and consider a random variable X, on some probability space $(\Omega, \mathcal{F}, P)$, with distribution μ, i.e., $\mu(B) := P(X \in B),\ B \in \mathcal{B}(\mathbb{R})$. We define

$$\alpha_+ = \alpha_+(\mu) := \sup\{x \in \mathbb{R}\colon \mu([x, \infty)) > 0\}) \quad \text{and}$$

$$\alpha_- = \alpha_-(\mu) := \sup\{x \in \mathbb{R}\colon \mu((-\infty, -x])) > 0\}).$$

Thus, $-\alpha_-, \alpha_+$ are the endpoints of the support of μ. Note that we may have $\alpha_\pm = +\infty$. We define $I_\mu = (-\alpha_-, \alpha_+)$. Recall that

$$\Lambda^*_\mu(x) := \sup\{tx - \Lambda_\mu(t)\colon t \in \mathbb{R}\}, \qquad x \in \mathbb{R}.$$

In fact, since $tx - \Lambda_\mu(t) < 0$ for $t < 0$ when $x \in [0, \alpha_+)$, we have that $\Lambda^*_\mu(x) = \sup\{tx - \Lambda_\mu(t)\colon t \geqslant 0\}$ in this case, and similarly $\Lambda^*_\mu(x) := \sup\{tx - \Lambda_\mu(t)\colon t \leqslant 0\}$ when $x \in (-\alpha_-, 0]$. One can also check that $\Lambda^*_\mu(\alpha_\pm) = +\infty$. See [GG09, Lemma 2.8] for the case $\alpha_\pm < +\infty$. In the case $\alpha_\pm = \pm\infty$, the convexity and monotonicity properties of Λ^*_μ imply again that $\lim_{t\to\pm\infty} \Lambda^*_\mu(t) = +\infty$.

Proposition 2.6.6 *Let μ be a centered probability measure on $\mathbb{R}$ which is absolutely continuous with respect to Lebesgue measure and let $I_\mu = \mathrm{supp}(\mu)$. Then,*

$$\int_{I_\mu} \exp(\Lambda^*_\mu(x)/2)\, d\mu(x) \leqslant 4.$$

Proof Let $F(x) = \mu(-\infty, x]$. For any $x \in [0, \alpha_+)$ and $t \geqslant 0$ we have

$$\min\{F(x), 1 - F(x)\} = \varphi_\mu(x) \leqslant e^{-\Lambda_\mu^*(x)}.$$

It follows that

$$\begin{aligned}\int_{I_\mu} e^{\Lambda_\mu^*(x)/2} d\mu(x) &\leqslant \int_{I_\mu} \frac{1}{\sqrt{\min\{F(x), 1 - F(x)\}}} f(x)\, dx \\ &\leqslant \int_{I_\mu} \frac{1}{\sqrt{F(x)}} f(x)\, dx + \int_{I_\mu} \frac{1}{\sqrt{1 - F(x)}} f(x)\, dx.\end{aligned} \tag{2.30}$$

Write f for the density of μ with respect to Lebesgue measure. Then, $(1 - F)'(x) = -f(x)$, which implies that

$$\int_0^{\alpha_+} \frac{1}{\sqrt{1 - F(x)}} f(x)\, dx \leqslant -\int_0^{\alpha_+} \frac{1}{\sqrt{1 - F(x)}} (1 - F)'(x) dx =$$

$$= -2\sqrt{1 - F(x)}\Big|_0^{\alpha_+} = 2\sqrt{1 - F(0)}$$

since $F(\alpha_+) = 1$. In the same way we check that

$$\int_{-\alpha_-}^0 \frac{1}{\sqrt{1 - F(x)}} f(x)\, dx \leqslant -\int_{-\alpha_-}^0 \frac{1}{\sqrt{1 - F(x)}} (1 - F)'(x) dx =$$

$$= -2\sqrt{1 - F(x)}\Big|_{-\alpha_-}^0 = 2 - 2\sqrt{1 - F(0)}.$$

This shows that

$$\int_{I_\mu} \frac{1}{\sqrt{1 - F(x)}} f(x)\, dx \leqslant 2.$$

In a similar way we obtain the same upper bound for the second summand in (2.30) and the result follows. □

Proposition 2.6.6 can be extended to products. Let μ_i, $1 \leqslant i \leqslant n$ be centered probability measures on $\mathbb{R}$, all of them absolutely continuous with respect to Lebesgue measure. If $\overline{\mu} = \mu_1 \otimes \cdots \otimes \mu_n$ then $I_{\overline{\mu}} = \prod_{i=1}^n I_{\mu_i}$ and we can easily check that

$$\Lambda_{\overline{\mu}}^*(x) = \sum_{i=1}^n \Lambda_{\mu_i}^*(x_i)$$

for all $x = (x_1, \dots, x_n) \in I_{\overline{\mu}}$, which implies that

$$\int_{I_{\overline{\mu}}} e^{\Lambda^*_{\overline{\mu}}(x)/2} d\overline{\mu}(x) = \prod_{i=1}^{n} \left(\int_{I_{\mu_i}} e^{\Lambda^*_{\mu_i}(x_i)/2} d\mu_i(x_i) \right) \leqslant 4^n.$$

In particular, for all $p \geqslant 1$ we have that

$$\int_{I_{\overline{\mu}}} (\Lambda^*_{\overline{\mu}}(x))^p \, d\overline{\mu}(x) < +\infty.$$

We close this section with one more case where we can establish that Λ^*_{μ} has finite moments of all orders. We consider an arbitrary centered log-concave probability measure on $\mathbb{R}^n$ but we have to assume that it satisfies some additional condition on the growth of its one-sided L_t-centroid bodies $Z_t^+(\mu)$; namely, that the family of the one-sided L_t-centroid bodies grows with some mild rate as $t \to \infty$ (note that the assumption in the next proposition can be satisfied only for log-concave probability measures μ with support $\mathrm{supp}(\mu) = \mathbb{R}^n$).

Proposition 2.6.7 *Let μ be a centered log-concave probability measure on $\mathbb{R}^n$. Assume that there exists an increasing function $g : [1, \infty) \to [1, \infty)$ with $\lim_{t\to\infty} g(t)/\ln(t+1) = +\infty$ such that $Z_t^+(\mu) \supseteq g(t) Z_2^+(\mu)$ for all $t \geqslant 2$. Then,*

$$\int_{\mathbb{R}^n} |\Lambda^*_{\mu}(x)|^p d\mu(x) < +\infty$$

for every $p \geqslant 1$.

Proof In Lemma 2.3.4 we saw that if $t \geqslant 1$ then for every $x \in \frac{1}{2} Z_t^+(\mu)$ we have

$$\varphi_{\mu}(x) \geqslant e^{-c_1 t},$$

where $c_1 > 1$ is an absolute constant. Since $\Lambda^*_{\mu}(x) \leqslant \ln \frac{1}{\varphi_{\mu}(x)}$, this shows that $\Lambda^*_{\mu}(x) \leqslant c_1 t$ for all $x \in \frac{1}{2} Z_t^+(\mu)$. In other words,

$$\frac{1}{2} Z^+_{t/c_1}(\mu) \subseteq B_t(\mu), \qquad t \geqslant c_1. \tag{2.31}$$

Since $\lim_{t\to\infty} g(t) = +\infty$, there exists $t_0 \geqslant c_1$ such that $\mu\left(\frac{g(t_0/c_1)}{2} Z_2^+(\mu)\right) \geqslant 2/3$. From Borell's lemma [BGVV14, Lemma 2.4.5] we know that, for all $t \geqslant t_0$,

$$1 - \mu\left(\frac{g(t/c_1)}{2} Z_2^+(\mu)\right) \leqslant e^{-c_2 g(t/c_1)/g(t_0/c_1)},$$

where $c_2 > 0$ is an absolute constant. We write

$$\int_{\mathbb{R}^n} |\Lambda_\mu^*(x)|^p d\mu(x) = \int_0^\infty pt^{p-1}\mu(\{x : \Lambda_\mu^*(x) \geqslant t\})\, dt = p\int_0^\infty t^{p-1}(1 - \mu(B_t(\mu)))\, dt.$$

From (2.31) it follows that

$$1 - \mu(B_t(\mu)) \leqslant 1 - \mu\left(\frac{1}{2}Z_{t/c_1}^+(\mu)\right) \leqslant 1 - \mu\left(\frac{g(t/c_1)}{2}Z_2^+(\mu)\right) \leqslant e^{-c_2 g(t/c_1)/g(t_0/c_1)}$$

for all $t \geqslant t_0$. Since $\lim_{t\to\infty} g(t)/\ln(t+1) = +\infty$, there exists $t_p \geqslant t_0$ such that

$$(p-1)\ln(t) \leqslant \frac{c_2}{2g(t_0/c_1)} g(t/c_1)$$

for all $t \geqslant t_p$. Assume that $p > 2$. Then, from the previous observations we get

$$\begin{aligned} p\int_{t_p}^\infty t^{p-1}(1 - \mu(B_t(\mu)))\, dt &\leqslant p\int_{t_p}^\infty t^{p-1}\left(1 - \mu\left(\tfrac{g(t/c_1)}{2}Z_2^+(\mu)\right)\right) dt \\ &\leqslant p\int_{t_p}^\infty t^{p-1}t^{-2(p-1)}\, dt = p\int_{t_p}^\infty t^{-(p-1)}\, dt < \infty. \end{aligned}$$

This proves the result for $p > 2$ and then from Hölder's inequality it is clear that the assertion of the proposition is also true for all $p \geqslant 1$. □

Note It is not hard to construct examples of log-concave probability measures, even on the real line, for which $\mathrm{supp}(\mu) = \mathbb{R}^n$ but the assumption of Proposition 2.6.7 is not satisfied. Consider for example a measure μ on $\mathbb{R}$ with density $f(x) = c \cdot \exp(-\psi)$ where ψ is an even convex function rapidly increasing to infinity, e.g. $\psi(t) = e^{t^2}$.

However, this does not exclude the possibility that for every centered log-concave probability measure μ on $\mathbb{R}^n$ the function Λ_μ^* has finite second or higher moments.

2.7 Threshold for the Measure: The Approach and Examples

Let μ be a log-concave probability measure on $\mathbb{R}^n$ such that $\|\Lambda_\mu^*\|_{L^2(\mu)} < \infty$. We define the parameter

$$\beta(\mu) = \frac{\mathrm{Var}_\mu(\Lambda_\mu^*)}{(\mathbb{E}_\mu(\Lambda_\mu^*))^2} \tag{2.32}$$

We shall give a general estimate for the upper threshold $\varrho_1(\mu, \delta)$ in terms of $\beta(\mu)$.

Theorem 2.7.1 *Let $\beta, \delta > 0$ with $8\beta < \delta < 1$. If μ is a log-concave probability measure on $\mathbb{R}^n$ with $\beta(\mu) = \beta$ and $n/L_\mu^2 \geqslant c_2 \ln(2/\delta)\sqrt{\delta/\beta(\mu)}$, then*

$$\varrho_1(\mu, \delta) \geqslant \left(1 - \sqrt{8\beta(\mu)/\delta}\right) \frac{\mathbb{E}_\mu(\Lambda_\mu^*)}{n}.$$

Proof Recall that $B_t(\mu) = \{x \in \mathbb{R}^n : \Lambda_\mu^*(x) \leqslant t\}$. We use Lemma 2.3.7 in the following way. Let $m := \mathbb{E}_\mu(\Lambda_\mu^*)$. Then, for all $\varepsilon \in (0, 1)$, from Chebyshev's inequality we have that

$$\mu(\{\Lambda_\mu^* \leqslant m - \varepsilon m\}) \leqslant \mu(\{|\Lambda_\mu^* - m| \geqslant \varepsilon m\}) \leqslant \frac{\mathbb{E}_\mu|\Lambda_\mu^* - m|^2}{\varepsilon^2 m^2} = \frac{\beta(\mu)}{\varepsilon^2}.$$

Equivalently,

$$\mu(B_{(1-\varepsilon)m}(\mu)) \leqslant \frac{\beta(\mu)}{\varepsilon^2}.$$

Let $\delta \in (\beta(\mu), 1)$. Since $8\beta(\mu) < \delta < 1$, choosing $\varepsilon = \sqrt{2\beta(\mu)/\delta}$ we have that

$$\mu(B_{(1-\varepsilon)m}(\mu)) \leqslant \frac{\delta}{2}.$$

Then, from Lemma 2.3.7 we see that

$$\begin{aligned}\sup\{\mathbb{E}_{\mu^N}(\mu(K_N)) : N \leqslant e^{(1-2\varepsilon)m}\} &\leqslant \mu(B_{(1-\varepsilon)m}(\mu)) + e^{(1-2\varepsilon)m}e^{-(1-\varepsilon)m} \\ &\leqslant \frac{\delta}{2} + e^{-\varepsilon m} \leqslant \delta,\end{aligned}$$

provided that $\varepsilon m \geqslant \ln(2/\delta)$. Since $m \geqslant c_1 n/L_\mu^2$, this condition is satisfied if $n/L_\mu^2 \geqslant c_2 \ln(2/\delta)\sqrt{\delta/\beta(\mu)}$. By the choice of ε we conclude that $\varrho_1(\mu, \delta) \geqslant \left(1 - \sqrt{8\beta(\mu)/\delta}\right) \frac{\mathbb{E}_\mu(\Lambda_\mu^*)}{n}$. □

For the proof of the lower threshold we work in a similar way, using Lemma 2.3.8. In the case of the uniform measure on a convex body we obtain the next theorem.

Theorem 2.7.2 *Let $\beta, \delta > 0$ with $2\beta < \delta < 1$. If K is a centered convex body of volume 1 in $\mathbb{R}^n$ with $\beta(\mu_K) = \beta$ and $n/L_{\mu_K}^2 \geqslant c_2 \ln(2/\delta)\sqrt{\delta/\beta}$ then*

$$\varrho_2(\mu_K, \delta) \leqslant \left(1 + \sqrt{8\beta/\delta}\right) \frac{\mathbb{E}_{\mu_K}(\Lambda_{\mu_K}^*)}{n}.$$

Proof Note that if $m := \mathbb{E}_\mu(\Lambda_\mu^*)$ then as before, for all $\varepsilon \in (0, 1)$, from Chebyshev's inequality we have that

$$\mu(\{\Lambda_\mu^* \geqslant m + \varepsilon m\}) \leqslant \mu(\{|\Lambda_\mu^* - m| \geqslant \varepsilon m\}) \leqslant \frac{\beta(\mu)}{\varepsilon^2}.$$

If $\beta(\mu) < 1/2$ and $2\beta(\mu) < \delta < 1$ then, choosing $\varepsilon = \sqrt{2\beta(\mu)/\delta}$ we have that

$$\mu(B_{(1+\varepsilon)m}(\mu)) \geqslant 1 - \frac{\delta}{2}.$$

Therefore, we will have that

$$\varrho_2(\mu, \delta) \leqslant (1 + 2\varepsilon)m/n$$

if our lower bound for $\inf_{x\in B_{(1+\varepsilon)m}(\mu)} \varphi_\mu(x)$ gives

$$2\binom{N}{n}\left(1 - \inf_{x\in B_{(1+\varepsilon)m}(\mu)} \varphi_\mu(x)\right)^{N-n} \leqslant \frac{\delta}{2} \tag{2.33}$$

for all $N \geqslant N_0 := \exp((1 + 2\varepsilon)m)$. Recall that in the case of the uniform measure on a centered convex body of volume 1, Theorem 2.5.1 shows that

$$\inf_{x\in B_{(1+\varepsilon)m}(\mu_K)} \varphi_{\mu_K}(x) \geqslant \frac{1}{10}\exp(-(1 + \varepsilon)m - 2\sqrt{n}).$$

We require that n and m are large enough so that $1/2^n < \delta/2$ and $2\sqrt{n} \leqslant \frac{\varepsilon m}{2}$. Using also the fact that $\binom{N}{n} \leqslant e^{-1}\left(\frac{eN}{n}\right)^n$ we see that (2.33) will be satisfied if we also have

$$\left(\frac{2eN}{n}\right)^n e^{-\frac{N-n}{10}e^{-(1+3\varepsilon/2)m}} < 1.$$

Setting $x := N/n$ we see that this last inequality is equivalent to

$$e^{(1+3\varepsilon/2)m} < \frac{x-1}{10\ln(2ex)}.$$

One can now check that if $N \geqslant \exp((1 + 2\varepsilon)m)$ then all the restrictions are satisfied if we assume that $n/L_{\mu_K}^2 \geqslant c_2\ln(2/\delta)\sqrt{\delta/\beta(\mu_K)}$. □

For an arbitrary log-concave probability measure we can use the recent estimate (2.27) of Brazitikos and Chasapis. Using this lower bound for $\varphi_-(B_t(\mu))$ instead of Theorem 2.5.1 we arrive at a similar conclusion.

Theorem 2.7.3 *Let $\beta, \delta > 0$ with $128\beta < \delta < 1$. If $n \geqslant n_0(\beta, \delta)$ then for any log-concave probability measure μ on $\mathbb{R}^n$ with $\beta(\mu) = \beta$ we have that*

$$\varrho_2(\mu, \delta) \leqslant \left(1 + c\sqrt{\beta(\mu)/\delta}\right)\frac{\mathbb{E}_\mu(\Lambda_\mu^*)}{n},$$

where $c > 0$ is an absolute constant.

From the discussion in this section it is clear that our approach is able to provide good bounds for the threshold $\varrho(\mu,\delta)$ if the parameter $\beta(\mu)$ is small, especially if $\beta(\mu)=o_n(1)$ as the dimension increases. We illustrate this with a number of examples, starting with the uniform measure on the cube. Let μ_{C_n} be the uniform measure on $C_n=\left[-\frac{1}{2},\frac{1}{2}\right]^n$. Since $\mu_{C_n}=\mu_{C_1}\otimes\cdots\otimes\mu_{C_1}$ we have

$$\mathrm{Var}_{\mu_{C_n}}(\Lambda^*_{\mu_{C_n}})=n\mathrm{Var}_{\mu_{C_1}}(\Lambda^*_{\mu_{C_1}})\qquad\text{and}\qquad\mathbb{E}_{\mu_{C_n}}(\Lambda^*_{\mu_{C_n}})=n\mathbb{E}_{\mu_{C_1}}(\Lambda^*_{\mu_{C_1}}).$$

Therefore,

$$\beta(\mu_{C_n})=\frac{\mathrm{Var}_{\mu_{C_n}}(\Lambda^*_{\mu_{C_n}})}{(\mathbb{E}_{\mu_{C_n}}(\Lambda^*_{\mu_{C_n}}))^2}=\frac{\beta(\mu_{C_1})}{n}\longrightarrow 0.$$

as $n\to\infty$. Then, Theorems 2.7.1 and 2.7.2 show that for any $\delta\in(0,1)$ there exists $n_0(\delta)$ such that, for any $n\geqslant n_0$,

$$\varrho(\mu_{C_n},\delta)\leqslant\frac{c}{\sqrt{\delta n}},$$

where $c>0$ is an absolute constant. This estimate provides a sharp threshold for the measure of a random polytope K_N with independent vertices uniformly distributed in C_n. It provides a direct proof of the result of Dyer et al. in [DFM92] with a stronger estimate for the "width of the threshold".

Next, let us consider the example of the standard n-dimensional Gaussian measure γ_n with density $f_{\gamma_n}(x)=(2\pi)^{-n/2}e^{-|x|^2/2}$, $x\in\mathbb{R}^n$. Note that $\gamma_n=\gamma_1\otimes\cdots\otimes\gamma_1$, and hence we may argue as in the previous example. We may also use direct computation to see that $\Lambda_{\gamma_n}(\xi)=|\xi|^2/2$ for all $\xi\in\mathbb{R}^n$ and $\Lambda^*_{\gamma_n}(x)=|x|^2/2$ for all $x\in\mathbb{R}^n$. It follows that $B_t(\gamma_n)=\sqrt{2t}B_2^n$. We check that if $x\in\partial(B_t(\gamma_n))$ then

$$\varphi_{\gamma_n}(x)=\frac{1}{\sqrt{2\pi}}\int_{\sqrt{2t}}^{\infty}e^{-u^2/2}du\geqslant\frac{c}{\sqrt{t}}e^{-t}$$

for all $t\geqslant 1$ (see [IM65, p. 17] for a refined form of the lower bound that we use). By the standard concentration estimate for the Euclidean norm with respect to γ_n (see [Ver18, Theorem 3.1.1]), for all $s>0$ we have that

$$\gamma_n(\{x\in\mathbb{R}^n:|\,|x|-\sqrt{n}\,|\geqslant s\sqrt{n}\})\leqslant 2\exp(-cs^2n),$$

where $c>0$ is an absolute constant. This shows that

$$\max\{\gamma_n((1-s)\sqrt{n}B_2^n),\,1-\gamma_n((1+s)\sqrt{n}B_2^n)\}\leqslant 2\exp(-cs^2n)$$

for every $s \in (0,1)$. From Lemma 2.3.7 we see that $\varrho_1(\gamma_n,\delta) \geqslant \frac{1}{2} - \frac{c_1\sqrt{\ln(4/\delta)}}{\sqrt{n}}$, and applying Lemma 2.3.8 we get $\varrho_2(\gamma_n,\delta) \leqslant \frac{1}{2} + \frac{c_2\sqrt{\ln(4/\delta)}}{\sqrt{n}}$. Combining the above we get

$$\varrho(\gamma_n,\delta) \leqslant \frac{C\sqrt{\ln(4/\delta)}}{\sqrt{n}},$$

where $C > 0$ is an absolute constant.

Finally, we discuss the example of the uniform measure on the Euclidean ball. It was proved in [BCG+19] that if $\varepsilon \in (0,1)$ and $K_N = \mathrm{conv}\{x_1,\dots,x_N\}$ where $x_1,\dots,x_N$ are random points independently and uniformly chosen from B_2^n then

$$\lim_{n\to\infty} \sup\left\{\frac{\mathbb{E}|K_N|}{|B_2^n|} : N \leqslant \exp\left((1-\varepsilon)\left(\frac{n+1}{2}\right)\ln n\right)\right\} = 0$$

and

$$\lim_{n\to\infty} \inf\left\{\frac{\mathbb{E}|K_N|}{|B_2^n|} : N \geqslant \exp\left((1+\varepsilon)\left(\frac{n+1}{2}\right)\ln n\right)\right\} = 1.$$

We shall obtain a similar conclusion with the approach of this work.

For any centered convex body K of volume 1 in $\mathbb{R}^n$ recall the definition of $\omega_{\mu_K} = \ln(1/\varphi_{\mu_K})$ in (2.25) and consider the parameter

$$\tau(\mu_K) = \frac{\mathrm{Var}_{\mu_K}(\omega_{\mu_K})}{(\mathbb{E}_{\mu_K}(\omega_{\mu_K}))^2}. \tag{2.34}$$

From (2.26) we know that $\omega_{\mu_K}(x) - 5\sqrt{n} \leqslant \Lambda^*_{\mu_K}(x) \leqslant \omega_{\mu_K}(x)$ for every $x \in \mathrm{int}(K)$. This allows us to compare $\beta(\mu_K)$ with $\tau(\mu_K)$; some simple calculations show that

$$\beta(\mu_K) = \left(\tau(\mu_K) + O(L^2_{\mu_K}/\sqrt{n})\right)\left(1 + O(L^2_{\mu_K}/\sqrt{n})\right) \tag{2.35}$$

Note also that if K is a centered convex body in $\mathbb{R}^n$ and $r > 0$ then $\Lambda^*_{\mu_{rK}}(x) = \Lambda^*_{\mu_K}(x/r)$ for all $x \in \mathbb{R}^n$, where μ_{rK} is the uniform measure on rK. It follows that

$$\frac{1}{|rK|}\int_{rK}[\Lambda^*_{\mu_{rK}}(x)]^p dx = \frac{1}{|K|}\int_K [\Lambda^*_{\mu_K}(x)]^p dx$$

for every $p > 0$ and $r > 0$. This shows that if D_n is the centered Euclidean ball of volume 1 in $\mathbb{R}^n$ then in order to compute $\beta(\mu_{D_n})$ it suffices to compute the ratio

$$\beta(\mu_{D_n}) + 1 = \frac{\frac{1}{|B_2^n|}\int_{B_2^n}\Lambda^*(x)^2 dx}{\left(\frac{1}{|B_2^n|}\int_{B_2^n}\Lambda^*(x) dx\right)^2}$$

where $\Lambda^* := \Lambda^*_{\mu_{B_2^n}}$ and, because of (2.35), it is enough to compute $\tau(\mu_{B_2^n})$. Set $\omega := \omega_{\mu_{B_2^n}}$. Then, $\omega(x) = \ln(1/\varphi(x))$ where $\varphi(x) = F(|x|)$,

$$F(r) = c_n \int_r^1 (1-t^2)^{\frac{n-1}{2}} dt, \qquad r \in [0,1]$$

and $c_n = \pi^{-1/2}\Gamma(\frac{n}{2}+1)/\Gamma(\frac{n+1}{2})$. From [BCG+19, Lemma 2.2] we know that

$$F(r) = (1-r^2)^{\frac{n+1}{2}} h(r,n),$$

where

$$\frac{1}{\sqrt{2\pi(n+2)}} \leqslant h(r,n) \leqslant \frac{1}{r\sqrt{2\pi n}} \tag{2.36}$$

for all $r \in (0,1]$. We assume that n is even (the case where n is odd can be treated in a similar way). Using polar coordinates we compute

$$\frac{1}{|B_2^n|}\int_{B_2^n} \omega(x)\,dx = \frac{n+1}{2} H_{\frac{n}{2}} + O(\ln n) \tag{2.37}$$

and

$$\frac{1}{|B_2^n|}\int_{B_2^n} (\omega(x))^2\,dx = \frac{(n+1)^2}{4} H_{\frac{n}{2}}^2 + O(n^2). \tag{2.38}$$

From (2.37) and (2.38) we finally get

$$\tau(\mu_{B_2^n}) = O(1/(\ln n)^2).$$

Then, (2.35) and a simple computation show that

$$\beta(\mu_{D_n}) = \left(\tau(\mu_{B_2^n}) + O(L^2_{\mu_{B_2^n}}/\sqrt{n})\right)\left(1 + O(L^2_{\mu_{B_2^n}}/\sqrt{n})\right) = O(1/(\ln n)^2),$$

because $L_{\mu_{B_2^n}} \approx 1$. Finally, note that by the estimate (2.26) in Corollary 2.5.3 we have

$$\mathbb{E}_{\mu_n}(\Lambda^*_{\mu_n}) = \frac{1}{|B_2^n|}\int_{B_2^n} \omega(x)\,dx + O(\sqrt{n}) = \frac{(n+1)}{2} H_{\frac{n}{2}} + O(\sqrt{n})$$

as $n \to \infty$. Combining the above we see that $\beta(\mu_{D_n}) = o_n(1)$ but $\varrho(\mu_{D_n}, \delta) = O(1/\sqrt{\delta})$.

Note The above discussion leaves open the following question: to estimate

$$\beta_n^* := \sup\{\beta(\mu_K) : K \text{ is a centered convex body of volume 1 in } \mathbb{R}^n\}$$

or, more generally,

$$\beta_n := \sup\{\beta(\mu) : \mu \text{ is a centered log-concave probability measure on } \mathbb{R}^n\}.$$

2.8 Product Measures

In [GG09] a threshold for $\mathbb{E}_{\mu^N}|K_N|/(2\alpha)^n$ was established for the case where X_i have independent identically distributed coordinates supported on a bounded interval, under some mild additional assumptions (see below for a more precise description). This result was generalized by Pafis in [Paf24] as follows. Let μ be an even Borel probability measure on the real line and let $X_1, \dots, X_n$ be independent and identically distributed random variables, defined on some probability space $(\Omega, \mathcal{F}, P)$, each with distribution μ. Consider the random vector $\mathbf{X} = (X_1, \dots, X_n)$ and, for a fixed N satisfying $N > n$, consider N independent copies $\mathbf{X}_1, \dots, \mathbf{X}_N$ of $\mathbf{X}$. The distribution of $\mathbf{X}$ is $\mu_n := \mu \otimes \cdots \otimes \mu$ (n times) and the distribution of $(\mathbf{X}_1, \dots, \mathbf{X}_N)$ is $\mu_n^N := \mu_n \otimes \cdots \otimes \mu_n$ (N times). The goal is to obtain a sharp threshold for the expected μ_n-measure of the random polytope

$$K_N := \mathrm{conv}\{\mathbf{X}_1, \dots, \mathbf{X}_N\}.$$

Assume that μ is non-degenerate, i.e. $\mathrm{Var}(X) > 0$. Let

$$x^* = x^*(\mu) := \sup\{x \in \mathbb{R} : \mu([x, \infty)) > 0\}$$

be the right endpoint of the support of μ and set $I_\mu = (-x^*, x^*)$. Note that since μ is non-degenerate and even, we have that $x^* > 0$. As usual, let

$$M_\mu(t) := \mathbb{E}(e^{tX}) := \int_{\mathbb{R}} e^{tx}\, d\mu(x), \qquad t \in \mathbb{R}$$

denote the moment generating function of X, and let $\Lambda_\mu(t) := \ln M_\mu(t)$ be its logarithmic moment generating function. Finally, consider the Legendre transform $\Lambda_\mu^* : I_\mu \to \mathbb{R}$ of Λ_μ.

We say that μ is *admissible* if it is non-degenerate, i.e. $\mathrm{Var}_\mu(X) > 0$, and satisfies the following conditions:

(i) There exists $r > 0$ such that $\mathbb{E}(e^{tX}) < \infty$ for all $t \in (-r, r)$; in particular, X has finite moments of all orders.
(ii) One of the following holds: (1) $x^* < +\infty$ and $P(X = x^*) = 0$, or (2) $x^* = +\infty$ and $\{\Lambda_\mu < \infty\} = \mathbb{R}$, or (3) $x^* = +\infty$, $\{\Lambda_\mu < \infty\}$ is bounded and μ is log-concave.

Finally, we say that μ satisfies *the Λ^*-condition* if

$$\lim_{x\uparrow x^*}\frac{-\ln\mu([x,\infty))}{\Lambda_\mu^*(x)}=1.$$

Theorem 2.8.1 *Let μ be an admissible even probability measure on $\mathbb{R}$ that satisfies the Λ^*-condition. Then, for any $\delta\in\left(0,\frac{1}{2}\right)$ and any $\varepsilon\in(0,1)$ there exists $n_0(\mu,\delta,\varepsilon)$ such that*

$$\varrho_1(\mu_n,\delta)\geqslant(1-\varepsilon)\mathbb{E}_\mu(\Lambda_\mu^*)\quad\textit{and}\quad\varrho_2(\mu_n,\delta)\leqslant(1+\varepsilon)\mathbb{E}_\mu(\Lambda_\mu^*)$$

for every $n\geqslant n_0(\mu,\delta,\varepsilon)$. In particular, $\{\mu_n\}_{n=1}^\infty$ exhibits a sharp threshold, i.e. $\lim_{n\to\infty}\varrho(\mu_n,\delta)=0$, with "threshold constant" $\mathbb{E}_\mu(\Lambda_\mu^*)$.

An application of Theorem 2.8.1 is also given to the case of the product p-measure $\nu_p^n:=\nu_p^{\otimes n}$. For any $p\geqslant 1$ we denote by ν_p the probability distribution on $\mathbb{R}$ with density $(2\gamma_p)^{-1}\exp(-|x|^p)$, where $\gamma_p=\Gamma(1+1/p)$. It is proved in [Paf24] that ν_p satisfies the Λ^*-condition.

Theorem 2.8.2 *For any $p\geqslant 1$ we have that*

$$\lim_{x\to\infty}\frac{-\ln(\nu_p[x,\infty))}{\Lambda_{\nu_p}^*(x)}=1.$$

Note that the measure ν_p is admissible for all $1\leqslant p<\infty$; it satisfies condition (ii-3) if $p=1$ and condition (ii-2) for all $1<p<\infty$. Therefore, Theorem 2.8.2 implies that if K_N is the convex hull of $N>n$ independent random vectors $\mathbf{X}_1,\dots,\mathbf{X}_N$ with distribution ν_p^n then the expected measure $\mathbb{E}_{(\nu_p^n)^N}(\nu_p^n(K_N))$ exhibits a sharp threshold at $N=\exp((1\pm\varepsilon)\mathbb{E}_{\nu_p}(\Lambda_{\nu_p}^*)n)$.

The variant of this question that was studied in [GG09] dealt with the case where μ is an even, compactly supported, Borel probability measure on the real line, $\mu_n(K_N)$ is replaced by the volume of K_N, and

$$\kappa=\kappa(\mu):=\frac{1}{2x^*}\int_{-x^*}^{x^*}\Lambda_\mu^*(x)dx.$$

If $0<\kappa(\mu)<\infty$ then one has that, for every $\varepsilon\in(0,\kappa)$,

$$\lim_{n\to\infty}\sup\left\{(2x^*)^{-n}\mathbb{E}(|K_N|)\colon N\leqslant\exp((\kappa-\varepsilon)n)\right\}=0\tag{2.39}$$

and if the distribution μ satisfies the Λ^*-condition then one also has

$$\lim_{n\to\infty}\inf\left\{(2x^*)^{-n}\mathbb{E}(|K_N|)\colon N\geqslant\exp((\kappa+\varepsilon)n)\right\}=1.\tag{2.40}$$

References

[AAGM15] S. Artstein-Avidan, A. Giannopoulos, and V. D. Milman. *Asymptotic Geometric Analysis, Vol. I*, volume 202 of *Mathematical Surveys and Monographs*. American Mathematical Society, Providence, RI, 2015.

[Bal88] K. M. Ball. Logarithmically concave functions and sections of convex sets in $\mathbb{R}^n$. *Studia Math.*, 88:69–84, 1988.

[BC24] S. Brazitikos and G. Chasapis. Sharp estimates for the cramér transform of log-concave measures and geometric applications. *Adv. Math.* 478(2025), Article 110407. arXiv:2405.07253, 2024.

[BCG+19] G. Bonnet, G. Chasapis, J. Grote, D. Temesvari, and N. Turchi. Threshold phenomena for high-dimensional random polytopes. *Commun. Contemp. Math.*, 21(5):1850038, 30 pp., 2019.

[BGP24a] S. Brazitikos, A. Giannopoulos, and M. Pafis. Half-space depth of log-concave probability measures. *Probab. Theory Related Fields*, 188:309–336, 2024.

[BGP24b] S. Brazitikos, A. Giannopoulos, and M. Pafis. Threshold for the expected measure of random polytopes. *Math. Ann.*, 388(3):2991–3024, 2024.

[BGVV14] S. Brazitikos, A. Giannopoulos, P. Valettas, and B-H. Vritsiou. *Geometry of isotropic convex bodies*, volume 196 of *Mathematical Surveys and Monographs*. American Mathematical Society, Providence, RI, 2014.

[BKT21] G. Bonnet, Z. Kabluchko, and N. Turchi. Phase transition for the volume of high-dimensional random polytopes. *Random Structures Algorithms*, 58(4):648–663, 2021.

[BLW18] F. Besau, M. Ludwig, and E. Werner. Weighted floating bodies and polytopal approximation. *Trans. Amer. Math. Soc.*, 370:7129–7148, 2018.

[Bor74] C. Borell. Convex measures on locally convex spaces. *Ark. Mat.*, 12:239–252, 1974.

[Bor75] C. Borell. Convex set functions in d-space. *Period. Math. Hungar.*, 6:111–136, 1975.

[Bou91] J. Bourgain. On the distribution of polynomials on high dimensional convex sets. In *Lecture Notes in Mathematics*, volume 1469, pages 127–137, Berlin, 1991. Springer.

[Che21] Y. Chen. An almost constant lower bound of the isoperimetric coefficient in the kls conjecture. *Geom. Funct. Anal.*, 31:34–61, 2021.

[CTV21] D. Chakraborti, T. Tkocz, and B-H. Vritsiou. A note on volume thresholds for random polytopes. *Geom. Dedicata*, 213:423–431, 2021.

[DFM92] M. E. Dyer, Z. Füredi, and C. McDiarmid. Volumes spanned by random points in the hypercube. *Random Structures Algorithms*, 3:91–106, 1992.

[DZ98] A. Dembo and O. Zeitouni. *Large Deviations Techniques and Applications*. Springer, New York, 1998.

[FMW16] M. Fradelizi, M. Madiman, and L. Wang. Optimal concentration of information content for log-concave densities. In *High dimensional probability VII*, volume 71 of *Progr. Probab.*, pages 45–60. Springer, 2016.

[FPT20] A. Frieze, W. Pegden, and T. Tkocz. Random volumes in d-dimensional polytopes. *Discrete Anal.*, pages Paper No. 15, 17 pp., 2020.

[Fra97] M. Fradelizi. Sections of convex bodies through their centroid. *Arch. Math. (Basel)*, 69(6):515–522, 1997.

[GG09] D. Gatzouras and A. Giannopoulos. Threshold for the volume spanned by random points with independent coordinates. *Israel J. Math.*, 169:125–153, 2009.

[GGT23] A. Gorban, B. Grechuk, and I. Tyukin. Stochastic separation theorems: how geometry may help to correct ai errors. *Notices Amer. Math. Soc.*, 70(1):25–33, 2023.

[GM11] O. Guédon and E. Milman. Interpolating thin-shell and sharp large-deviation estimates for isotropic log-concave measures. *Geom. Funct. Anal.*, 21:1043–1068, 2011.

[IM65] K. Itô and H. P. McKean. *Diffusion processes and their sample paths*. Springer-Verlag, 1965.

[H55] J. L. Hodges Jr. A bivariate sign test. *Ann. Math. Statist.*, 26:523–527, 1955.

[KL22] B. Klartag and J. Lehec. Bourgain's slicing problem and kls isoperimetry up to polylog. *Geom. Funct. Anal.*, 32(5):1134–1159, 2022.

[Kla04] B. Klartag. A geometric inequality and a low m-estimate. *Proc. Amer. Math. Soc.*, 132:2619–2628, 2004.

[Kla06] B. Klartag. On convex perturbations with a bounded isotropic constant. *Geom. Funct. Anal.*, 16:1274–1290, 2006.

[Kla07] B. Klartag. A central limit theorem for convex sets. *Invent. Math.*, 168:91–131, 2007.

[Kla23] B. Klartag. Logarithmic bounds for isoperimetry and slices of convex sets. *Ars Inven. Anal.*, pages Paper No. 4, 17 pp., 2023.

[Mat] Mathoverflow. Expected value of tukey's half-space depth for log-concave measures. https://mathoverflow.net/questions/386972/. Question asked.

[Ngu14] V. H. Nguyen. Dimensional variance inequalities of brascamp-lieb type and a local approach to dimensional prékopa's theorem. *J. Funct. Anal.*, 266:931–955, 2014.

[NR03] A. Naor and D. Romik. Projecting the surface measure of the sphere of ℓ_p^n. *Ann. Inst. H. Poincaré Probab. Statist.*, 39:241–261, 2003.

[NSW19] S. Nagy, C. Schütt, and E. M. Werner. Halfspace depth and floating body. *Stat. Surv.*, 13:52–118, 2019.

[Paf24] M. Pafis. Threshold for the expected measure of the convex hull of random points with independent coordinates. *Bull. Lond. Math. Soc.*, 56(1):240–262, 2024.

[Piv07] P. Pivovarov. Volume thresholds for gaussian and spherical random polytopes and their duals. *Studia Math.*, 183(1):15–34, 2007.

[Roc70] R. T. Rockafellar. *Convex analysis*, volume 28 of *Princeton Mathematical Series*. Princeton University Press, Princeton, NJ, 1970.

[Sch14] R. Schneider. *Convex Bodies: The Brunn-Minkowski Theory*, volume 151 of *Encyclopedia of Mathematics and its Applications*. Cambridge University Press, Cambridge, second expanded edition, 2014.

[SW90] C. Schütt and E. M. Werner. The convex floating body. *Math. Scand.*, 66(2):275–290, 1990.

[Tuk75] J. W. Tukey. Mathematics and the picturing of data. In *Proceedings of the International Congress of Mathematicians (Vancouver, B. C., 1974)*, volume 2, pages 523–531, Montreal, Que., 1975. Canad. Math. Congress.

[Ver18] R. Vershynin. *High-dimensional probability. An introduction with applications in data science*, volume 47 of *Cambridge Series in Statistical and Probabilistic Mathematics*. Cambridge University Press, Cambridge, 2018.

[Wan14] L. Wang. *Heat capacity bound, energy fluctuations and convexity*. PhD thesis, Yale University, 2014.

Chapter 3
Approximation in Geometry

Márton Naszódi

This chapter covers three areas in the broad field of geometric approximation. First, we consider a classical problem in convexity and computer science, the *approximation of a convex body* in high dimensional Euclidean space *by a convex polytope* of small complexity. Our second topic is *quantitative Helly-type results.* Helly's classical theorem states that if the intersection of any $d + 1$ members of a finite family of convex sets in $\mathbb{R}^d$

Introduction

This chapter contains the lecture notes of the course 'Approximation in Geometry' given in 2023 at the Convex and Discrete Geometry Summer School in Budapest at the Erdős Center, an affiliate of the Rényi Institute of Mathematics. It covers three areas in the broad field of geometric approximation. First, in Sect. 3.1, we consider a classical problem in convexity and computer science, the **approximation of a convex body** in high dimensional Euclidean space **by a convex polytope** of small complexity. One obtains problems of very distinct nature by choosing the measure of the quality of approximation (that is, a metric on the space of convex bodies), and by setting the range for the complexity of the polytope, for example, considering polytopes with vertex number that is polynomial or exponential in the dimension.

The two most studied metrics on the space of convex bodies are the *Hausdorff distance* and the *Banach–Mazur distance*. The first is defined as the minimal radius $\rho > 0$ such that the ρ-neighborhood of each set contains the other set. This definition is applicable, when the approximation is expected to respect the Euclidean structure on $\mathbb{R}^d$.

The *Banach–Mazur distance*, which despite its name is not a metric, is a natural choice when dealing with an affine invariant problem. For convex bodies K and L,

M. Naszódi (✉)
HUN-REN Alfréd Rényi Institute of Math. and Eötvös Loránd Univ., Budapest, Hungary
e-mail: marton.naszodi@renyi.hu

J. Pach and G. Tóth (eds.), *Courses in Discrete and Computational Geometry*,
Bolyai Society Mathematical Studies 31,
https://doi.org/10.1007/978-3-032-10503-5_3

it is defined as the smallest R such that $K - v \subseteq \mathcal{A}L \subseteq R(K - v)$ for some affine transformation $\mathcal{A}$ of $\mathbb{R}^d$ and translation vector $v \in \mathbb{R}^d$. We will use a simpler, more intuitive notion, the *geometric distance*, defined as

$$d(K, L) = \inf\{\alpha/\beta : \alpha, \beta > 0, \beta K \subseteq L \subseteq \alpha K\},$$

since results stated for this distance are easily translated to the language of the Banach–Mazur distance.

Our second topic, considered in Sect. 3.2, is **quantitative Helly-type results**. Helly's classical theorem states that if the intersection of any $d + 1$ members of a finite family of convex sets in $\mathbb{R}^d$ is not empty, then the sets have a point in common. This fundamental result has grown into a heavily researched area of discrete geometry, convexity and combinatorics with applications in computer science. Quantitative questions in the area are those, where instead of looking for a point in the intersection of some sets, we want to see if the intersection is large in some sense. The following result by Bárány, Katchalski, and Pach from 1982 gives a taste of it. *Assume that the intersection of any* $2d$ *members of a finite family* $\mathcal{F}$ *of convex sets in* $\mathbb{R}^d$ *is of volume at least one. Then the intersection of all members of the family is of volume at least* $c(d)$, *a constant depending only on* d.

In the discussion of recent bounds on $c(d)$, *John's decomposition of the identity*, a linear algebraic notion concerning rank one operators that sum to the identity operator, will play an important role. It motivates our third topic, **sums of matrices**, which can be introduced by the following question. Given a matrix A (often simply $A = I$) as a positive linear combination of some matrices. Can we select a small subset of the matrices whose positive linear combination (with possibly new coefficients) yields a matrix close to A in some norm?

Before stating a classical result of Rudelson from 1999, we recall the following definition from probability theory, a notion very closely related to John's decomposition of the identity. A random vector v in $\mathbb{R}^d$ is called *isotropic*, if $\mathbb{E}[v \otimes v] = I$, where I is the identity operator on $\mathbb{R}^d$, $\mathbb{E}$ denotes the expectation of a random variable, and $v \otimes v$ is the rank one linear operator on $\mathbb{R}^d$ represented by the $d \times d$ matrix vv^T.

Theorem *Let* $\gamma > 0$ *be given, and* $y_1, \ldots, y_k$ *be independent copies of an isotropic random vector* y *in* $\mathbb{R}^d$ *for which* $|y|^2 \leq \gamma$ *almost surely, with*

$$k = \left\lceil \frac{c\gamma \ln d}{\varepsilon^2} \right\rceil,$$

where $c > 0$ *is a universal constant. Then*

$$\mathbb{E}\left\| \frac{1}{k} \sum_{i=1}^{k} y_i \otimes y_i - I \right\| \leq \varepsilon$$

holds, where $\|A\|$ *denotes the operator norm of* A.

In Sect. 3.3, we discuss the proof of this theorem as well its generalizations, followed by an application to a quantitative Helly-type problem.

The main goal of these notes is to convey some ideas, teach techniques. We do not attempt to give a full account of the literature, but provide some pointers for those willing to go deeper.

Exercises constitute an integral part of the text, they are used in subsequent proofs. Thus, for most, a hint or a full solution is given at the end of the chapter, marked by a ♣ sign.

Definitions and notation.

The *Minkowski sum* of two sets A, B in $\mathbb{R}^d$ is denoted by $A + B = \{a + b : a \in A, b \in B\}$. The origin is denoted by o, the closed Euclidean ball centered at $x \in \mathbb{R}^d$ of radius ρ is $\mathbf{B}^d(x, \rho)$. The boundary of a convex set K is denoted by $\mathrm{bd}(K)$. A *convex body* is a compact, convex set with non-empty interior. The *center of mass* (or, centroid) of a bounded convex set K is the point obtained as the following integral

$$\frac{1}{\mathrm{vol}(K)} \int_K x \mathrm{d}x,$$

where $\mathrm{d}x$ denotes integration with respect to the Lebesgue measure in the affine hull of K, and $\mathrm{vol}(K)$ is the $\dim K$-dimensional volume (Lebesgue measure) of K. We denote the volume of the unit ball by $\omega_d = \mathrm{vol}\left(\mathbf{B}^d(o, 1)\right)$. Recall that the *polar* of a set K in $\mathbb{R}^d$ is defined as $K^\circ = \left\{y \in \mathbb{R}^d : \langle x, y\rangle \le 1 \text{ for all } x \in K\right\}$. In general, we mostly follow the notation in Schneider's book [Sch14].

3.1 Approximation of Convex Bodies by Polytopes

In order to define the problem of approximating a *convex body* (a compact convex set with nonempty interior) by a *convex polytope* (the convex hull of finitely many points in $\mathbb{R}^d$), we need to fix some notion of distance between convex sets. We will use two such notions. The *Hausdorff distance* of two convex sets K and L in $\mathbb{R}^d$ is defined as

$$\delta_H(K, L) = \inf\{\delta > 0 : K + \mathbf{B}^d(o, \delta) \supseteq L, L + \mathbf{B}^d(o, \delta) \supseteq K\}.$$

The *geometric distance* of K and L is

$$d(K, L) = \inf\{\alpha/\beta : \alpha, \beta > 0, \beta K \subseteq L \subseteq \alpha K\}.$$

Note that this quantity is sensitive to the choice of the origin, in other words, it is not translation invariant.

Exercise 3.1.1 Clearly, the geometric distance is not a metric. Consider the class of compact, convex sets containing o in the interior, and let two sets K_1 and K_2 be equivalent, if they are positive homothets of each other, that is, if $K_2 = \lambda K_1$ for some $\lambda > 0$. Show that $\ln(d(\cdot, \cdot))$ is a metric on the set of equivalence classes. ♣

In what follows, K is a given convex body in $\mathbb{R}^d$, and our goal is to find a convex polytope which is close to K in one of the two distances defined above. The following exercise is of fundamental importance, as it shows that finding a convex polytope that approximates K is essentially equivalent to finding a set X that hits every cap of K of a certain size.

Exercise 3.1.2 Let K be a convex body containing the origin in its interior and $\varepsilon \in (0, 1)$. Let $X \subset K$ be a finite set. Show that X intersects every *cap of metric depth* ε, that is, every set of the form $\operatorname{cap}(x, \varepsilon) = \{y \in K : \langle y, \nu\rangle \geq (1-\varepsilon)\langle x, \nu\rangle\}$, where x is an arbitrary boundary point of K and ν is any outer unit normal vector of K at x, if, and only if, the convex polytope $P = \operatorname{conv}(X)$ satisfies $d(K, P) \leq \frac{1}{1-\varepsilon}$. ♣

3.1.1 The Packing Bound

In this section, our goal is to find a convex polytope P with few vertices such that $\delta_H(P, K) \leq \varepsilon$ for a given convex body K in $\mathbb{R}^d$. Arguably, the simplest strategy is the following. Let Λ be the set of centers of a saturated (in other words, maximal with respect to containment) packing of balls of radius $\varepsilon/2$ in K. Then the balls of radius ε centered at points of Λ may cover K. One minor technical difficulty arises: if K has a 'sharp' point on its boundary, then these ε balls may not cover K. To overcome this, we will consider a packing of $\varepsilon/2$ balls in a slight extension of K.

Exercise 3.1.3 Show that if $\Lambda \subset \mathbb{R}^d$ is such that $\{\mathbf{B}^d(v, \varepsilon/2) : v \in \Lambda\}$ is a maximal packing of $\varepsilon/2$ radius balls in $K + \mathbf{B}^d(o, \varepsilon/2)$, then $P = \operatorname{conv}(\Lambda)$ is a convex polytope satisfying $\delta_H(P, K) \leq \varepsilon$. ♣

Exercise 3.1.4 Prove that there is a Λ with $\Lambda + \mathbf{B}^d(o, \varepsilon) \supseteq K$ of cardinality at most $\frac{\operatorname{vol}(K+\mathbf{B}^d(o,\varepsilon/2))}{\operatorname{vol}(\mathbf{B}^d(o,\varepsilon/2))}$. ♣

These two exercises provide approximation in the Hausdorff distance. Next, we turn to geometric distance.

Exercise 3.1.5 Let K be an origin symmetric convex body in $\mathbb{R}^d$, that is, $K = -K$. Prove that for any $0 < \varepsilon < 1$, there is a convex polytope P with no more than $\left(\frac{3}{\varepsilon}\right)^d$ vertices that is $(1+\varepsilon)$-close to K in the geometric distance. ♣

The following theorem extends the statement of the previous exercise to non-symmetric convex bodies.

Theorem 3.1.6 (Approximation in the geometric distance via packing) *In any dimension $d > 0$, for any $0 < \varepsilon < 1$ and convex body K in $\mathbb{R}^d$, there is a convex polytope P with no more than $\left(\frac{6}{\varepsilon}\right)^d$ vertices that is $(1+\varepsilon)$-close to K in the geometric distance with an appropriate translation vector $t \in \mathbb{R}^d$.*

The proof relies on the following fundamental fact due to Milman and Pajor [MP00] (see also [AAGM15, Theorem 4.1.20]).

Lemma 3.1.7 (Milman, Pajor: Large centrally symmetric body in any convex body) *If the center of mass of a convex body K in $\mathbb{R}^d$ is the origin, then* $\mathrm{vol}(K \cap (-K)) \geq 2^{-d}\mathrm{vol}(K)$.

Proof of Theorem 3.1.6 By translating K, we may assume that the center of mass of K is the origin. Let $\Lambda \subset \mathbb{R}^d$ be the set of translation vectors of a *maximal packing* of $(\varepsilon/2)(K \cap (-K))$ in $(1+\varepsilon/2)K$, that is, the members of the family $\{(\varepsilon/2)(K \cap (-K)) + u : u \in \Lambda\}$ are contained in $(1+\varepsilon/2)K$, and they have pairwise disjoint interiors, moreover, for any $v \in \mathbb{R}^d$ with the property $(\varepsilon/2)(K \cap (-K)) + v \subseteq (1+\varepsilon/2)K$, there is a $u \in \Lambda$ such that the interiors of $(\varepsilon/2)(K \cap (-K)) + u$ and $(\varepsilon/2)(K \cap (-K)) + v$ intersect.

The following standard volume argument, which uses Lemma 3.1.7, bounds the size of Λ from above.

$$\mathrm{card}(\Lambda) \leq \frac{\mathrm{vol}(K + (\varepsilon/2)K)}{\mathrm{vol}((\varepsilon/2)(K \cap (-K)))} \leq \frac{\mathrm{vol}((1+\varepsilon/2)K)}{2^{-d}\mathrm{vol}((\varepsilon/2)K)} \leq \left(\frac{6}{\varepsilon}\right)^d .$$

On the other hand, it easily follows from the maximality of Λ that $\Lambda + \varepsilon K \supseteq \Lambda + \varepsilon(K \cap (-K)) \supseteq K$, completing the proof of Theorem 3.1.6. □

References

Theorem 3.1.6 is a classical, standard argument whereby approximation by a convex polytope is obtained from a covering, which in turn, is derived from a packing. For results of this flavor, cf. [Nas18] and references therein.

3.1.2 The Bronshtein–Ivanov Net

In this section we show how a more sophisticated argument, which still relies on a packing, enables us to replace the ε^d factor by roughly $\varepsilon^{d/2}$ for what we will call '*nice*' convex bodies.

We call a convex body K in $\mathbb{R}^d$ a *nice convex body*, if it has smooth boundary (that is, at every boundary point, there is a unique supporting hyperplane to K), $\mathbf{B}^d(o, 1) \subset K \subset \mathbf{B}^d(o, R)$, and for every boundary point $x \in \mathrm{bd}(K)$, there exists a ball of radius Θ containing K whose boundary sphere touches K at x. Note that the last condition is a quantitative form of requiring that K is strictly convex, that is, it contains no non-degenerate line segment on the boundary. The definition of 'nice' depends on R and Θ.

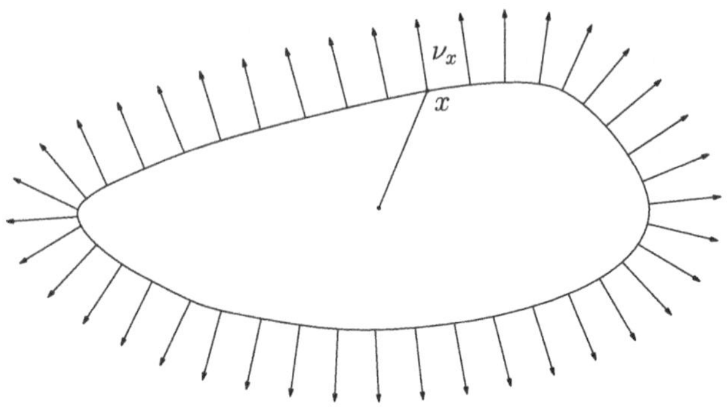

Fig. 3.1 The Bronshtein–Ivanov net

The goal of the present section is to prove the following result due to Bronshtein and Ivanov [BI75].

Theorem 3.1.8 (Bronshtein–Ivanov) *If K is a nice convex body with $R = d^2$ and $\Theta = d^5$, then there is a convex polytope P with no more than $d^{100d}\varepsilon^{-\frac{d-1}{2}}$ vertices satisfying $P \subseteq K \subseteq (1+\varepsilon)P$.*

As it turns out, the condition that K is a nice convex body is not too restrictive, in fact, every convex body has an affine image that can be well approximated by a nice convex body with $R = d^2$ and $\Theta = d^5$, see [NNR20].

Proof of Theorem 3.1.8 Let K be a convex body with smooth boundary containing the origin and contained in $\mathbf{B}^d(o, R)$, and let $\rho \in \left(0, \frac{1}{2}\right)$. Consider the set S of points $\{x + \rho\nu_x : x \in \mathrm{bd}(K)\}$, where ν_x is the outer unit normal to $\mathrm{bd}(K)$ at x. Let $\Lambda = \{x_j + \rho\nu_{x_j} : 1 \leq j \leq N\}$ be a maximal ρ *-separated set* in S, i.e., a set such that any two of its members are at Euclidean distance at least ρ (see Fig. 3.1). We call the corresponding set $\{x_j : 1 \leq j \leq N\}$ a *Bronshtein–Ivanov net* of mesh ρ for the body K.

Exercise 3.1.9 In the construction above, for every $x \in \mathrm{bd}(K)$, we can find j such that $|x - x_j|^2 + |\nu_x - \nu_{x_j}|^2 \leq \rho^2$. ♣

Lemma 3.1.10 (Upper bound on the size of a Bronshtein–Ivanov net) *We have* $N \leq 2^d(R+3)^d\rho^{-d+1}$.

Proof Assume that $s', s'' \geq 0$. Write

$$|x' + \nu' + s'\nu' - x'' - \nu'' - s''\nu''|^2 = |x' + \nu' - x'' - \nu''|^2 + |s'\nu' - s''\nu''|^2 + 2s'\langle \nu', x' - x''\rangle$$
$$+ 2s''\langle \nu'', x'' - x'\rangle + 2(s' + s'')(1 - \langle \nu', \nu''\rangle) \geq |x' + \nu' - x'' - \nu''|^2,$$

where, in the last inequality, we used the fact that ν' and ν'' are the outer unit normals of support hyperplanes of K at x' and x'' respectively, and hence, $\langle \nu', x' - x''\rangle, \langle \nu'', x'' - x'\rangle \geq 0$.

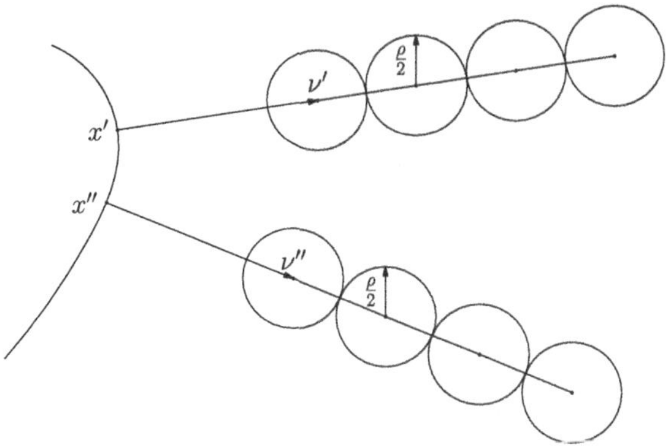

Fig. 3.2 The disjoint balls

Thus, if the balls of radius $\frac{\rho}{2}$ centered at $x' + \nu'$ and $x'' + \nu''$ are disjoint, so are the balls of radius $\frac{\rho}{2}$ centered at $x' + (1 + s')\nu'$ and $x'' + (1 + s'')\nu''$. From here we conclude that the balls of radius $\frac{\rho}{2}$ centered at the points $x_j + (1 + k\rho)\nu_{x_j}$ for all integer k with $0 \leq k \leq \frac{1}{\rho}$, are all pairwise disjoint (see Fig. 3.2) and contained in $\mathbf{B}^d\,(0, R + 3)$.

The total number of these balls is at least $\frac{N}{\rho}$ (for every point x_j in the net, there is a sequence of at least $\frac{1}{\rho}$ balls corresponding to different values of k), whence $\frac{N}{\rho} \leq \left(\frac{R+3}{\frac{\rho}{2}}\right)^d$ and the desired bound on N follows. □

Lemma 3.1.11 (Caps of nice bodies are of small diameter) *Let* $\varepsilon \in \left(0, \frac{1}{2}\right)$. *Assume that* K *is a nice convex body,* $x \in \mathrm{bd}(K)$*, and* ν *is the outer normal to* $\mathrm{bd}(K)$ *at* x*. If* $y \in K$ *and* $\langle y, \nu\rangle \geq (1 - \varepsilon)\langle x, \nu\rangle$*, then* $|y - x| \leq \sqrt{2\Theta\, R\, \varepsilon}$.

Proof Let Q be the ball of radius Θ containing K whose boundary sphere touches K at x. Then $y \in Q$ and ν is the outer unit normal to Q at x, so Q is centered at $x - \Theta\nu$. Note also that, since $o \in K \subset \mathbf{B}^d\,(0, R)$, we have $0 \leq \langle x, \nu\rangle \leq R$. Now we have

$$\Theta^2 \geq |y - x + \Theta\nu|^2 = |y - x|^2 + 2\Theta\langle y - x, \nu\rangle + \Theta^2,$$

so

$$|y - x|^2 \leq 2\Theta\langle x - y, \nu\rangle \leq 2\Theta\varepsilon\langle x, \nu\rangle \leq 2\Theta R\varepsilon,$$

as required. □

Lemma 3.1.12 (The cap about x may be replaced by the cap about x') *Fix* $\varepsilon, \rho \in \left(0, \frac{1}{2}\right)$. *Let* K *be a nice convex body,* $x, x', y \in \mathrm{bd}(K)$*; and* ν *and* ν' *the outer unit normals to* $\mathrm{bd}(K)$ *at* x *and* x' *respectively. Assume that* $|x - x'|^2 + |\nu - \nu'|^2 \leq \rho^2$ *and* $\langle y, \nu\rangle \geq \left(1 - \frac{\varepsilon}{2}\right)\langle x, \nu\rangle$*. Then*

$$\langle y, \nu'\rangle \geq \left(1 - \frac{\varepsilon}{2} - 2\rho(\rho + \varepsilon R + |y - x|)\right)\langle x', \nu'\rangle.$$

Proof We have

$$\begin{aligned}\langle y, v'\rangle = \langle x, v'\rangle + \langle y - x, v'\rangle = \\ \langle x', v'\rangle + \langle x - x', v'\rangle + \langle y - x, v\rangle + \langle y - x, v' - v\rangle \geq \\ \langle x', v'\rangle + \langle x - x', v' - v\rangle + \langle y - x, v\rangle + \langle y - x, v' - v\rangle \geq \\ \langle x', v'\rangle - \rho^2 - \frac{\varepsilon}{2}\langle x, v\rangle - \rho|y - x|.\end{aligned}$$

Here, when passing from the second line to the third one, we used the inequality $\langle x - x', v\rangle \geq 0$. In the last inequality, we used the assumptions of the theorem, in particular, the estimates $|x - x'|, |v - v'| \leq \rho$. Note that, since $\mathbf{B}^d(o, 1) \subset K \subset \mathbf{B}^d(o, R)$, we have

$$\langle x, v\rangle = \langle x, v'\rangle + \langle x, v - v'\rangle \leq \langle x', v'\rangle + \rho R$$

and $\langle x', v'\rangle \geq 1 > \frac{1}{2}$. So

$$\langle y, v'\rangle \geq \left(1 - \frac{\varepsilon}{2}\right)\langle x', v'\rangle - \rho\left(\rho + \frac{\varepsilon R}{2} + |y - x|\right) \geq \left(1 - \frac{\varepsilon}{2} - 2\rho(\rho + \varepsilon R + |y - x|)\right)\langle x', v'\rangle.$$

□

Combining Lemmas 3.1.10, 3.1.11, and 3.1.12, with the choice of $\rho = \frac{\sqrt{\varepsilon}}{10\sqrt{\Theta R}}$ yields the proof of Theorem 3.1.8.

3.1.3 Economic Cap Covering

In this section, we present the **Economic Cap Covering Theorem** due to Bárány and Larman [BL88, Bár89], using the idea of *Macbeath regions* introduced by Macbeath [Mac52], and ideas of Ewald et al. [ELR70]. For a wonderful exposition, see [Bár07].

For a convex body K, we define the *depth function* (also referred to as the *Tukey depth* with respect to the uniform measure on K, when $\mathrm{vol}(K) = 1$) from K to $\mathbb{R}_{\geq 0}$ as

$$\mathrm{depth}_K(x) = \min\{\mathrm{vol}(K \cap H) : H \text{ is a half-space containing } x\}.$$

For a point $x \in K$, we say that a cap $C = K \cap H^+$ is a *minimal cap* of x, if it has minimum volume among caps given by a closed half-space H^+ containing x. Note that it may not be unique for a given $x \in K$, nevertheless, we denote (any) minimal cap of x as $\mathrm{mincap}(x)$, and call the hyperplane H bounding H^+ the *base hyperplane* of $\mathrm{mincap}(x)$.

A *magnified cap* is defined as follows. Denote the base hyperplane of a cap $C = K \cap H^+$ by H, and let H' denote the supporting hyperplane of C parallel to and distinct from H. We will call the center of mass of $C \cap H'$ the *center of the cap C*.

Note that if K is strictly convex (that is, its boundary contains no line segment), then $C \cap H'$ is a singleton. Finally, for a $\lambda > 0$, we denote by C^λ the image of C under the magnification centered at the center of C by factor λ.

Magnified caps will play a similar role as radius ε Euclidean balls or translates of εK did in previous sections. We will pack them, use them to cover K, or $\mathrm{bd}(K)$, and thus obtain a convex polytope close to K in some metric.

The *floating body* (respectively, the *wet part*) of K for $t \geq 0$ is defined as the deep (resp., shallow) points of K, i.e.,

$$K_{\geq t} = \{x \in K : \mathrm{depth}_K(x) \geq t\}, \text{ and}$$

$$K_{\leq t} = \{x \in K : \mathrm{depth}_K(x) \leq t\}.$$

Exercise 3.1.13 Show that $K_{\geq t}$ is a convex set for any $t > 0$. ♣

Our goal is to cover the wet part by small caps.

Theorem 3.1.14 (Bárány, Larman: Economic Cap Covering) *There are constants $c(d), C(d) > 0$ depending only on the dimension d such that the following hold. Assume that K is a convex body in $\mathbb{R}^d$ with $\mathrm{vol}(K) = 1$, and $0 < \varepsilon < (2d)^{-2d}$. Then there are caps $C_1, \ldots, C_m$ and pairwise disjoint convex sets $C'_1, \ldots, C'_m$ such that $C'_i \subseteq C_i$, for each i, and*

1. $\bigcup_{i=1}^m C'_i \subseteq K_{\leq\varepsilon} \subseteq \bigcup_{i=1}^m C_i$,
2. $\mathrm{vol}(C'_i) > c(d)\varepsilon$ *and* $\mathrm{vol}(C_i) < C(d)\varepsilon$ *for each i,*
3. *for each cap C with $C \cap K_{\geq\varepsilon} = \emptyset$ there is a C_i containing C.*

The power of the Economic Cap Covering Theorem is that it provides a very fine description of the boundary of K. The downside is the same: it may be too fine for certain applications, as it assumes $\varepsilon < (2d)^{-2d}$.

The central objects underpinning the Economic Cap Covering Theorem are Macbeath regions. The *Macbeath region* of K at $x \in K$ with parameter $\lambda > 0$ is the centrally symmetric convex set

$$M_K(x, \lambda) = x + \lambda[(K - x) \cap (x - K)].$$

When omitted, λ is one, and we will often omit K as well. We define the *Macbeath depth* of a point $x \in K$ as

$$\text{M-depth}_K(x) = \mathrm{vol}(M_K(x)).$$

The advantage of Macbeath depth over depth is that in order to compute it, no minimization is necessary.

The rest of this section is devoted to an outline of the proof of the Economic Cap Covering Theorem following [Bár07] very closely. Throughout the section, K is a fixed convex body in $\mathbb{R}^d$, of volume 1; caps, minimal caps, and Macbeath regions are all defined with respect to K.

The following exercise (due to Ewald et al. [ELR70]) shows that Macbeath regions behave in a similar manner as balls: a saturated packing of Macbeath regions of parameter $\frac{1}{2}$ yields a covering by Macbeath regions of parameter 5.

Exercise 3.1.15 Let x, y be points in K. Show that if $M\left(x, \frac{1}{2}\right) \cap M\left(y, \frac{1}{2}\right) \neq \emptyset$, then $M(x, 5) \supseteq M(y, 1)$. ♣

The most intricate part of the proof of Theorem 3.1.14 is to show that when x is close to the boundary of K, we have $\text{M-depth}_K(x) \approx \text{depth}_K(x)$. The following exercises and lemmas will yield it.

Exercise 3.1.16 Show that $\text{M-depth}_K(x) \leq 2\text{depth}_K(x)$ for all $x \in K$. ♣

The proof of the converse statement (from [ELR70, Bár07]) requires a sophisticated argument, which we omit, cf. [Bár07, Lemma 6.4].

Lemma 3.1.17 *Set* $\varepsilon_0 = 3^{-d}/d$. *If* $\text{depth}_K(x) \leq \varepsilon_0$, *then* $\text{mincap}(x) \subseteq M(x, 2d)$.

We immediately obtain the following.

$$\text{If } \text{depth}_K(x) \leq 3^{-d}/d, \text{ then } \text{depth}_K(x) \leq (2d)^d \text{M-depth}_K(x). \tag{3.1}$$

Next, we show that the floating body is *strictly convex*, that is, its boundary contains no line segment, a fact notable on its own right.

Lemma 3.1.18 (Strict convexity of the floating body) *For any convex body K in $\mathbb{R}^d$ and any $0 < \varepsilon < \text{vol}(K)$, we have that $K_{\geq \varepsilon}$ is strictly convex.*

For the proof of Lemma 3.1.18, we need the following fundamental observation.

Lemma 3.1.19 (The center of mass of the base of a minimal cap) *Let x be an interior point of K, with its minimal cap given as* $\text{mincap}(x) = K \cap H^+$, *where the closed half-space H^+ is bounded by the hyperplane H containing x. Then x is the center of mass of the $(d-1)$-dimensional convex body $K \cap H$.*

We sketch the proof, a standard variation argument, which is omitted in many sources.

Proof of Lemma 3.1.19 Denote by c the center of mass of $K \cap H$, and let e be an arbitrary unit vector parallel to H. First, by the definition of the center of mass, we have

$$\langle c, e \rangle = \frac{1}{\text{vol}_{d-1}(K \cap H)} \int_{K \cap H} \langle s, e \rangle \text{dvol}_{d-1}(s).$$

Second, let L denote the $(d-2)$-dimensional affine subspace in H passing through x with normal vector e. For any angle $\vartheta \in (-\pi, \pi)$, consider the volume $f(\vartheta) = \text{vol}\big(K \cap H^+(\vartheta)\big)$, where $H^+(\vartheta)$ denotes the rotation of H^+ about L by angle ϑ. On the one hand, it is not difficult to see that

$$\left.\frac{\mathrm{d}}{\mathrm{d}\vartheta}\right|_{\vartheta=0} f(\vartheta) = \left.\frac{\mathrm{d}}{\mathrm{d}\vartheta}\right|_{\vartheta=0} \left(\int_{K\cap H} \tan(\vartheta) \left(\langle s, e\rangle - \langle x, e\rangle\right) \mathrm{d}\mathrm{vol}_{d-1}(s) \right) =$$

$$\int_{K\cap H} \left(\langle s, e\rangle - \langle x, e\rangle\right) \mathrm{d}\mathrm{vol}_{d-1}(s)\,.$$

On the other hand, since $\mathrm{mincap}(x)$ is a minimal cap of x, and hence, 0 is a critical point of the function f, we have that $\left.\frac{\mathrm{d}}{\mathrm{d}\vartheta}\right|_{\vartheta=0} f(\vartheta) = 0$. Combining our equations, we obtain that $\langle x, e\rangle = \langle c, e\rangle$ for any unit vector e parallel to H, and hence, $x = c$, completing the proof of Lemma 3.1.19. □

Several much deeper results on $K \cap H$ and $K \cap H^+$ can be found in Ryabogin's article [Rya23, Sects. 2 and 3], in particular, Dupin's theorems with accessible proofs in English.

Exercise 3.1.20 Let x be a point on the boundary of the floating body $K_{\geq\varepsilon}$. Show that no minimal cap of x intersects the interior of $K_{\geq\varepsilon}$. ♣

Proof of Lemma 3.1.18 Suppose for a contradiction that the line segment $[x, y]$ is on the boundary of the floating body $K_{\geq\varepsilon}$. By Exercise 3.1.20, any minimal cap $\mathrm{mincap}(z)$ of $z = (x + y)/2$ supports $K_{\geq\varepsilon}$, and hence, its bounding hyperplane H contains x and y. Thus, $\mathrm{mincap}(z)$ is a minimal cap of x as well. However, $K \cap H$ has a unique center of mass contradicting Lemma 3.1.19. □

Lemma 3.1.21 (Minimal caps cover the wet part) *For every $\varepsilon > 0$, the wet part $K_{\leq\varepsilon}$ is covered by the minimal caps of points of* $\mathrm{bd}(K_{\geq\varepsilon})$.

Proof Let y be a point in $K_{\leq\varepsilon}$, and assume that $\mathrm{depth}_K(y) < \varepsilon$, as otherwise there is nothing to prove. Let H_1^+ denote the half-space defining a minimal cap of y. Let H^+ be the translate of H_1^+ which supports $K_{\geq\varepsilon}$ from outside, that is, which intersects the boundary, but not the interior of $K_{\geq\varepsilon}$. Clearly, $y \in H^+$.

By Lemma 3.1.18, the bounding hyperplane H of H^+ intersects $K_{\geq\varepsilon}$ at a unique point, call it x. If $K_{\geq\varepsilon}$ has a unique supporting hyperplane at x, then we are done, as that hyperplane must bound a minimal çap of x according to Exercise 3.1.20.

Consider the difficult case, when $K_{\geq\varepsilon}$ has multiple supporting hyperplanes at x, and let N denote the set of their outer normal vectors. It is well known that N is a convex cone with apex at the origin, and thus, it is the convex hull of its extreme rays, cf. [Sch14, Chap. 1]. By taking the polar of the normal cone, it is clear, that $K \cap H^+$ is contained in the union of the caps whose base hyperplane passes through x and whose normal rays are the extreme rays of N (check it!). We will show that these caps are minimal caps of x.

Fix an outer normal ray ℓ of $K_{\geq\varepsilon}$ at x. By a theorem of Alexandrov (cf. [Sch14, Theorem 2.2.5]), the set of points of any convex body, in particular, of $K_{\geq\varepsilon}$ having multiple supporting hyperplanes is of measure zero in the boundary, and hence, there is a sequence $x_1, x_2, \ldots$ of boundary points converging to x, with a unique supporting hyperplane at each. We may assume more, namely, that the sequence of the unique outer normal rays $\ell_1, \ell_2, \ldots$ of $x_1, x_2, \ldots$ converge to ℓ. From the considerations above, we know that the caps whose bases pass through x_i and have outer normal ray ℓ_i are minimal caps, and they clearly converge to the cap whose base passes through x and has outer normal ray ℓ, completing the proof of the difficult case, and hence, that of Lemma 3.1.21. □

A somewhat similar argument yields the following lemma, cf. [Bár07, Lemma 6.8].

Lemma 3.1.22 (A cap tangent to the floating body is contained in a Macbeath region) *Let $\varepsilon < \varepsilon_0$, and assume that a cap C intersects the floating body $K_{\geq\varepsilon}$ at a single point, x. Then* $\operatorname{vol}(C) \leq d\varepsilon$ *and* $C \subseteq M(x, 2d)$.

Sketch of the proof of the Economic Cap Covering Theorem

Finally, we combine the lemmas above to obtain Theorem 3.1.14. The plan is to take a packing of Macbeath regions, and show that their magnified copies cover the wet part. Then, we replace these Macbeath regions by caps that will satisfy the conclusion of the theorem.

In this spirit, let $x_1, x_2, \dots, x_m$ be a maximal set of points on the boundary of $K_{\geq\varepsilon}$ with the property that $M(x_i, 1/2) \cap M(x_j, 1/2) = \emptyset$ for all $i \neq j$. Note that such maximal set is finite, since, by (3.1), we have $\operatorname{vol}(M(x_i, 1/2)) = 2^{-d}\text{M-depth}_K(x_i) \geq (4d)^{-d}\text{depth}_K(x_i) = (4d)^{-d}\varepsilon$.

We claim that

$$K_{\leq\varepsilon} \subseteq \bigcup_{x\in \operatorname{bd}(K_{\geq\varepsilon})} M(x). \tag{3.2}$$

Indeed, let $y \in K$ be such that $\text{depth}_K(y) \leq \varepsilon$, and suppose for a contradiction that $y \notin M(x) = K \cap (2x - K)$ for every $x \in \operatorname{bd}(K_{\geq\varepsilon})$. That is, $\operatorname{bd}(K_{\geq\varepsilon})$ is disjoint from $\frac{K+y}{2}$. Since y is not in $K_{\geq\varepsilon}$, it means that $K_{\geq\varepsilon}$ and $\frac{K+y}{2}$ can be separated by a hyperplane H. We may assume that H is tangent to $K_{\geq\varepsilon}$, and $\frac{K+y}{2}$ is contained in the other open half-space bounded by H. Let C denote the cap of K bounded by H containing $\frac{K+y}{2}$. By Lemma 3.1.22, $\operatorname{vol}(C) \leq d\varepsilon$. On the other had, C contains $\frac{K+y}{2}$, and hence,

$$d\varepsilon \geq \operatorname{vol}(C) \geq 2^d,$$

contradicting the choice of ε, and hence, proving (3.2).

Exercise 3.1.23 Show that

$$K_{\leq\varepsilon} \subseteq \bigcup_{i=1}^{m} M(x_i, 5). \tag{3.3}$$

♣

We will turn the covering of the wet part by Macbeath regions given in (3.3) into a covering by caps. We start with some notation. We may write a hyperplane (resp., a half-space) as $[a = t] = \{x \in \mathbb{R}^d : a(x) = t\}$ (resp., $[a \leq t] = \{x \in \mathbb{R}^d : a(x) \leq t\}$), where a is linear functional on $\mathbb{R}^d$ and $t \in \mathbb{R}$ is a scalar. For each $i =$

$1, \dots, m$, we write a minimal cap of x_i as $\mathrm{mincap}(x_i) = K \cap [a_i \le t_i]$. We may take a magnified copy of $\mathrm{mincap}(x_i)$ in (at least) two natural manners. First, we may consider $C^{10d+1}(x_i)$ defined as the magnification of $\mathrm{mincap}(x_i)$ by factor $(10d+1)$ from the center of $\mathrm{mincap}(x_i)$, as introduced on "Economic Cap Covering". Second, we may consider $C_i = K \cap [a_i \le t_i + 10dw_i]$, where w_i is the width of the minimal cap $\mathrm{mincap}(x_i)$ in the direction a_i. By the convexity of K, we have

$$C_i \subseteq \mathrm{mincap}^{10d+1}(x_i)\,. \tag{3.4}$$

We set $C_i' = M(x_i, 1/2) \cap [a_i \le t_i]$.

Exercise 3.1.24 Show that C_i' and C_i satisfy 1 and 2 in Theorem 3.1.14. ♣

Finally, in order to prove 3 in Theorem 3.1.14, let C be a cap disjoint from the interior of $K_{\ge\varepsilon}$. We may clearly assume that C intersects $\mathrm{bd}(K_{\ge\varepsilon})$. The intersection, by Lemma 3.1.18, is a single point, say x, and, by Lemma 3.1.22, $C \subseteq M(x, 2d)$. By Exercise 3.1.15 and the maximality of $x_1, \dots, x_m$, there is an i with $M(x) \subseteq M(x_i, 5)$.

Exercise 3.1.25 Show that $M(x, 2d) \subseteq M(x_i, 10d)$. ♣

Since $C \subseteq M(x, 2d)$, and $M(x_i, 10d) \subseteq C_i$ (check it!), the proof of Theorem 3.1.14 is complete.

3.1.3.1 An Application of the Economic Cap Covering Theorem

As an application, we sketch the proof of the following theorem, which guarantees the existence of a convex polytope with few facets (that is, $(d-1)$-dimensional faces) between the floating body of K and K.

Theorem 3.1.26 (Bárány) *Let K be a convex body with* $\mathrm{vol}(K) = 1$ *and $t \in (0, t_0)$ (where t_0 depends only on d). Then there is a convex polytope P with $K_{\ge t} \subseteq P \subseteq K$ with no more than*

$$C(d)\frac{\mathrm{vol}\big(K_{\ge t}\big)}{t}$$

facets, where $C(d) > 0$ depends only on d.

Proof Set $\tau = 6^{-d}t$, and choose a set of points $\{x_1, \dots, x_m\}$ from $\mathrm{bd}\big(K_{\ge\tau}\big)$ maximal with respect to the property that the Macbeath regions $M(x_i, 1/2)$ are pairwise disjoint. One can show that

$$c(d)m < \frac{\mathrm{vol}\big(K_{\le\tau}\big)}{\tau} < C(d)\frac{\mathrm{vol}\big(K_{\le t}\big)}{t},$$

for some $c(d), C(d) > 0$ depending only on d. Now, we remove the magnified (by factor 6) minimal caps from K to obtain

$$P = K \setminus \bigcup_{i=1}^{m} \operatorname{mincap}^6(x_i)\,.$$

It can be shown that (1) no $z \in \operatorname{bd}(K)$ belongs to P, and (2) $K_{\geq t} \subseteq P$. □

3.1.4 VC-Dimension and Epsilon-Nets

Exercise 3.1.2 shows that in order to obtain a convex polytope that approximates a convex body K well, we need to find a (preferably small) set X that intersects every cap of K of a certain depth. In a more combinatorial language: X is a *transversal* to the family of caps of K. In this section, we will see that it is easy to find a transversal for set families that are 'simple' in a purely combinatorial sense. Caps are 'simple', thus, the combinatorial machinery that we introduce here will yield an approximating polytope.

The *Vapnik–Chervonenkis dimension* (VC-dimension, in short) of a set family $\mathcal{F} \subset 2^V$ on a (finite or infinite) set V is the size of the largest subset A of V such that $\mathcal{F}|_A = \{F \cap A : F \in \mathcal{F}\}$ is the power set 2^A of A.

Exercise 3.1.27 Show that if V is a (finite or infinite) subset of $\mathbb{R}^d$ and $\mathcal{G}$ is some (finite or infinite) set of half-spaces in $\mathbb{R}^d$, then the VC-dimension of $\mathcal{F} = \{G \cap V : G \in \mathcal{G}\}$ is at most $d + 1$. ♣

The main advantage of a set family $\mathcal{F}$ with low VC-dimension is that it is easy to find a small subset U of V that intersects all members of $\mathcal{F}$, provided that there is a measure on V according to which each set in $\mathcal{F}$ is of not too small measure. More precisely, we have the following fundamental result.

Theorem 3.1.28 (Epsilon-net Theorem) *Let $0 < \varepsilon < 1/e$, and let D be a positive integer. Let $\mathcal{F}$ be a family of some measurable subsets of a probability space (U, μ), where the probability of each member F of $\mathcal{F}$ is $\mu(F) \geq \varepsilon$. Assume that the VC-dimension of $\mathcal{F}$ is at most D. Set*

$$t := \left\lceil 3\frac{D}{\varepsilon} \ln \frac{1}{\varepsilon} \right\rceil.$$

Choose t elements $X_1, \ldots, X_t$ of U randomly, independently according to μ. Then $\{X_1, \ldots, X_t\}$ is a transversal of $\mathcal{F}$ with probability at least $1 - \delta$, where

$$\delta := (200\varepsilon)^D.$$

In order to use it for approximation of a convex body by a convex polytope, we need to find a measure according to which no relevant cap is too small. First, we recall a classical result of Grünbaum [Grü60] as a non-obvious exercise.

Exercise 3.1.29 (*Grünbaum: Half-spaces containing the center of mass are of big volume*) Let K be a convex body in $\mathbb{R}^d$ with center of mass at the origin, and let F be a half-space containing o. Then

$$\mathrm{vol}(K \cap F) \geq \left(\frac{d}{d+1}\right)^d \mathrm{vol}(K)\,. \tag{3.5}$$

Note that the right hand side is greater than $\frac{\mathrm{vol}(K)}{e}$. ♣

Exercise 3.1.30 (*Stability of Grünbaum's theorem*) Let K be a convex body in $\mathbb{R}^d$ with center of mass at the origin. Let $0 < \vartheta < 1$, and F be a half-space that supports ϑK from outside. Show that

$$\mathrm{vol}(K \cap F) \geq \mathrm{vol}(K)\, \frac{(1-\vartheta)^d}{e}. \tag{3.6}$$

♣

Theorem 3.1.31 (A polytope obtained by uniform sampling from K) *Let* $\vartheta \in (0, 1)$. *Set*

$$t = \left\lceil 3\frac{(d+1)e}{(1-\vartheta)^d} \ln \frac{e}{(1-\vartheta)^d} \right\rceil.$$

Then for any centered convex body K in $\mathbb{R}^d$, if t points $X_1, \ldots, X_t$ are chosen randomly, independently and uniformly from K, then

$$\vartheta K \subseteq \mathrm{conv}(X_1, \ldots, X_t) \subseteq K$$

with probability at least $1 - \delta$, *where*

$$\delta = \left[200\left(\frac{(1-\vartheta)^d}{e}\right)\right]^{d+1}.$$

Exercise 3.1.32 Prove Theorem 3.1.31. ♣

References

The Epsilon-net theorem was found by Haussler and Welzl [HW87] based on ideas of Vapnik and Cervonenkis [VC68], and further developed to its sharpest form by Komlós et al. [KPW92]; cf. the classical books of Pach and Agarwal [PA95] and of Matoušek [Mat02], and the monograph [Mus22] by Mustafa, where a deep analysis of this result shows that fundamentally, it is, again, achieved exploiting a (combinatorial) packing phenomenon. Theorem 3.1.31 is from [Nas19].

3.1.5 A Measure Obtained Using the Polar Body

In all of the preceding arguments, we used volume as a measure for our convex sets—caps, balls, etc. In this section we present another measure that depends on K, which yields a better approximation of K by a convex polytope. Recall that the *polar* of a set K in $\mathbb{R}^d$ is defined as $K^\circ = \{y \in \mathbb{R}^d : \langle x, y\rangle \leq 1 \text{ for all } x \in K\}$. It is well known that if K is a convex body containing the origin in its interior, then so is K°. If, furthermore, K is smooth (that is, it has a unique supporting hyperplane at each boundary point) and strictly convex (that is, its boundary contains no line segment), then so is K°, and in this case, there is a bijection between $\mathrm{bd}(K)$ and $\mathrm{bd}(K^\circ)$, $x \mapsto x^*$, where $x^* \in \mathrm{bd}(K^\circ)$ is the unique point with $\langle x, x^*\rangle = 1$.

For a Borel set $C \subset \mathrm{bd}(K)$, let $C^* = \{x^* \in \mathrm{bd}(K)^\circ : x \in C\}$ denote the corresponding points of the polar of K. Consider the cone-like sets $\mathrm{cone}(C) = \{rx : x \in C, 0 \leq r \leq 1\}$ and $\mathrm{cone}(C^*) = \{ry : y \in C^*, 0 \leq r \leq 1\}$. We set the measure of C to be

$$\mu(C) = \frac{1}{2}\left(\frac{\mathrm{vol}_d(\mathrm{cone}(C))}{\mathrm{vol}_d(K)} + \frac{\mathrm{vol}_d(\mathrm{cone}(C^*))}{\mathrm{vol}_d(K^\circ)}\right).$$

Using this measure, it is shown in [NNR20] that each cap $\mathrm{cap}\,(x, \varepsilon)$ (see Exercise 3.1.2 for the definition) is at least of measure

$$\mu(\mathrm{cap}\,(x, \varepsilon)) \geq C^d \varepsilon^{\frac{d-1}{2}}.$$

The definition of the measure is motivated by the following philosophy. If C contains 'very smooth, flat' boundary points of K, then $\mathrm{vol}_d(\mathrm{cone}(C))$ will be large. On the other hand, if K is 'sharp', that is, it has 'large curvature' at C, then $\mathrm{cone}(C^*)$ will have large volume. In particular, the following is shown in [NNR20].

Lemma 3.1.33 *Let K be a strictly convex body with smooth boundary. Assume that K contains o in its interior and satisfies the Santaló bound* $\mathrm{vol}(K)\,\mathrm{vol}(K^\circ) \leq e^{O(d)} d^{-d}$. *Then μ is a probability measure on* $\mathrm{bd}(K)$ *invariant under linear automorphisms of $\mathbb{R}^d$ and* $\mu(\mathrm{cap}\,(x, \varepsilon)) \geq e^{O(d)} \varepsilon^{\frac{d-1}{2}}$ *for all $x \in \mathrm{bd}(K)$ and all $\varepsilon \in (0, \frac{1}{2})$.*

If one combines Lemma 3.1.33 with the method of Sect. 3.1.4, we obtain a convex polytope P of $e^{O(d)} \varepsilon^{-\frac{d-1}{2}} \ln \frac{1}{\varepsilon}$ vertices approximating K by ε error in the geometric distance, that is, $(1 - \varepsilon)K \subset P \subset K$. The main result of [NNR20] shows that the $\ln \frac{1}{\varepsilon}$ factor may be removed:

Theorem 3.1.34 *Let K be a convex body in $\mathbb{R}^d$ with the center of mass at the origin, and let $\varepsilon \in \left(0, \frac{1}{2}\right)$. Then there exists a convex polytope P with at most $e^{O(d)} \varepsilon^{-\frac{d-1}{2}}$ vertices such that* $(1 - \varepsilon)K \subset P \subset K$.

The quantitative strength of this theorem is the dependence on ε, as well as the factor $e^{O(d)}$, which is better then what is obtained from the Bronshtein–Ivanov net.

A somewhat weaker result was obtained by Barvinok [Bar14] by a completely different method using the 'tensorization trick' [Bar18]. For further developments, see [AdFM23, AM23] by Arya, Fonseca, and Mount, where a refinement is given combining this approach with the Economic Cap Covering Theorem, and an efficient algorithm is presented to find an approximating convex polytope.

3.2 Quantitative Helly-Type Questions

Helly's theorem states the following. *If the intersection of any $d + 1$ members of a finite family of at least $d + 1$ convex sets in $\mathbb{R}^d$ is not empty, then the intersection of the entire family is not empty.* The study of quantitative versions, whereby instead of looking for a point in the intersection, we want to ensure that the intersection is not too small in some sense, was initiated by Bárány et al. in [BKP82].

The **Quantitative Volume Theorem** from [BKP82] states the following. *Assume that the intersection of any $2d$ members of a finite family $\mathcal{F}$ of at least $2d$ convex sets in $\mathbb{R}^d$ is of volume at least one. Then the volume of the intersection of all members of the family is of volume at least $c(d)$, a constant depending only on d.*

In Sect. 3.2.1, we discuss the Quantitative Volume Theorem, in Sect. 3.2.2, we present a colorful generalization. Finally, in Sect. 3.2.3, we consider the dual question: how to select $2d$ vertices of a convex polytope L whose convex hull L' is not too small—a positive homothet of L' of factor roughly d^2 will contain L.

3.2.1 Rough Approximation of the Volume

In [BKP82], it is proved that in the Quantitative Volume Theorem, one may take $c(d) = d^{-2d^2}$ and conjectured that it should hold with $c(d) = d^{-cd}$ for an absolute constant $c > 0$. It was confirmed with $c(d) \approx d^{-2d}$ in [Nas16], whose argument was refined by Brazitikos [Bra17], who showed that one may take $c(d) \approx d^{-3d/2}$. For more on quantitative Helly-type results, see the surveys [HW18, DLGMM19]. For recent quantitative variants of the Fractional Helly Theorem and the (p, q)-Theorem cf. [RS17, DLLHRS17, SXS21, JN22], and [FJT24]. In what follows, we present the proof from [Nas16].

3.2.1.1 Preliminaries for the Proof of the Quantitative Volume Theorem

One of our tools will be John's fundamental theorem, Theorem 3.2.3, which can often be deployed in the study of affine invariant problems. For a non-zero vector $w \in \mathbb{R}^d$, we denote by $w \otimes w$ the linear operator on $\mathbb{R}^d$ which sends any vector $x \in \mathbb{R}^d$ to $w \otimes w(x) = \langle w, x\rangle w$. In matrix form, $w \otimes w = w^\top w$, if w is a column vector. In particular, if w is a unit vector, then $w \otimes w$ is the rank one orthogonal projection of $\mathbb{R}^d$ onto the line spanned by w.

Definition 3.2.1 We say that a set of vectors $w_1, \dots, w_m \in \mathbb{R}^d$ with weights $c_1, \dots, c_m > 0$ form a *John's decomposition of the identity*, if

$$\sum_{i=1}^{m} c_i w_i = o \quad \text{and} \quad \sum_{i=1}^{m} c_i w_i \otimes w_i = I, \tag{3.7}$$

where I is the identity operator on $\mathbb{R}^d$.

Exercise 3.2.2 Show that (3.7) yields $\sum_{i=1}^{m} c_i = d$. ♣

John's theorem [Joh48] (see also Ball's instructive lecture notes [Bal97]) states that once we take an appropriate affine image of a convex body K, we obtain a set of contact points of K and $\mathbf{B}^d(o, 1)$ that has a very special property: these points (unit vectors) yield a John's decomposition of the identity.

Theorem 3.2.3 (John's theorem) *For any convex body K in $\mathbb{R}^d$, there is a unique ellipsoid of maximal volume in K. Furthermore, this ellipsoid is $\mathbf{B}^d(o, 1)$ if, and only if, there are points $w_1, \dots, w_m \in \mathrm{bd}\left(\mathbf{B}^d(o, 1)\right) \cap \mathrm{bd}(K)$ (called* contact points*) and corresponding weights $c_1, \dots, c_m > 0$ that form a John's decomposition of the identity.*

The 'only if' direction was proved by John, while Ball observed in [Bal92] that the 'if' direction holds as well. We say that a convex body K is in *John's position*, if the largest volume ellipsoid contained in K is $\mathbf{B}^d(o, 1)$. A basic application of John's theorem, which we will need later, follows. For the solution, cf. [Bal97], or [GLM+04, Theorem 5.1].

Exercise 3.2.4 Show that if $w_1, \dots, w_m \in \mathrm{bd}\left(\mathbf{B}^d(o, 1)\right)$ and corresponding weights $c_1, \dots, c_m > 0$ form a John's decomposition of the identity, then $\{w_1, \dots, w_m\}^\circ \subset \mathbf{B}^d(o, d)$. Moreover, $\mathbf{B}^d\left(o, \frac{1}{d}\right) \subset \mathrm{conv}(w_1, \dots, w_m)$. Deduce that if K is in John's position, then $K \subset \mathbf{B}^d(o, d)$. ♣

A finer application of John's decomposition is the Dvoretzky-Rogers lemma [DR50], according to which, when a set of unit vectors form a John's decomposition of the identity, they contain a basis whose d vectors are 'well spread'.

Lemma 3.2.5 (Dvoretzky-Rogers lemma) *Let $w_1, \dots, w_m \in \mathrm{bd}\left(\mathbf{B}^d(o, 1)\right)$ and $c_1, \dots, c_m > 0$ form a John's decomposition of the identity. Then there is an orthonormal basis $z_1, \dots, z_d$ of $\mathbb{R}^d$, and a subset $\{v_1, \dots, v_d\}$ of $\{w_1, \dots, w_m\}$ such that*

$$v_i \in \mathrm{span}\{z_1, \dots, z_i\}, \quad \textit{and} \quad \sqrt{\frac{d-i+1}{d}} \leq \langle v_i, z_i \rangle \leq 1, \quad \textit{for all } i = 1, \dots, d. \tag{3.8}$$

Exercise 3.2.6 Prove Lemma 3.2.5. Note that one may drop the condition $\sum_{i=1}^{m} c_i w_i = o$, and the conclusion of the Dvoretzky-Rogers lemma will still hold. ♣

Exercise 3.2.7 Let Δ denote a regular simplex in $\mathbb{R}^d$ such that the ball $\mathbf{B}^d(o, 1)$ is the largest volume ellipsoid in Δ. Prove that

$$\mathrm{vol}(\Delta) = \frac{d^{d/2}(d+1)^{(d+1)/2}}{d!}. \tag{3.9}$$

♣

3.2.1.2 Proof of the Quantitative Volume Theorem

Without loss of generality, we may assume that $\mathcal{F}$ consists of closed half-spaces, and $0 < \mathrm{vol}(\cap\mathcal{F}) < \infty$ (that is, $\cap\mathcal{F}$ is a convex body in $\mathbb{R}^d$), and $\mathcal{F}$ is a finite family. In short, we will assume that $P = \cap\mathcal{F}$ is a d-dimensional convex polyhedron. The problem is clearly affine invariant, so we may further assume that $\mathbf{B}^d(o, 1) \subset P$ is the ellipsoid of maximal volume in P.

By John's theorem (Theorem 3.2.3), there are contact points $w_1, \ldots, w_m \in \mathrm{bd}\big(\mathbf{B}^d(o, 1)\big) \cap \mathrm{bd}(P)$ (and weights $c_1, \ldots, c_m > 0$) that form a John's decomposition of the identity. We denote their convex hull by $Q = \mathrm{conv}(w_1 \ldots, w_m)$. Lemma 3.2.5 yields that there is an orthonormal basis $z_1, \ldots, z_d$ of $\mathbb{R}^d$, and a subset $\{v_1, \ldots, v_d\}$ of the contact points $\{w_1, \ldots, w_m\}$ such that (3.8) holds.

Let $S_1 = \mathrm{conv}(o, v_1, v_2, \ldots, v_d)$ be the simplex spanned by these contact points, and let E_1 be the largest volume ellipsoid contained in S_1. We denote the center of E_1 by u. Let ℓ be the ray emanating from the origin in the direction of the vector $-u$. Clearly, the origin is in the interior of Q. In fact, by Exercise 3.2.4, $\mathbf{B}^d\left(o, \frac{1}{d}\right) \subset Q$. Let w be the point of intersection of the ray ℓ with $\mathrm{bd}(Q)$. Then $|w| \geq 1/d$. Let S_2 denote the simplex $S_2 = \mathrm{conv}(w, v_1, v_2, \ldots, v_d)$. See Fig. 3.3.

We apply a contraction with center w and ratio $\lambda = \frac{|w|}{|w-u|}$ on E_1 to obtain the ellipsoid E_2. Clearly, E_2 is centered at the origin and is contained in S_2. Furthermore,

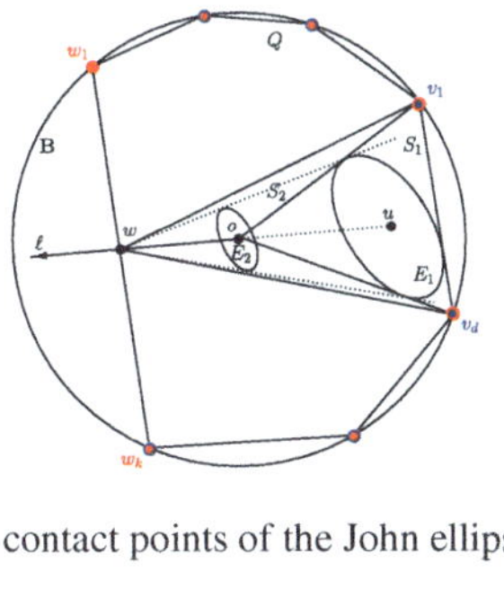

Fig. 3.3 Proof of the QVT using contact points of the John ellipsoid

$$\lambda = \frac{|w|}{|u| + |w|} \geq \frac{|w|}{1 + |w|} \geq \frac{1}{d+1}. \tag{3.10}$$

Since w is on $\mathrm{bd}(Q)$, by Carathéodory's theorem, w is in the convex hull of some set of at most d vertices of Q. By re-indexing the vertices, we may assume that $w \in \mathrm{conv}(w_1, \ldots, w_k)$ with $k \leq d$. Now,

$$E_2 \subset S_2 \subset \mathrm{conv}(w_1, \ldots, w_k, v_1, \ldots, v_d). \tag{3.11}$$

Let $X = \{w_1, \ldots, w_k, v_1, \ldots, v_d\}$ be the set of these unit vectors, and let $\mathcal{G}$ denote the family of those half-spaces which support $\mathbf{B}^d(o, 1)$ at the points of X. Clearly, $|\mathcal{G}| \leq 2d$. Since the points of X are contact points of P and $\mathbf{B}^d(o, 1)$, we have that $\mathcal{G} \subseteq \mathcal{F}$. By (3.11),

$$\cap \mathcal{G} = X^* \subset E_2^*. \tag{3.12}$$

By (3.8),

$$\mathrm{vol}(S_1) \geq \frac{1}{d!} \cdot \frac{\sqrt{d!}}{d^{d/2}} = \frac{1}{\sqrt{d!}d^{d/2}}. \tag{3.13}$$

Since $\mathbf{B}^d(o, 1) \subset \cap\mathcal{F}$, by (3.12) and (3.10), (3.9), (3.13) we have

$$\frac{\mathrm{vol}(\cap\mathcal{G})}{\mathrm{vol}(\cap\mathcal{F})} \leq \frac{\mathrm{vol}\left(E_2^*\right)}{\mathrm{vol}\left(\mathbf{B}^d(o, 1)\right)} = \frac{\mathrm{vol}\left(\mathbf{B}^d(o, 1)\right)}{\mathrm{vol}(E_2)} \leq (d+1)^d \frac{\mathrm{vol}\left(\mathbf{B}^d(o, 1)\right)}{\mathrm{vol}(E_1)} \tag{3.14}$$

$$= (d+1)^d \frac{\mathrm{vol}(\Delta)}{\mathrm{vol}(S_1)} = \frac{d^{d/2}(d+1)^{(3d+1)/2}}{d!\mathrm{vol}(S_1)} \leq \frac{d^d d^{3d/2} e^{3/2} (d+1)^{1/2}}{(d!)^{1/2}} \leq e^{d+1} d^{2d+\frac{1}{2}},$$

where Δ is as defined above (3.9). This completes the proof.

We note that Brazitikos' improvement [Bra17], whereby it is shown that one may take $c(d) \approx d^{-3d/2}$ in the Quantitative Volume Theorem, relies on the following idea. Instead of considering the ellipsoid E_1 contained in the simplex S_1, one sets u as the centroid of S_1, that is, $u = \frac{1}{d+1}\sum_{i=1}^{d} v_i$, and sets E_1 to be the intersection of S_1 with $2u - S_1$, that is, with the reflection of S_1 about u. This intersection is then obviously centrally symmetric, and by Lemma 3.1.7, its volume is the same (up to an inconsequential 2^{-d} factor) as the volume of S_1. The rest of the proof follows as the one presented above, with E_2 being the contraction of this E_1.

3.2.2 Colorful Quantitative Volume Theorem

The **Colorful Helly Theorem**, proved by Lovász, and presented by Bárány [Bár82] states the following. *If* $\mathbf{C}_1, \ldots, \mathbf{C}_{d+1}$ *are finite families (color classes) of convex sets in* $\mathbb{R}^d$*, such that for any colorful selection* $C_1 \in \mathbf{C}_1, \ldots, C_{d+1} \in \mathbf{C}_{d+1}$*, the intersec-*

tion $\bigcap_{i=1}^{d+1} C_i$ *is non-empty, then for some* j*, the intersection* $\bigcap_{C\in\mathbf{C}_j} C$ *is also non-empty.* A quantitative variant appeared in [DFN21], see [SXS21] for further results.

Theorem 3.2.8 (Colorful Quantitative Volume Theorem) *Let* $\mathbf{C}_1, \ldots, \mathbf{C}_{3d}$ *be finite families of convex bodies in* $\mathbb{R}^d$*. Assume that for any colorful selection,* $C_i \in \mathbf{C}_i$ *for each* $i \in [3d]$*, the intersection* $\bigcap_{k=1}^{3d} C_i$ *contains an ellipsoid of volume at least 1. Then, there is an* $i \in [3d]$ *such that* $\bigcap_{C\in\mathbf{C}_i} C$ *contains an ellipsoid of volume at least* $c^{d^2} d^{-5d^2/2}$ *with an absolute constant* $c \geq 0$*.*

Exercise 3.2.9 Deduce the following result. Let $\mathbf{C}_1, \ldots, \mathbf{C}_{3d}$ be finite families of convex bodies in $\mathbb{R}^d$. Assume that for any colorful selection of **2d sets**, $C_{i_k} \in \mathbf{C}_{i_k}$ for each $k \in [2d]$ with $1 \leq i_1 < \cdots < i_{2d} \leq 3d$, the intersection $\bigcap_{k=1}^{2d} C_{i_k}$ is of volume at least 1. Then, there is an $i \in [3d]$ such that $\bigcap_{C\in\mathbf{C}_i} C$ is of volume at least $c^{d^2} d^{(-5d^2/2)-d}$ with an absolute constant $c \geq 0$. ♣

Exercise 3.2.10 (*Quantitative Volume Theorem with Ellipsoids*) Let $C_1, \ldots, C_n$ be convex sets in $\mathbb{R}^d$. Assume that the intersection of any $2d$ of them contains an ellipsoid of volume at least 1. Then $\bigcap_{i=1}^{n} C_i$ contains an ellipsoid of volume at least $c^d d^{-5d/2}$ with an absolute constant $c > 0$. ♣

Proof of Theorem 3.2.8

Preliminaries The proof of Theorem 3.2.8 to some extent follows the proof of the (non-quantitative) Colorful Helly theorem hence, we sketch the latter.

For each colorful choice of d-tuples $C_1 \in \mathbf{C}_{i_1}, \ldots, C_d \in \mathbf{C}_{i_d}$, where $1 \leq i_1 < i_2 < \cdots < i_d \leq d+1$, take the lexicographically lowest point of the intersection $\bigcap_{j=1}^{d} C_{i_j}$. Now, select that colorful d-tuple, for which this lexicographically lowest point is the highest. Using the following exercise, it is easy to see that this point p belongs to all members of the color class which does not appear in the colorful choice.

Exercise 3.2.11 Let $K_1, \ldots, K_n$ be convex sets in $\mathbb{R}^d$, and assume that p is the lexicographically lowest point of $\bigcap_{i=1}^{n} K_i$. Then, there is a subset $\sigma \subseteq \{1, \ldots, n\}$ of indices of size at most d such that p is the lexicographically lowest point of $\bigcap_{i\in\sigma} K_i$. ♣

Next, we turn to the proof of the quantitative result. Recall that we denote the volume of the unit ball by $\omega_d = \operatorname{vol}\left(\mathbf{B}^d(o, 1)\right)$.

Exercise 3.2.12 Assume that $\mathbf{B}^d(o, 1)$ is the largest volume ellipsoid contained in the convex set C in $\mathbb{R}^d$. Let $\mathcal{E}$ be another ellipsoid in C of volume at least $\delta\omega_d$ with $0 < \delta < 1$. Prove that there is a translate of $\frac{\delta}{d^{d-1}}\mathbf{B}^d(o, 1)$ which is contained in $\mathcal{E}$. ♣

We will follow Lovász' idea of the proof of the Colorful Helly Theorem. The first step is to fix an ordering of the objects of study. This time, we are looking for an ellipsoid and not a point in the intersection, therefore we need an *ordering on the ellipsoids*.

For an ellipsoid $\mathcal{E}$, we define its *height* as the largest value of the orthogonal projection of $\mathcal{E}$ on the last coordinate axis, that is, $\max\{\langle x, e_d\rangle : x \in \mathcal{E}\}$, where $e_d = (0, 0, \dots, 0, 1)$.

Lemma 3.2.13 *Let C be a convex body that contains an ellipsoid of volume ω_d. Then there is a unique ellipsoid of volume ω_d such that every other ellipsoid of volume ω_d in C has larger height. Furthermore, if $\tau \in \mathbb{R}$ denotes the height of this ellipsoid, then the largest volume ellipsoid of the convex body $H_\tau \cap C$ is this ellipsoid, where H_τ denotes the closed half-space $H_\tau = \{x \in \mathbb{R}^d : \langle x, e_d\rangle \le \tau\}$.*

We call this ellipsoid the *lowest ellipsoid* in C.

***Proof of Lemma* 3.2.13** It is not difficult to see that $H_\tau \cap C$ does not contain any ellipsoid of volume larger than ω_d. Indeed, otherwise for a sufficiently small $\varepsilon > 0$, the set $H_{\tau-\varepsilon} \cap C$ would contain an ellipsoid of volume equal to ω_d, where $H_{\tau-\varepsilon}$ denotes the closed half-space $H_{\tau-\varepsilon} = \{x \in \mathbb{R}^d : \langle x, e_d\rangle \le \tau - \varepsilon\}$. By Theorem 3.2.3, the largest volume ellipsoid of $H_\tau \cap C$ is unique denote it by E. It follows that E is the unique lowest ellipsoid of C. □

Exercise 3.2.14 Let $\mathbf{C}_1, \dots, \mathbf{C}_{d+1}$ be finite families of convex bodies, and L a convex body in $\mathbb{R}^d$. Assume that for any colorful selection $C_1 \in \mathbf{C}_1, \dots, C_{d+1} \in \mathbf{C}_{d+1}$, the intersection $\bigcap_{i=1}^{d+1} C_i$ contains a translate of L. Then for some j, the intersection $\bigcap_{C \in \mathbf{C}_j} C$ contains a translate of L. ♣

***The proof of Theorem* 3.2.8** Consider the lowest ellipsoid in the intersection of each colorful selections of $2d-1$ sets. We may assume that the highest one of these ellipsoids is $\mathbf{B}^d(o, 1)$. By possibly changing the indices of the families, we may assume that the selection is $C_1 \in \mathbf{C}_1, \dots, C_{2d-1} \in \mathbf{C}_{2d-1}$. We call $\mathbf{C}_{2d}, \mathbf{C}_{2d+1}, \dots, \mathbf{C}_{3d}$ the *remaining families.*

Consider the half-space $H_1 = \{x \in \mathbb{R}^d : \langle x, e_d\rangle \le 1\} \supset \mathbf{B}^d(o, 1)$. By Lemma 3.2.13, $\mathbf{B}^d(o, 1)$ is the largest volume ellipsoid contained in $M := C_1 \cap \dots \cap C_{2d-1} \cap H_1$.

Next, take an arbitrary colorful selection $C_{2d} \in \mathbf{C}_{2d}, C_{2d+1} \in \mathbf{C}_{2d+1}, \dots, C_{3d} \in \mathbf{C}_{3d}$ of the remaining families. We claim that the intersection of any $2d$ sets of

$$C_1, \dots, C_{2d-1}, H_1, C_{2d}, \dots, C_{3d}$$

contains an ellipsoid of volume at least ω_d. Indeed, if H_1 is not among those $2d$ sets, then our assumption ensures this. If H_1 is among them, then by the choice of H_1, the claim holds. Therefore, by Exercise 3.2.10, the intersection

$$\bigcap_{i=1}^{3d} C_i \cap H_1$$

contains an ellipsoid $\mathcal{E}$ of volume at least $\delta\omega_d$, where $\delta := c^d d^{-3d/2}$. Clearly, $\mathcal{E} \subset M$. Since $\mathbf{B}^d(o, 1)$ is the maximum volume ellipsoid contained in M, by Exercise 3.2.12, we have that there is a translate of $\frac{\delta}{d^{d-1}}\mathbf{B}^d(o, 1)$ which is contained in $\mathcal{E}$ and thus in $\bigcap_{i=2d}^{3d} C_i$. Thus, we have shown that any colorful selection $C_{2d} \in \mathbf{C}_{2d}, C_{2d+1} \in \mathbf{C}_{2d+1}, \ldots, C_{3d} \in \mathbf{C}_{3d}$ of the remaining $d+1$ families, $\bigcap_{i=2d}^{3d} C_i$ contains a translate of the same convex body $c^d d^{-5d/2}\mathbf{B}^d(o, 1)$. It follows from Exercise 3.2.14 that there is an index $2d \le i \le 3d$ such that $\bigcap_{C\in\mathbf{C}_i} C$ contains a translate of $c^d d^{-5d/2}\mathbf{B}^d(o, 1)$, which is an ellipsoid of volume $c^{d^2} d^{-5d^2/2}\omega_d$, finishing the proof of Theorem 3.2.8.

3.2.3 Back to Geometric Distance

Next, we turn to the problem of (a rough) approximation of a convex polytope as the convex hull of a well chosen small subset of its vertices. The following theorem is due to Almendra-Hernández et al. [AHAK23], improving a result in [IN22].

Theorem 3.2.15 *Let $\lambda > 0$, and $L \subset \mathbb{R}^d$ be a convex polytope such that $L \subset -\lambda L$. Then there is a set of at most $2d$ vertices of L, whose convex hull L' satisfies*

$$L \subset -(\lambda + 2)d \cdot L'.$$

According to the next exercise, with the right choice of the origin, the assumption of the theorem always holds with $\lambda = d$.

Exercise 3.2.16 Show that for any convex set K in $\mathbb{R}^d$, there is a translation $t \in \mathbb{R}^d$ such that $K - t \subseteq d(t - K)$. ♣

In [AHAK23], Theorem 3.2.15 is used to show the following quantitative Helly-type result for the diameter: *For any finite family $\mathcal{F}$ of closed convex sets, one can select a subfamily of at most $2d$ members such that their intersection is of diameter at most $2d^2 D$, where D denotes the diameter of the intersection of $\mathcal{F}$.*

Proof of Theorem 3.2.15 The condition $L \subseteq -\lambda L$ ensures that the origin belongs to the interior of L. Among all simplices with d vertices from the set of vertices of L and one vertex at the origin, consider a simplex $S = \operatorname{conv}(0, v_1, \ldots, v_d)$ with maximal volume. The simplex S can be represented as

$$S = \left\{ x \in \mathbb{R}^d : x = \alpha_1 v_1 + \ldots + \alpha_d v_d \quad \text{for } \alpha_i \ge 0 \text{ and } \sum_{i=1}^{d} \alpha_i \le 1 \right\}. \quad (3.15)$$

Define $P = \sum_{i \in [d]} [-v_i, v_i]$. It is easy to see that P is a parallelotope that can be represented as

$$P = \{x \in \mathbb{R}^d : x = \beta_1 v_1 + \ldots + \beta_d v_d \quad \text{for } \beta_i \in [-1, 1]\}. \tag{3.16}$$

Since S is chosen maximally, Eq. (3.16) shows that for any vertex v of L, $v \in P$. By convexity,

$$L \subset P. \tag{3.17}$$

Let $S' = -2dS + (v_1 + \ldots + v_d)$. By (3.15),

$$S' = \left\{ x \in \mathbb{R}^d : x = \gamma_1 v_1 + \ldots + \gamma_d v_d \quad \text{for } \gamma_i \leq 1 \text{ and } \sum_{i=1}^{d} \gamma_i \geq -d \right\},$$

which, together with (3.16), yields

$$P \subseteq S'. \tag{3.18}$$

Let y be the intersection of the ray emanating from 0 in the direction $-(v_1 + \cdots + v_d)$ and the boundary of Q. By Carathéodory's theorem, we can choose $k \leq d$ vertices $\{v_1', \ldots, v_k'\}$ of L such that $y \in \operatorname{conv}(v_1', \ldots, v_k')$. Set $L' = \operatorname{conv}(v_1, \ldots, v_d, v_1', \ldots, v_k')$. Clearly, $\frac{v_1 + \cdots + v_d}{d} \in S \subset L$. Thus, $0 \in L'$, and consequently,

$$S \subseteq Q'. \tag{3.19}$$

Since $L \subset -\lambda L$, we also have that

$$\frac{v_1 + \cdots + v_d}{d} \in -\lambda[y, 0] \subset -\lambda L'.$$

Combining it with (3.17), (3.18), (3.19), we obtain

$$L \subset P \subset S' = -2dS + (v_1 + \cdots + v_d) \subset -2d\, L' - \lambda d\, L' = -(\lambda + 2)d\, L', \tag{3.20}$$

Completing the proof of Theorem 3.2.15. □

Recall Exercise 3.2.4, according to which if K is in John's position, then $K \subset \mathbf{B}^d(o, d)$.

Exercise 3.2.17 Let Q be a convex polytope in $\mathbb{R}^d$ in John's position. Then there is a subset of at most $2d$ vertices of Q whose convex hull Q' satisfies

$$Q \subseteq -2d^2 Q'.$$

♣

Steinitz's theorem states that, if the origin belongs to the interior of the convex hull of a set $Q \subset \mathbb{R}^d$, then there is a subset of Q of at most $2d$ points whose convex hull contains the origin in the interior. The first quantitative version was given by Bárány et al. in [BKP82]. The current best bound on the radius of the ball appeared in [IN24], and it reads as follows.

Theorem 3.2.18 *Let Q be a subset of $\mathbb{R}^d$ whose convex hull contains the Euclidean unit ball $\mathbf{B}^d(o, 1)$. Then there exists a subset of Q of at most $2d$ points whose convex hull Q' satisfies*

$$\mathbf{B}^d\left(o, \frac{1}{6d^2}\right) \subset Q'.$$

3.3 Approximation of Sums of Matrices

In the previous lecture, we saw how taking a John's decomposition of the identity and selecting a small subset of the unit vectors yields geometric results. In this lecture, we consider the following general question. Given a matrix A (often, we simply have $A = I$) as a (positive) linear combination of some other matrices. Can we select a small subset of the matrices whose linear combination (with possibly new coefficients) yields a matrix *close to* A in some norm.

In particular, for geometric applications, our main question is the following. Given a John's decomposition of the identity as in (3.7). By Carathéodory's theorem applied in the vector space of $d \times d$ real matrices, we may assume that the sum contains roughly d^2 summands which, by observing that the matrices in question are symmetric, drops to roughly $d^2/2$, for the details, see [GS05] or [FNZ19]. Can we decrease this number further? The answer is no. However, if we relax our goal, and require the new convex combination to be *close to* the identity operator, as opposed to *equal to it*, we can. In this section, we present a method that achieves a convex combination with $c_1 d \ln d$ summands (Rudelson's theorem), and then another one that reaches $c_2 d$ summands, where the constants $c_1, c_2 > 0$ depend on the precision of the approximation (how close we get to the identity operator), and do not depend on d.

As a warm up, we consider a model problem in $\mathbb{R}^d$ in place of the space of (symmetric, positive definite, etc.) matrices. According to Carathéodory's theorem (cf. [Mat02]), if o is in the convex hull of a set of vectors, than it is in the convex hull of a subset of them of size $d + 1$. When getting close to o is sufficient, the size of the subset is much smaller: it is independent of d. For more, cf. [ABMT20] by Adiprasito, Bárány, Mustafa, and Terpai.

Exercise 3.3.1 Let P be a set of n points in $\mathbb{R}^d$, $k \le d$ and $o \in \operatorname{conv}(P)$. Then there is a subset Q of P of size at most k with

$$\operatorname{dist}(o, \operatorname{conv}(Q)) \le \frac{\operatorname{diam}(P)}{\sqrt{k}},$$

where dist denotes the distance of the closest point of $\mathrm{conv}(Q)$ to o, and $\mathrm{diam}(P) = \sup\{|x-y| : x, y \in P\}$ is the diameter of P. ♣

3.3.1 Rudelson's Theorem: A Probabilistic Approach

In this section, we consider random positive semi-definite (PSD) matrices. First, we recall some basic facts.

Exercise 3.3.2 Show the following.

1. The set $\mathcal{P}^d$ of positive semi-definite $d \times d$ matrices (with real entries) forms a convex cone with apex at the origin in the vector space $\mathbb{R}^{d(d+1)/2}$ of symmetric matrices.
2. Matrices of trace 1 form a hyperplane H_1 containing $\frac{1}{d}I$ in $\mathbb{R}^{d(d+1)/2}$.
3. The set $\mathcal{P}^d \cap H_1$ is a convex body in H_1.

♣

Note that if $v \in \mathbb{R}^d$ is a non-zero vector, then $v \otimes v$ is a rank one PSD matrix of trace $|v|^2$. One can show that the extremal rays of the PSD cone $\mathcal{P}^d$ are exactly the matrices obtained this way. We will make use of the following *partial ordering defined on the space of symmetric matrices*. For A and B symmetric matrices, we say that $A \preceq B$ if $B - A$ is positive semi-definite.

A random vector v in $\mathbb{R}^d$ is called *isotropic*, if $\mathbb{E} v \otimes v = I$, where $\mathbb{E}$ denotes the expectation of a random variable, and I is the identity operator on $\mathbb{R}^d$. According to Rudelson's theorem [Rud99], if we take k independent copies $y_1, \ldots, y_k$ of an isotropic random vector y in $\mathbb{R}^d$ for which $|y|^2 \le \gamma$ almost surely, with

$$k = \left\lceil \frac{c\gamma \ln d}{\varepsilon^2} \right\rceil, \text{ then } \mathbb{E} \left\| \frac{1}{k} \sum_{i=1}^{k} y_i \otimes y_i - I \right\| \le \varepsilon,$$

where $\|A\| = \max\{\langle Ax, Ax\rangle^{1/2} : x \in \mathbb{R}^d, \langle x, x\rangle = 1\}$ denotes the *operator norm* of the matrix A.

Rudelson's result applies in the setting of a John's decomposition of the identity, see (3.7): if we take $\alpha_i = c_i/d$, then by Exercise 3.2.2, we may interpret α_i as the probability of a random unit vector taking the value u_i.

Let $\sigma = \{i_1, \ldots, i_k\}$ be a multiset obtained by k independent draws from $\{1, \ldots, m\}$ according to the distribution where $\mathrm{nex}(i \text{ is drawn}) = \alpha_i$, and consider the following average of matrices $\frac{1}{k}\sum_{i\in\sigma} \sqrt{d}u_i \otimes \sqrt{d}u_i$. It follows that, in expectation, this average is not farther than ε from I in the operator norm, provided that k is at least $\frac{cd \ln d}{\varepsilon^2}$, where c is some constant. We state it in a slightly more general form as presented in [INP20].

Theorem 3.3.3 (Rudelson's theorem) *Let $0 < \varepsilon < 1$ and $Q_1, \ldots, Q_k$ be independent random matrices distributed according to (not necessarily identical) probability distributions $\mathcal{P}_1, \ldots, \mathcal{P}_k$ on the set $\mathcal{P}^d$ of $d \times d$ real positive semi-definite matrices such that $\mathbb{E}Q_i = A$ for some $A \in \mathcal{P}^d$ and all $i \in \{1, \ldots, k\}$. Set $\gamma = \mathbb{E}(\max_{i \in \{1,\ldots,k\}} \|Q_i\|)$, and assume that*

$$k \geq \frac{c\gamma(1 + \|A\|)\ln d}{\varepsilon^2},$$

where c is an absolute constant. Then

$$\mathbb{E}\left\|\frac{1}{k}\sum_{i \in \{1,\ldots,k\}} Q_i - A\right\| \leq \varepsilon. \tag{3.21}$$

3.3.1.1 Proof of Rudelson's Theorem

We give Rudelson's proof from [Rud99] as presented in [INP20]. Let $\mathcal{P}^d$ denote the cone of positive semi-definite symmetric matrices in $\mathbb{R}^{d \times d}$. The *Schatten p-norm* of a real $d \times d$ matrix A is defined as

$$\|A\|_{C_p^d} := \left(\sum_{i=1}^{d}(s_i(A))^p\right)^{1/p},$$

where $s_1(A), \ldots, s_d(A)$ is the sequence of eigenvalues of the positive semi-definite matrix $\hat{A} = \sqrt{A^\top A}$. Clearly, $\|A\| = \left\|\hat{A}\right\|$. Now, $\left\|\hat{A}\right\|$ is a diagonalizable matrix, and the largest absolute value of its eigenvalues appears in the sum in the definition of $\|A\|_{C_p^d}$. Thus for any $p \geq 1$, we have

$$\|A\| \leq \|A\|_{C_p^d}.$$

Exercise 3.3.4 Show that for $p = \ln d$,

$$\|A\| \leq \|A\|_{C_p^d} \leq e\,\|A\| \tag{3.22}$$

holds, where ln denotes the natural logarithm, and e denotes its base. ♣

From this point on, $\mathbf{r}$ denotes a sequence of k *Rademacher variables*, that is, $\mathbf{r} = (r_1, \ldots, r_k)$, where the r_i are random variables uniformly distributed on $\{1, -1\}$, independent of each other and all other random variables in the context. We state the following inequality due to Lust–Piquard and Pisier [LP86, LPP91], essentially in the form as it appears in the book [Pis98, Theorem 8.4.1].

Theorem 3.3.5 (Lust–Piquard) $2 \le p < \infty$. *For any d and any $Q_1, \dots, Q_k$ (not necessarily positive definite) square matrices of size d we have*

$$\left[\mathop{\mathbb{E}}_{\mathbf{r}}\left\|\sum_{j=1}^{k} r_j Q_j\right\|_{C_p^d}^{p}\right]^{1/p} \le c\sqrt{p}\max\left\{\left\|\left(\sum_{j=1}^{k} Q_j Q_j^{\top}\right)^{1/2}\right\|_{C_p^d}, \left\|\left(\sum_{j=1}^{k} Q_j^{\top} Q_j\right)^{1/2}\right\|_{C_p^d}\right\}$$

for a universal constant $c > 0$.

Note that for any $d \times d$ matrix Q, the product $Q^{\top}Q$ is positive semi-definite. Since, by Weyl's inequality, the Schatten p-norm is monotone on the cone of positive semi-definite matrices, we may deduce from the theorem of Lust–Piquard the following inequality

$$\left[\mathop{\mathbb{E}}_{\mathbf{r}}\left\|\sum_{j=1}^{k} r_j Q_j\right\|_{C_p^d}^{p}\right]^{1/p} \le c\sqrt{p}\left\|\left(\sum_{j=1}^{k} Q_j Q_j^{\top} + Q_j^{\top} Q_j\right)^{1/2}\right\|_{C_p^d}. \tag{3.23}$$

Lemma 3.3.6 (Symmetrization by Rademacher variables) *Let $q_1, \dots, q_k$ be independent random vectors distributed according to (not necessarily identical) probability distributions $\mathcal{P}_1, \dots, \mathcal{P}_k$ on a normed space X with $\mathbb{E}q_i = q$ for all $i \in \{1, \dots, k\}$. Then*

$$\mathop{\mathbb{E}}_{q_1,\dots,q_k}\left\|\frac{1}{k}\sum_{\ell=1}^{k} q_\ell - q\right\| \le \frac{2}{k}\mathop{\mathbb{E}}_{q_1,\dots,q_k}\mathop{\mathbb{E}}_{\mathbf{r}}\left\|\sum_{\ell=1}^{k} r_\ell q_\ell\right\|.$$

Proof of Lemma 3.3.6 Let $\bar{q}_1, \dots, \bar{q}_k$ be independent random vectors chosen according to $\mathcal{P}_1, \dots, \mathcal{P}_k$, respectively.

$$\mathop{\mathbb{E}}_{q_1,\dots,q_k}\left\|\frac{1}{k}\sum_{\ell=1}^{k} q_\ell - q\right\| = \frac{1}{k}\mathop{\mathbb{E}}_{q_1,\dots,q_k}\left\|\sum_{\ell=1}^{k} q_\ell - \mathop{\mathbb{E}}_{\bar{q}_1,\dots,\bar{q}_k}\sum_{\ell=1}^{k} \bar{q}_\ell\right\|$$

$$\le \frac{1}{k}\mathop{\mathbb{E}}_{\mathbf{r}}\mathop{\mathbb{E}}_{q_1,\dots,q_k}\mathop{\mathbb{E}}_{\bar{q}_1,\dots,\bar{q}_k}\left\|\sum_{\ell=1}^{k} r_\ell(q_\ell - \bar{q}_\ell)\right\| \le \frac{2}{k}\mathop{\mathbb{E}}_{q_1,\dots,q_k}\mathop{\mathbb{E}}_{\mathbf{r}}\left\|\sum_{\ell=1}^{k} r_\ell q_\ell\right\|,$$

where we used that $r_\ell(q_\ell - \bar{q}_\ell)$ and $q_\ell - \bar{q}_\ell$ have the same distribution. □

Proof of Theorem 3.3.3 The argument follows very closely Rudelson's. The philosophy is a standard route in probability theory: once we have a bound on the p-th moment of some random variable (in our case, this bound is provided by the Lust-Picard inequality), we turn it into a bound on the expected deviation of the average of an independent sample from the expected value of the distribution. Note that here, the Lust-Picard inequality is not a bound directly on the p-th moment, but on a symmetrized version it. This, however is well treated by Lemma 3.3.6. Denote by $D = \frac{1}{k}\sum_{i\in\{1,\dots,k\}} Q_i - A$, and $p = \ln d$. Then

$$\begin{aligned}
\mathop{\mathbb{E}}_{Q_1,\dots,Q_k} \|D\| &\le \mathop{\mathbb{E}}_{Q_1,\dots,Q_k} \|D\|_{C_p^d} \overset{(S)}{\le} \frac{2}{k} \mathop{\mathbb{E}}_{Q_1,\dots,Q_k} \mathop{\mathbb{E}}_{\mathbf{r}} \left\| \sum_{\ell=1}^{k} r_\ell Q_\ell \right\|_{C_p^d} \\
&\overset{(H)}{\le} \frac{2}{k} \mathop{\mathbb{E}}_{Q_1,\dots,Q_k} \left[\mathop{\mathbb{E}}_{\mathbf{r}} \left\| \sum_{\ell=1}^{k} r_\ell Q_\ell \right\|_{C_p^d}^{p} \right]^{1/p} \overset{(L\text{-}P)}{\le} \frac{c_0\sqrt{p}}{k} \mathop{\mathbb{E}}_{Q_1,\dots,Q_k} \left\| \left(\sum_{\ell=1}^{k} Q_\ell^2 \right)^{1/2} \right\|_{C_p^d} \\
&\overset{(PSD)}{\le} \frac{c_0\sqrt{p}}{k} \mathop{\mathbb{E}}_{Q_1,\dots,Q_k} \left[\max_{\ell\in\{1,\dots,k\}} \|Q_\ell\|^{1/2} \cdot \left\| \left(\sum_{\ell=1}^{k} Q_\ell \right)^{1/2} \right\|_{C_p^d} \right] \\
&\le \frac{c_1\sqrt{p}}{k} \mathop{\mathbb{E}}_{Q_1,\dots,Q_k} \left[\max_{\ell\in\{1,\dots,k\}} \|Q_\ell\|^{1/2} \cdot \left\| \left(\sum_{\ell=1}^{k} Q_\ell \right) \right\|^{1/2} \right] \\
&\overset{(B)}{\le} \frac{c_1\sqrt{\gamma p}}{k} \left[\mathop{\mathbb{E}}_{Q_1,\dots,Q_k} \left\| \left(\sum_{\ell=1}^{k} Q_\ell \right) \right\|^{1/2} \right] \overset{(H)}{\le} \frac{c_1\sqrt{\gamma p}}{k} \left[\mathop{\mathbb{E}}_{Q_1,\dots,Q_k} \left\| \left(\sum_{\ell=1}^{k} Q_\ell \right) \right\| \right]^{1/2} \le \\
&\frac{c_1\sqrt{\gamma p}}{\sqrt{k}} \left[\mathop{\mathbb{E}}_{Q_1,\dots,Q_k} \|D\| + \|A\| \right]^{1/2},
\end{aligned}$$

where c_0 and c_1 are positive constants. Here, we use Lemma 3.3.6 in step (S) and the inequality (3.23) in step (L-P). The inequality (PSD) relies on the fact that the matrices Q_i are positive semi-definite. Inequality (B) follows from the assumed bound on the norm of Q_ℓ, and (H) follows from Hölder's inequality (or, the Cauchy–Schwarz inequality). Thus, we obtain

$$\mathbb{E}\|D\| \le \frac{c_1\sqrt{\gamma \ln d}}{\sqrt{k}} \sqrt{\mathbb{E}\|D\| + \|A\|}.$$

Denoting by $\alpha = \left(\frac{c_1\sqrt{\gamma \ln d}}{\sqrt{k}} \right)^2$, we have

$$(\mathbb{E}\|D\|)^2 - \alpha\mathbb{E}\|D\| - \alpha\|A\| \le 0.$$

Therefore, we get $\mathbb{E}\,\|D\| \le \alpha + \sqrt{\alpha \,\|A\|}$, and thus the inequality

$$\mathbb{E}\,\|D\| \le \frac{c_1^2 \gamma \ln d}{k} + \frac{c_1 \sqrt{\gamma \,\|A\| \ln d}}{\sqrt{k}} \le \varepsilon$$

holds for $k \ge \frac{c\gamma(1+\|A\|)\ln d}{\varepsilon^2}$ with sufficiently large c. Theorem 3.3.3 is proved. $\square$

3.3.2 An Algorithmic Approach

If instead of considering the *expectation* of the average of randomly chosen $u_i \otimes u_i$, we want to show the *existence* of a small subset of the set of $u_i \otimes u_i$ whose average is close to I, then the picture changes, as was shown by a completely different approach introduced in the seminal paper of Batson et al. [BSS14]. It was developed further by Marcus et al. [MSS15] (see also [Sri12]), and by Friedland and Youssef [FY17]. In [FY17], the following is shown.

Theorem 3.3.7 *Let $u_1, \dots, u_m$ be unit vectors in $\mathbb{R}^d$ that yield a John's decomposition of I. Then there is a multi-subset σ of $\{1, \dots, m\}$ of size $|\sigma| = \frac{cd}{\varepsilon^2}$ with $\left\| \frac{1}{|\sigma|} \sum_{i\in\sigma} \sqrt{d}u_i \otimes \sqrt{d}u_i - I \right\| < \varepsilon$. Furthermore, σ can be found by a deterministic algorithm.*

3.3.3 How It All Comes Together

In this concluding section, we present a result of Brazitikos [Bra17, Theorem 1.4], to show how an approximate John's decomposition, similar to the one given in Theorem 3.3.3 with $A = I$, or the one in Theorem 3.3.7, yields a quantitative volume theorem with a higher Helly number than the Quantitative Volume Theorem given at the top of Sect. 3.2.1 (constant times d in place of $2d$) but better approximation in the volume (roughly of order d^d as opposed to $d^{3d/2}$).

Theorem 3.3.8 (Brazitikos) *There is an absolute constant $\alpha > 1$, such that the following holds in any dimension d. Let $F_1, \dots, F_n$ be a family of closed half-spaces in $\mathbb{R}^d$ such that $P = \bigcap_{i=1}^n F_i$ is of positive volume. Then there is an index set $\sigma \subset \{1, \dots, n\}$ of size at most αd with*

$$\operatorname{vol}\left(\bigcap_{i\in\sigma} F_i \right) \le (Cd)^d \operatorname{vol}(P),$$

where $C > 0$ is a universal constant.

The first main tool in the proof of Theorem 3.3.8 is the following approximate John's decomposition result of Srivastava [Sri12].

Lemma 3.3.9 (Srivastava: Approximate John's decomposition) *For any $\varepsilon > 0$ there is a $\vartheta_\varepsilon > 0$ with the following property. If the unit vectors $w_1, \dots, w_m$ satisfy* (3.7) *with weights $c_1, \dots, c_m > 0$, then there is a $\sigma \subset \{1, \dots, m\}$ of size at most $\vartheta_\varepsilon d$, positive scalars α_i for $i \in \sigma$ and a vector v such that*

$$\|v\|^2 \le \frac{\varepsilon}{\sum_{i\in\sigma}\alpha_i}, \tag{3.24}$$

$$I \preceq \sum_{i\in\sigma} \alpha_i (w_i + v) \otimes (w_i + v) \preceq (4+\varepsilon) I \tag{3.25}$$

and

$$\sum_{i\in\sigma} \alpha_i (w_i + v) = o. \tag{3.26}$$

The second main tool is a stability version of the Brascamp–Lieb inequality.

Lemma 3.3.10 (Brazitikos: Approximate Brascamp–Lieb inequality) *Let $\gamma > 1$, and $u_1, \dots, u_k \in \mathbb{R}^d$ unit vectors satisfying*

$$I \preceq A := \sum_{i=1}^{k} \beta_i u_i \otimes u_i \preceq \gamma I$$

with appropriate $\beta_1, \dots, \beta_k > 0$. Set $\kappa_i = \beta_i \langle A^{-1} u_i, u_i \rangle > 0$ for $i = 1, \dots, k$. Then for any measurable functions $f_1, \dots, f_k : \mathbb{R} \to \mathbb{R}_+$, we have

$$\int_{\mathbb{R}^d} \prod_{i=1}^{k} f_i^{\kappa_i}(\langle x, u_i\rangle)\mathrm{d}x \le \gamma^{\frac{d}{2}} \prod_{i=1}^{k} \left(\int_{\mathbb{R}^d} f_i(\langle x, u_i\rangle)\mathrm{d}x \right)^{\kappa_i}.$$

Sketch of the proof of Theorem 3.3.8 The proof is strongly influenced by Ball's proof of the reverse isoperimetric inequality [Bal91], where the exact (not an approximate) form of the Brascamp–Lieb inequality is used. For the interested reader, we recommend reading that proof (beautifully presented in Lecture 4 of [Bal97]) alongside the one presented here to gain more intuition.

The problem is affine invariant, so we may assume that P is in John's position, that is, $\mathbf{B}^d(o, 1)$ is the largest volume ellipsoid contained in it. John's theorem (Theorem 3.2.3) yields the existence of vectors w_i and positive scalars c_i satisfying (3.7), which by Lemma 3.3.9 yields a subset σ of indices of size $k \le \vartheta_\varepsilon d$, positive scalars α_i for $i \in \sigma$ and a vector v satisfying (3.24), (3.25), and (3.26). It is not difficult to deduce from (3.24), (3.25), and (3.26) that $\|v\|^2 \le \frac{\varepsilon}{d}$. (Check it!)

Setting $w = \frac{v}{\sqrt{\varepsilon d}}$, Exercise 3.2.4 yields that $w \in \mathrm{conv}(w_1, \dots, w_m)$, and thus, by Carathéodory's theorem, there is an index set $\tau \subset \{1, \dots, n\}$ of size $d+1$ such that

$$w = \sum_{i \in \tau} \rho_i w_i, \text{ where } \rho_i \geq 0 \text{ and } \sum_{i \in \tau} \rho_i = 1.$$

We set

$$Q = \{x \in \mathbb{R}^d : \langle x, w_i \rangle \leq 1 \text{ for } i \in \sigma \cup \tau\},$$

and will prove that $\mathrm{vol}(Q) \leq (Cd)^d \mathrm{vol}(P)$. By (3.26), we have $\sum_{i \in \sigma} \alpha_i(-v) = \sum_{i \in \sigma} \alpha_i w_i$, and hence, $-v \in \mathrm{conv}(w_i : i \in \sigma)$. We obtain that $\left[-v, \frac{v}{\sqrt{\varepsilon d}}\right] \subset \mathrm{conv}(w_i : i \in \sigma \cup \tau)$.

Next, we lift our vectors into $\mathbb{R}^{d+1}$ by setting $v_j = \sqrt{\frac{d}{d+1}}\left(-w_j, \frac{1}{\sqrt{d}}\right)$. By computation, one obtains the following comparison of $(d+1) \times (d+1)$ PSD matrices.

$$I_{d+1} \preceq A := \sum_{i \in \sigma} \beta_i v_i \otimes v_i \preceq 10 I_{d+1} \tag{3.27}$$

for $\beta_i = \frac{2(d+1)}{d} c_i > 0$, where I_{d+1} denotes the identity operator on $\mathbb{R}^{d+1}$. We lift Q into $\mathbb{R}^{d+1}$ as a cone by setting

$$\hat{Q} = \left\{(x, r) \in \mathbb{R}^d \times \mathbb{R}_+ : x \in \frac{rQ}{\sqrt{d}}\right\}.$$

By definition, if $y \in \hat{Q}$, then $\langle y, v_i \rangle > 0$ for all $i \in \sigma$.
Note also that if $(x, r) \in \hat{Q}$, then

$$\left(\frac{1}{\sum_{i \in \sigma} c_i}\right) \langle \sum_{i \in \sigma} c_i u_i, x \rangle = \langle -u, x \rangle \geq -\sqrt{\varepsilon} r.$$

We set $\varepsilon = 10^{-3}$, and $f_i(t) = e^{-\frac{\beta_i}{\kappa_i}} \mathbf{1}_{[0,\infty)}(t)$ for $i \in \sigma$, where $\mathbf{1}_{[0,\infty)}(\cdot)$ denotes the characteristic function of non-negative reals. It follows that if $y = (x, r) \in \hat{Q}$, then

$$\prod_{i \in \sigma} f_i^{\kappa_i}(\langle x, v_i \rangle) \geq \exp\left(-10r(d+1)\right).$$

We apply Lemma 3.3.10, to obtain

$$\int_{\mathbb{R}^{d+1}} \prod_{i \in \sigma} f_i^{\kappa_i}(\langle y, v_i \rangle) \mathrm{d}y \leq 10^{\frac{d+1}{2}} \prod_{i \in \sigma} \left(\int_{\mathbb{R}^{d+1}} f_i(\langle y, v_i \rangle) \mathrm{d}y\right)^{\kappa_i},$$

with $\kappa_i = \beta_i \langle A^{-1}v_i, v_i\rangle > 0$ for $i \in \sigma$. Note that by (3.27), $\kappa_i/\beta_i = \langle A^{-1}v_i, v_i\rangle \le 1$. Combined with the preceding equations, we have

$$\frac{\mathrm{vol}(Q)}{d^{d/2}}\int_0^\infty r^d e^{-10r(d+1)}\mathrm{d}r = \int_0^\infty \int_{\frac{rQ}{\sqrt{d}}} e^{-10r(d+1)}\mathrm{d}x\mathrm{d}r =$$

$$\int_{y=(r,x)\in\hat{Q}} e^{-10r(d+1)}\mathrm{d}y \le \int_{\mathbb{R}^{d+1}} \prod_{i\in\sigma} f_i^{\kappa_i}(\langle y, v_i\rangle)\mathrm{d}y \le$$

$$10^{\frac{d+1}{2}} \prod_{i\in\sigma}\left(\int_{\mathbb{R}^{d+1}} f_i(\langle y, v_i\rangle)\mathrm{d}y\right)^{\kappa_i} = 10^{\frac{d+1}{2}}\prod_{i\in\sigma}\left(\frac{\kappa_i}{c_i}\right)^{\kappa_i} = 40^{\frac{d+1}{2}}.$$

Stirling's approximation now yields

$$\mathrm{vol}(Q) \le C^d d^{d/2}.$$

Since $\mathbf{B}^d(o, 1) \subseteq P \subseteq Q$, and $\mathrm{vol}\left(\mathbf{B}^d(o, 1)\right)$ is of order $d^{-d/2}$, the desired volume estimate follows, completing the sketch of the proof of Theorem 3.3.8. □

Acknowledgements The author of the chapter was supported by the National Research, Development and Innovation Fund (NRDI) grants K147544 and K131529, as well as the ELTE TKP 2021-NKTA-62 funding scheme, and by the NRDI EXCELLENCE-24 grant no. 151504 Combinatorics and Geometry.

References

[AAGM15] Shiri Artstein-Avidan, Apostolos Giannopoulos, and Vitali D. Milman. *Asymptotic geometric analysis, Part I*, volume 202. American Mathematical Soc., 2015.

[ABMT20] Karim Adiprasito, Imre Bárány, Nabil H. Mustafa, and Tamás Terpai. Theorems of Carathéodory, Helly, and Tverberg without dimension. *Discrete Comput. Geom.*, 64(2):233–258, 2020.

[AdFM23] Sunil Arya, Guilherme D. da Fonseca, and David M. Mount. Economical convex coverings and applications. In *Proceedings of the 2023 Annual ACM-SIAM Symposium on Discrete Algorithms (SODA)*, pages 1834–1861. SIAM, Philadelphia, PA, 2023.

[AHAK23] Víctor Hugo Almendra-HernÃndez, Gergely Ambrus, and Matthew Kendall. Quantitative Helly-type theorems via sparse approximation. *Discrete and Computational Geometry*, pages 1707–1714, 2023.

[AM23] Sunil Arya and David M. Mount. Optimal volume-sensitive bounds for polytope approximation. In *39th International Symposium on Computational Geometry*, volume 258 of *LIPIcs. Leibniz Int. Proc. Inform.*, pages Paper No. 9, 16. Schloss Dagstuhl. Leibniz-Zent. Inform., Wadern, 2023.

[Bal91] Keith Ball. Volume ratios and a reverse isoperimetric inequality. *Journal of the London Mathematical Society*, 2(2):351–359, 1991.

[Bal92] Keith Ball. Ellipsoids of maximal volume in convex bodies. *Geometriae Dedicata*, 41(2):241–250, 1992.

[Bal97] Keith Ball. An elementary introduction to modern convex geometry. In *Flavors of geometry*, volume 31 of *Math. Sci. Res. Inst. Publ.*, pages 1–58. Cambridge Univ. Press, Cambridge, 1997.

[Bár82] Imre Bárány. A generalization of Carathéodory's theorem. *Discrete Math.*, 40(2-3):141–152, 1982.

[Bár89] Imre Bárány. Intrinsic volumes and f-vectors of random polytopes. *Math. Ann.*, 285(4):671–699, 1989.

[Bár07] Imre Bárány. Random polytopes, convex bodies, and approximation. In *Stochastic geometry*, volume 1892 of *Lecture Notes in Math.*, pages 77–118. Springer, Berlin, 2007.

[Bar14] Alexander Barvinok. Thrifty approximations of convex bodies by polytopes. *International Mathematics Research Notices*, 2014(16):4341–4356, 2014.

[Bar18] Alexander Barvinok. The tensorization trick in convex geometry. In Gergely Ambrus, Imre Bárány, Károly J. Böröczky, Gábor Fejes Tóth, and János Pach, editors, *New Trends in Intuitive Geometry*, pages 1–23. Springer Berlin Heidelberg, Berlin, Heidelberg, 2018.

[BGVV14] Silouanos Brazitikos, Apostolos Giannopoulos, Valettas, and Beatrice-Helen Vritsiou. *Geometry of isotropic convex bodies*, volume 196 of *Mathematical Surveys and Monographs*. American Mathematical Society, Providence, RI, 2014.

[BI75] Efim M. Bronshtein and L. D. Ivanov. The approximation of convex sets by polyhedra. *Sibirsk. Mat. Ž.*, 16(5):1110–1112, 1132, 1975.

[BKP82] Imre Bárány, Meir Katchalski, and János Pach. Quantitative Helly-type theorems. *Proc. Amer. Math. Soc.*, 86(1):109–114, 1982.

[BL88] Imre Bárány and David G. Larman. Convex bodies, economic cap coverings, random polytopes. *Mathematika*, 35(2):274–291, 1988.

[Bra17] Silouanos Brazitikos. Brascamp-Lieb inequality and quantitative versions of Helly's theorem. *Mathematika*, 63(1):272–291, 2017.

[BSS14] Joshua Batson, Daniel A. Spielman, and Nikhil Srivastava. Twice-Ramanujan sparsifiers. *SIAM Rev.*, 56(2):315–334, 2014.

[DFN21] Gábor Damásdi, Viktória Földvári, and Márton Naszódi. Colorful Helly-type theorems for the volume of intersections of convex bodies. *Journal of Combinatorial Theory, Series A*, 178:105361, 2021.

[DLGMM19] Jesús A. De Loera, Xavier Goaoc, Frédéric Meunier, and Nabil H. Mustafa. The discrete yet ubiquitous theorems of Carathéodory, Helly, Sperner, Tucker, and Tverberg. *Bull. Amer. Math. Soc. (N.S.)*, 56(3):415–511, 2019.

[DLLHRS17] Jesús A. De Loera, Reuben N. La Haye, David Rolnick, and Pablo Soberón. Quantitative combinatorial geometry for continuous parameters. *Discrete Comput. Geom.*, 57(2):318–334, 2017.

[DR50] Aryeh Dvoretzky and C. Ambrose Rogers. Absolute and unconditional convergence in normed linear spaces. *Proc. Nat. Acad. Sci. U. S. A.*, 36:192–197, 1950.

[ELR70] Günther Ewald, David G. Larman, and C. Ambrose Rogers. The directions of the line segments and of the r-dimensional balls on the boundary of a convex body in Euclidean space. *Mathematika*, 17:1–20, 1970.

[FJT24] Nóra Frankl, Attila Jung, and István Tomon. The quantitative fractional Helly theorem, 2024.

[FNZ19] Ferenc Fodor, Márton Naszódi, and Tamás Zarnócz. On the volume bound in the Dvoretzky-Rogers lemma. *Pacific J. Math.*, 301(1):89–99, 2019.

[FY17] Omer Friedland and Pierre Youssef. Approximating matrices and convex bodies. *International Mathematics Research Notices*, 2019(8):2519–2537, 2017.

[GLM+04] Yehoram Gordon, Alexander E. Litvak, Mathieu Meyer, Alain Pajor, et al. John's decomposition in the general case and applications. *J. Differential Geom.*, 68(1):99–119, 09 2004.

[Grü60] Branko Grünbaum. Partitions of mass-distributions and of convex bodies by hyperplanes. *Pacific Journal of Mathematics*, 10(4):1257–1261, 1960.

[GS05] Peter M. Gruber and Franz E. Schuster. An arithmetic proof of John's ellipsoid theorem. *Arch. Math. (Basel)*, 85(1):82–88, 2005.

[HW87] David Haussler and Emo Welzl. ε-nets and simplex range queries. *Discrete Comput. Geom.*, 2(2):127–151, 1987.

[HW18] Andreas F. Holmsen and Rephael Wenger. Helly-type theorems and geometric transversals. In Jacob E. Goodman, Joseph O'Rourke, and Csaba D. Tóth, editors, *Handbook of discrete and computational geometry*, pages 91–123. CRC Press, Boca Raton, FL, 2018. Third edition.

[IN22] Grigory Ivanov and Márton Naszódi. A quantitative Helly-type theorem: containment in a homothet. *SIAM J. Discrete Math.*, 36(2):951–957, 2022.

[IN24] Grigory Ivanov and Márton Naszódi. Quantitative steinitz theorem: A polynomial bound. *Bulletin of the London Mathematical Society*, 56(2):796–802, 2024.

[INP20] Grigory Ivanov, Márton Naszódi, and Alexandr Polyanskii. Approximation of the average of some random matrices. *J. Funct. Anal.*, 279(7):108684, 18, 2020.

[JN22] Attila Jung and Márton Naszódi. Quantitative fractional Helly and (p, q)-theorems. *European J. Combin.*, 99:Paper No. 103424, 10, 2022.

[Joh48] Fritz John. Extremum problems with inequalities as subsidiary conditions. In *Studies and Essays Presented to R. Courant on his 60th Birthday, January 8, 1948*, pages 187–204. Interscience Publishers, Inc., New York, N. Y., 1948.

[KPW92] János Komlós, János Pach, and Gerhard Woeginger. Almost tight bounds for ε-nets. *Discrete Comput. Geom.*, 7(2):163–173, 1992.

[LP86] Françoise Lust-Piquard. Inégalités de Khintchine dans C_p $(1 < p < \infty)$. *C. R. Acad. Sci. Paris Sér. I Math.*, 303(7):289–292, 1986.

[LPP91] Françoise Lust-Piquard and Gilles Pisier. Noncommutative Khintchine and Paley inequalities. *Ark. Mat.*, 29(2):241–260, 1991.

[Mac52] A. Murray Macbeath. A theorem on non-homogeneous lattices. *Ann. of Math. (2)*, 56:269–293, 1952.

[Mat02] Jiří Matoušek. *Lectures on discrete geometry*, volume 212 of *Graduate Texts in Mathematics*. Springer-Verlag, New York, 2002.

[MP00] Vitali D. Milman and Alain Pajor. Entropy and asymptotic geometry of non-symmetric convex bodies. *Advances in Mathematics*, 152(2):314–335, 2000.

[MSS15] Adam W. Marcus, Daniel A. Spielman, and Nikhil Srivastava. Interlacing families II: Mixed characteristic polynomials and the Kadison-Singer problem. *Ann. of Math. (2)*, 182(1):327–350, 2015.

[Mus22] Nabil H. Mustafa. *Sampling in combinatorial and geometric set systems*, volume 265 of *Mathematical Surveys and Monographs*. American Mathematical Society, Providence, RI, [2022] 2022.

[Nas16] Márton Naszódi. Proof of a conjecture of Bárány, Katchalski and Pach. *Discrete Comput. Geom.*, 55(1):243–248, 2016.

[Nas18] Márton Naszódi. Flavors of translative coverings. In *New trends in intuitive geometry*, volume 27 of *Bolyai Soc. Math. Stud.*, pages 335–358. János Bolyai Math. Soc., Budapest, 2018.

[Nas19] Márton Naszódi. Approximating a convex body by a polytope using the epsilon-net theorem. *Discrete Comput. Geom.*, 61(3):686–693, 2019.

[NNR20] Márton Naszódi, Fedor Nazarov, and Dmitry Ryabogin. Fine approximation of convex bodies by polytopes. *Amer. J. Math.*, 142(3):809–820, 2020.

[PA95] János Pach and Pankaj K. Agarwal. *Combinatorial geometry.* New York, NY: John Wiley & Sons, 1995.

[Pis98] Gilles Pisier. *Non-commutative vector valued Lp-spaces and completely p-summing maps*. Société mathématique de France, Paris, France, 1998.

[RS17] David Rolnick and Pablo Soberón. Quantitative (p,q) theorems in combinatorial geometry. *Discrete Mathematics*, 340(10):2516 – 2527, 2017.

[Rud99] Mark Rudelson. Random vectors in the isotropic position. *J. Funct. Anal.*, 164(1):60–72, 1999.

[Rya23] Dmitry Ryabogin. On bodies floating in equilibrium in every orientation. *Geom. Dedicata*, 217(4):Paper No. 70, 17, 2023.

[Sch14] Rolf Schneider. *Convex bodies: the Brunn–Minkowski theory*. Number 151. Cambridge university press, 2014.

[Sri12] Nikhil Srivastava. On contact points of convex bodies. In *Geometric aspects of functional analysis*, volume 2050 of *Lecture Notes in Math.*, pages 393–412. Springer, Heidelberg, 2012.

[SXS21] Sherry Sarkar, Alexander Xue, and Pablo Soberón. Quantitative combinatorial geometry for concave functions. *J. Combin. Theory Ser. A*, 182:Paper No. 105465, 33, 2021.

[VC68] Vladimir N. Vapnik and Alexey Ja. Chervonenkis. On the uniform convergence of relative frequencies of events to their probabilities. *Dokl. Akad. Nauk SSSR, 181*, 4:781ff., 1968. in Russian; English translation in Theor. Probab. Appl., **16** (1971), 264–280.

Part II
The Existential Theory of the Reals

Chapter 4
Intersection Graphs of Segments and $\exists\mathbb{R}$

Jiří Matoušek

A graph G with vertex set $\{v_1, v_2, \ldots, v_n\}$ is an *intersection graph of segments* if there are segments $s_1, \ldots, s_n$ in the plane such that s_i and s_j have a common point if and only if $\{v_i, v_j\}$ is an edge of G. In this expository paper, we consider the algorithmic problem of testing whether a given abstract graph is an intersection graph of segments. It turned out that this problem is complete for an interesting recently introduced class of computational problems, denoted by $\exists\mathbb{R}$. This class consists of problems that can be reduced, in polynomial time, to solvability of a system of polynomial inequalities in several variables over the reals. We discuss some subtleties in the definition of $\exists\mathbb{R}$, and we provide a complete and streamlined account of a proof of the $\exists\mathbb{R}$-completeness of the recognition problem for segment intersection graphs. Along the way, we establish $\exists\mathbb{R}$-completeness of several other problems. We also present a decision algorithm, due to Muchnik, for the first-order theory of the reals.

4.1 Introduction

Let G be a graph with vertex set $\{v_1, \ldots, v_n\}$. We say that G is an *intersection graph of segments* if there are straight segments $s_1, \ldots, s_n$ in the plane such that, for every i, j, $1 \le i < j \le n$, the segments s_i and s_j have a common point if and only if $\{v_i, v_j\} \in E(G)$. Such segments $s_1, \ldots, s_n$ are called a *segment representation* of G. For brevity, we will often say "segment graph" instead of "intersection graph of segments," and we let SEG denote the class of all segment graphs.

Supported by the ERC Advanced Grant No. 267165 and by the project CE-ITI (GACR P202/12/G061).

J. Matoušek (✉)
Department of Applied Mathematics, Charles University, Praha 1, Czech Republic

Institute of Theoretical Computer Science, ETH Zurich, Zurich, Switzerland

J. Pach and G. Tóth (eds.), *Courses in Discrete and Computational Geometry*, Bolyai Society Mathematical Studies 31,
https://doi.org/10.1007/978-3-032-10503-5_4

Segment graphs constitute a difficult and much studied class. They turned out to have strong algebraic aspects, and these are the focus of the present paper. We should stress that there are also numerous interesting non-algebraic results; one recent highlight, settling a long-standing conjecture, is that every planar graph belongs to SEG [CG09], and another is a construction of triangle-free SEG graphs with arbitrarily large chromatic number [PKK+14].

The story we want to present here begins with the algorithmic question, given an abstract graph G, does it belong to SEG? We will refer to this as the *recognition problem for segment graphs*, abbreviated RECOG(SEG).

At first encounter, it is probably not obvious that there are any graphs at all not belonging to SEG. Here is one of the simplest examples, a K_5 with subdivided edges:

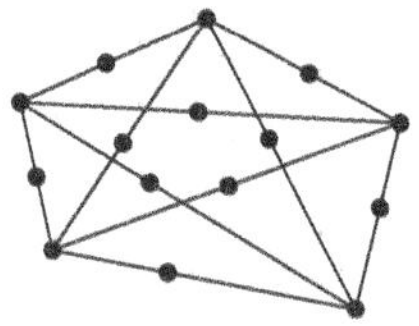

Without going into any details, we just mention the idea of the proof: assuming for contradiction that this graph is a segment graph, one obtains a planar drawing of K_5, which is impossible.

After enough understanding of combinatorial properties of segment graphs was accumulated, RECOG(SEG) was proved NP-hard in [KM94]. For many combinatorial problems, such as the existence of a Hamiltonian cycle in a graph, proving NP-hardness is more or less the end of the story, since *membership* in NP, i.e., a polynomial-time certificate of a YES answer, is obvious.

However, for RECOG(SEG) this was only a beginning, since membership in NP is not clear at all, and today it seems quite believable that RECOG(SEG) is actually not in NP.

Indeed, how should one certify that G is a segment graph? Some thought, which we leave to the reader, reveals that if G has a segment representation, then we may perturb and scale the segments so that all of their endpoints have integer coordinates. So if we knew that all the coordinates of the endpoints have at most polynomially many digits (in other words, that the absolute value of the coordinates is bounded by 2^{n^C}, where $n = |V(G)|$ and C is a constant), membership in NP would follow immediately.

Serious people seriously conjectured that the number of digits can be polynomially bounded—but it cannot.

Theorem 4.1.1 *For every sufficiently large n, there are n-vertex graphs in* SEG *for which every segment representation with integral endpoints has coordinates doubly exponential in n, that is, with* $2^{\Omega(n)}$ *digits.*

A first result of this kind, with the weaker bound of $2^{\Omega(\sqrt{n})}$, was proved in [KM94]. We will establish this weaker bound in Sect. 4.4. The stronger result in the theorem was obtained by Müller and McDiarmid [MM13]. A similar "large coordinates" phenomenon in the setting of line arrangements, which will be mentioned later, was observed earlier by Goodman, Pollack, and Sturmfels [GPS90], based on a fundamental work of Mnëv [Mne89].

Related to these developments, it was also gradually revealed that the recognition problem for segment graphs is computationally at least as hard as various other geometric and algebraic computational problems—most notably, the solvability of a system of strict polynomial inequalities in several variables over the reals.

More recently Schaefer and Štefankovič [SŠ11] (also see [Sch10]) introduced a new complexity class, denoted by $\exists\mathbb{R}$, and based on previous work, they proved many natural and well-known problems complete for this class, including RECOG(SEG). It is known that $\exists\mathbb{R}$ contains NP, and is contained in PSPACE (the computational problems solvable in polynomial space), but for all we know, both of these inclusions may be strict.

The purpose of the present expository paper is twofold. First, we aim at a complete and streamlined proof of the $\exists\mathbb{R}$-completeness of the recognition problem for segment graphs, which contains many nice ideas and has not been readily available in the literature in full. Second, and more significantly, we use the problem RECOG(SEG) as an opportunity to introduce the class $\exists\mathbb{R}$ in detail and to treat some subtleties in its definition, as well as several algebraic $\exists\mathbb{R}$-complete problems. Hopefully this may contribute to popularizing this class, which so far may not be as widely known as it would deserve.

The best known algorithms for $\exists\mathbb{R}$-complete problems work in exponential time and polynomial space, and they are quite sophisticated and technically demanding. We thus present another, suboptimal but much simpler algorithm due to Muchnik. It actually solves a more general problem, that of quantifier elimination in the first-order theory of the reals, a classical problem first solved by Tarski in the 1950s. Muchnik's algorithm can be explained in several pages, and it contains some of the ideas used in more sophisticated algorithms.

Since this paper started out as a course material, it contains a number of exercises. These may be useful for someone wishing to teach the material, and also for self-study—ideas that one discovers on his/her own are usually remembered much better. Readers not interested in solving exercises may take them just as a peculiar way of stating auxiliary results without proofs.

4.2 Recognizing Segment Graphs and the Existential Theory of $\mathbb{R}$

An algebraic re-formulation. It is not easy to find any algorithm at all to recognize segment graphs. In desperation we can turn to computational algebra and formulate the problem using polynomial equations and inequalities.

Suppose that a given graph $G = (V, E)$, $V = \{v_1, \ldots, v_n\}$, is a segment graph, represented by segments $s_1, \ldots, s_n$. By a suitable rotation of the coordinate system we can achieve that none of the segments is vertical. Then the segment s_i representing vertex i can be algebraically described as the set $\{(x, y) \in \mathbb{R}^2 : y = a_i x + b_i, c_i \leq x \leq d_i\}$ for some real numbers a_i, b_i, c_i, d_i.

Letting ℓ_i be the line containing s_i, we note that $s_i \cap s_j \neq \emptyset$ if either $\ell_i = \ell_j$ and the intervals $[c_i, d_i]$ and $[c_j, d_j]$ overlap, or ℓ_i and ℓ_j intersect in a single point whose x-coordinate lies in both of the intervals $[c_i, d_i]$ and $[c_j, d_j]$. As is easy to calculate, that x-coordinate equals $\frac{b_j - b_i}{a_i - a_j}$.

Let us introduce variables A_i, B_i, C_i, D_i representing the unknown quantities a_i, $b_i, c_i, d_i, i = 1, 2, \ldots, n$. (Following a convention in a part of the algebraic literature, we will denote variables by capital letters.) Then $s_i \cap s_j \neq \emptyset$ can be expressed by the following, somewhat cumbersome, predicate $\mathrm{INTS}(A_i, B_i, C_i, D_i, A_j, B_j, C_j, D_j)$, given by

$$\begin{aligned}
&\Big(A_i = A_j \wedge B_i = B_j \wedge \neg(D_i < C_j \vee D_j < C_i)\Big)\\
\vee &\Big(A_i > A_j \wedge C_i(A_i - A_j) \leq B_j - B_i \leq D_i(A_i - A_j)\\
&\quad \wedge C_j(A_i - A_j) \leq B_j - B_i \leq D_j(A_i - A_j)\Big)\\
\vee &\Big(A_i < A_j \wedge C_i(A_i - A_j) \geq B_j - B_i \geq D_i(A_i - A_j)\\
&\quad \wedge C_j(A_i - A_j) \geq B_j - B_i \geq D_j(A_i - A_j)\Big)
\end{aligned}$$

(this is only correct if we "globally" assume that $C_i \leq D_i$ for all i). Here, in accordance with a usual notation in logic, $\wedge$ stands for conjunction, $\vee$ for disjunction, and $\neg$ for negation.

The existence of a SEG-representation of G can then be expressed by the formula

$$\begin{aligned}
&(\exists A_1 B_1 C_1 D_1 \ldots A_n B_n C_n D_n)\Big(\bigwedge_{i=1}^{n} C_i \leq D_i\Big)\\
&\wedge \Big(\bigwedge_{\{i,j\} \in E} \mathrm{INTS}(A_i, B_i, B_i, D_i, A_j, B_j, C_j, D_j)\Big)\\
&\wedge \Big(\bigwedge_{\{i,j\} \notin E} \neg\,\mathrm{INTS}(A_i, B_i, B_i, D_i, A_j, B_j, C_j, D_j)\Big).
\end{aligned}$$

The existential quantifier in this formula quantifies all the variables written after it, and these variables range over $\mathbb{R}$, the real numbers.

Exercise 4.2.1 Prove that every segment graph has a representation in which no two segments lie on the same line (warning: this is not entirely easy). Therefore, the first part of the predicate INTS is not really necessary.

Exercise 4.2.2 Express the representability of a given graph G as an intersection graph of convex sets in the plane by a formula similar to the one above. Hint: disjoint compact convex sets can be strictly separated by a line.

First-order theory of the reals. It may seem that we have replaced the problem of recognizing segment graphs, which was at least geometrically intuitive, by a hopelessly complicated algebraic problem. However, the *only* known recognition algorithm relies exactly on such a conversion and on a general algorithm for testing the validity of this kind of formulas. What is more, recognizing segment graphs turns out to be *computationally equivalent* to validity testing for a fairly broad class of formulas.

First we define a still more general class of formulas. A **formula of the first-order theory of the reals** is made of the constants 0 and 1, the binary operator symbols $+$, $-$ and $\times$, the relation symbols $\leq$, $<$, $\geq$, $>$, $=$, $\neq$, the logical connectives $\wedge$, $\vee$, $\neg$, and $\Leftrightarrow$ (equivalence), the quantifiers $\forall$ and $\exists$, symbols for variables, and parentheses, using simple syntactic rules, which the reader could no doubt easily supply. In the rest of this paper, the word "formula" without other qualifications means a formula in the first order theory of the reals.

Including of "derived" symbols like $-$, $\neq$, $>$, $\Leftrightarrow$ etc. is slightly nonstandard, compared to the literature in logic or model theory, but for considering computational complexity it appears more convenient to regard them as primitive symbols.

Semantically, all variables range over the real numbers[1] and $+$, $-$, $\times$, $\leq$, etc. are to be interpreted as the usual operations and relations on $\mathbb{R}$.

A formula is called a **sentence** if it has no *free* variables; that is, every variable occurring in it is bound by a quantifier. Our formula above expressing the SEG-representability of a given graph is an example of a sentence.

The first-order theory of the reals is sufficiently powerful to express a wide variety of geometric problems. At the same time, it is *decidable*, which means that there is an algorithm that, given a sentence, decides whether it is true or false. We will present one such algorithm in Sect. 4.5. This is in contrast with the first-order theories in many other areas of mathematics, such as the Peano arithmetic, which are undecidable.

Prenex form. We will always assume, w.l.o.g., that the considered formulas are in **prenex form**; that is, of the form $(Q_1X_1)(Q_2X_2)\ldots(Q_kX_k)\Phi$, where $Q_1,\ldots,Q_k$ are quantifiers and Φ is a quantifier-free formula. An arbitrary formula is easily converted to a prenex form by pushing the quantifiers outside using simple rules, such as rewriting $\neg(\exists X)\Phi$ as $(\forall X)\neg\Phi$, rewriting $((\exists X)\Phi)\vee\Psi$ as $(\exists X)(\Phi\vee\Psi)$ (this assumes that the name of the quantified variable X does not collide with the name of any variable in Ψ), etc. Such a transformation changes the length of the formula by at most a constant factor.

---

[1] In a first-order theory, we can quantify over individual elements, in our case over the real numbers; in a second-order theory, quantification over *sets* of real numbers would be allowed as well.

A general quantifier-free formula Φ with variables $X_1, \ldots, X_n$ has the form

$$\Phi = \Phi(X_1, \ldots, X_n) = F(A_1, A_2, \ldots, A_m).$$

Here F is a Boolean formula and each A_i is an *atomic formula* of the form $p_i(X_1, \ldots, X_n)$ rel 0, where rel is one of $<, \leq, >, \geq, =, \neq$, and p_i is a polynomial in n variables with integer coefficients.

Size of formulas. When discussing the complexity of an algorithm accepting a formula as an input, we will measure the input size by the *length* of the formula, i.e., the number of symbols in it.[2]

In practice, one also writes integer constants, such as 13, and powers, such as X^5, in the formulas. An integer constant k can be formed with $O(\log k)$ symbols using the binary expansion of k; e.g., $13 = 2^3 + 2^2 + 1 = ((1 + 1) + 1) \times (1 + 1) \times (1 + 1) + 1$.

On the other hand, a power X^k is really an abbreviation for the k-fold product $X \times \cdots \times X$. We must be particularly careful with more complicated expressions involving powers: for example, $(\cdots((X^2)^2 \cdots)^2$, with n second powers, could be written using $O(n)$ symbols if we permitted the power operation—but we do not allow it, and so this expression needs a formula of length about 2^n.

Thus, in brief, integer constants are written in binary, but exponents must be taken as shorthands for multiple products.

Another thing to observe is that we cannot a priori assume polynomials to be written in the usual form, as a sum of monomials with integer coefficients. Indeed, the polynomial $(1 + X_1)(1 + X_2) \cdots (1 + X_n)$ has $O(n)$ symbols according to our definition, but if we multiplied out the parentheses, we would get 2^n distinct monomials, and thus a formula of exponential length. This kind of subtleties can usually be ignored in logic, but they start playing a role if we consider polynomial-time reductions.

Complexity of decision algorithms. The best known decision algorithms for the first-order theory of the reals can decide a sentence of length L in time at most $2^{2^{O(L)}}$, i.e., doubly exponential.

A more refined analysis of such algorithms uses several other parameters besides the formula length, such as the number of variables, the maximum degree of a polynomial appearing in the formula, or the number of quantifier alternations (e.g., the formula $(\exists X_1 X_2)(\forall X_3 X_4 X_5)(\exists X_6)\Phi$, where Φ is quantifier-free, has three blocks of alternating quantifiers). For example, some of the algorithms are doubly exponential only in the number of quantifier alternations, while the dependence on other parameters is much milder. We will discuss this somewhat more concretely at the end of Sect. 4.5.

---

[2] The length is basically proportional to the number of bits needed to encode the formula. An exceptional case are the symbols of variables: if the formula contains n variables, then encoding a variable symbol needs about $\log n$ bits, at least on the average. Thus, in the worst case, the encoding of a formula of length L may need up to $L \log L$ bits. But the complexity of the considered algorithms is so large that this subtlety can be safely ignored, as is done in most of the literature.

Semialgebraic sets. A set $S \subseteq \mathbb{R}^n$ is called **semialgebraic** if it can be described by a quantifier-free formula; i.e., if $S = \{x \in \mathbb{R}^n : \Phi(x)\}$ for some $\Phi = \Phi(X_1, \ldots, X_n)$ as above. Thus, a semialgebraic set is a set-theoretic combination of finitely many zero sets and nonnegativity sets of polynomials.

Exercise 4.2.3 (a) Describe semialgebraic sets in $\mathbb{R}^1$.

(b) Show that a semialgebraic set in $\mathbb{R}^1$ defined by a quantifier-free formula of length L has $O(L)$ connected components. More concretely, if the formula $\Phi(X)$ involves polynomials $p_1(X), \ldots, p_m(X)$, bound the number of components in terms of the degrees of the p_i.

The existential theory of the reals and ETR. Among the formulas of the first-order theory of the reals, the quantifier-free ones are, in a sense, the simplest. In particular, quantifier-free *sentences* are trivially decidable, since they involve only constants.

The next level of difficulty are the *existential formulas*, of the form

$$(\exists X_1 X_2 \ldots X_k)\Phi,$$

with Φ quantifier-free. For example, our formula above for recognizing segment graphs is existential, and many other questions about graph representations etc. can also be expressed by an existential formula.

Let ETR, abbreviating *existential theory of the reals*, stand for the decision problem whose input is an existential sentence Ψ of the first-order theory of the reals, and the output is the truth value of Ψ, TRUE or FALSE.

Exercise 4.2.4 Show that ETR is NP-hard.

The true complexity status of ETR is not known at present. The best available algorithms show that it belongs to PSPACE, i.e., it can be solved in polynomial space and exponential time (this result is due to Canny [Can88]; also see [BPR03] for more recent algorithms).

We have seen that an instance of RECOG(SEG) can be converted, in polynomial time, into an equivalent instance of ETR. This is usually expressed by saying that RECOG(SEG) *reduces* to ETR. A surprising and useful fact is that, conversely, ETR reduces to RECOG(SEG), and thus these two problems are, up to polynomial-time reductions, computationally equivalent. We will discuss this, and related result, in the subsequent sections.

As an executive summary, one may remember that the first-order theory of the reals can be decided in doubly-exponential time, and the existential theory in singly-exponential time.

Remark: real-closed fields. Let us mention in passing that the first-order theory does not fully describe the ordered field of the real numbers. Indeed, there are other ordered fields, some of them quite interesting and useful, that satisfy exactly the same set of sentences as the reals. Such fields are called the **real-closed fields**, and they can be described by a simple axiomatic system (for which we refer to the literature, e.g., [BPR03]). In the literature, instead of the first-order theory of the reals, one thus often speaks of the first-order theory of real-closed fields.

One example of a real-closed field consists of all the *algebraic* real numbers. Another consists of all real *Puiseux series*; these are formal series of fractional powers of the form $\sum_{i=k}^{\infty} a_i \varepsilon^{i/q}$, where $k \in \mathbb{Z}$, q is a natural number, ε is a formal variable, and the a_i are real coefficients. The real-closed field of Puiseux series - is non-Archimedean since it contains infinitesimals. This makes it quite unlike the real numbers.

Here we will not consider the Puiseux series in any detail. We just remark that they are widely used for *perturbation arguments*, where one needs to bring some semialgebraic sets into a suitably general position; then the variable ε plays the role of an infinitesimal quantity. A nice feature is that if we start with a real-closed field and make the perturbation, we are again in a (larger) real-closed field, and we can apply the same theory and even the same algorithms, provided that suitable computational primitives have been implemented.

4.3 The Complexity Class ∃ℝ

The class ∃ℝ consists of the computational decision problems that reduce to ETR in polynomial time. More precisely, the reduction is the usual many-to-one reduction also used in the definition of NP; that is, a decision problem A many-to-one reduces to a decision problem B if there is a polynomial-time mapping f such that x is a YES-instance of A iff $f(x)$ is a YES-instance of B.

Restricted versions of ETR **which are still complete for** ∃ℝ. The problem ETR is complete for ∃ℝ by definition. Here we will exhibit several restricted versions of ETR that also turn out to be ∃ℝ-complete, although in some cases this may be surprising. In the next section we will discuss geometric ∃ℝ-complete problems.

The significance of ∃ℝ-completeness is similar, on a smaller scale, to that of NP-completeness: it bundles together many seemingly very different problems into a single big complexity question.

The considered special cases of ETR all decide an existential sentence

$$(\exists X_1 \ldots X_n)\Phi$$

where Φ is quantifier-free and of the following special forms:

- For the problem INEQ, Φ is a conjunction of polynomial equations and inequalities. Moreover, we require that the polynomials in the equations and inequalities be written in the standard form, as sums of monomials with integer coefficients. Thus, for example, $(1 + X_1)(1 + X_2) \cdots (1 + X_n)$ is not allowed.
- STRICT-INEQ is the special case of INEQ using only strict inequalities and no equations.
- FEASIBLE is the special case of INEQ with a single polynomial equation $p(X_1, \ldots, X_n) = 0$.

Exercise 4.3.1 Formulate the problem of testing whether a given graph belongs to SEG as an instance of STRICT-INEQ. Use Exercise 4.2.1 (which says that all segments can be assumed to have distinct directions).

We begin with showing that INEQ, and even FEASIBLE, are ∃ℝ-complete. Even for INEQ this turns out to be trickier that one might think. One problem is with the exponential blow-up of formula size when converting polynomials into the standard form, as was mentioned in the preceding section. Another, similar problem appears when dealing with an arbitrary Boolean formulas.

Fortunately, both of these problems can be solved by the same idea, going back to Tseitin (in the context of Boolean satisfiability, one speaks of the **Tseitin transform** for converting Boolean formulas to *conjunctive normal form*). The main trick is to add new existentially quantified variables, which are used to "store" the values of subformulas.

Proposition 4.3.2 *The decision problems* FEASIBLE, *and consequently* INEQ, *are ∃ℝ-complete. In particular, solving a single polynomial equation in many variables over ℝ is, up to a polynomial-time reduction, as hard as solving a system of polynomial equations and inequalities over ℝ.*

Proof Let $(\exists X_1 \ldots X_n)\Phi$ be an existential sentence (in prenex form), of length L and with Φ quantifier-free. We transform it into an equivalent instance of FEASIBLE of length $O(L)$. The main trick is adding new real variables, which represent the values of the various subformulas in Φ.

We consider the way of how Φ is recursively built from subformulas. Here is an example, in which the subformulas are marked by lowercase Greek letters:

$$\underbrace{\Big(\underbrace{\underbrace{\underbrace{(X+Y)}_{\vartheta}\underbrace{(Z-Y)}_{\xi}}_{\zeta} > 0}_{\alpha} \vee \underbrace{X \le 0}_{\beta}\Big)}_{\gamma} \wedge \neg\underbrace{(\underbrace{Y = 0}_{\delta})}_{\varepsilon}$$

The subformulas ϑ, ξ, ζ are *arithmetic*, meaning that they yield a real number, while $\alpha, \beta, \ldots$ are Boolean, yielding true or false.

We process the subformulas one by one, starting from the innermost ones and proceeding outwards. During this process, we build a new formula Γ, which is a conjunction of polynomial equations. We start with Γ empty, and we add equations to it as we go along. We also add new existentially quantified real variables, one or several per subformula.

In the example above, we start with the subformula ϑ, we introduce a new existentially quantified real variable V_ϑ (V abbreviating *value*), and we add to Γ the equation $V_\vartheta = X + Y$. Similarly, for ξ we add $V_\xi = Z - Y$, and for ζ we add $V_\zeta = V_\vartheta V_\xi$. Thus, for each arithmetic subformula, the corresponding variable represents the value of the subformula.

For every Boolean subformula η, the plan is to introduce a corresponding real variable W_η, representing the truth value of η in the sense that $W_\eta = 1$ means η true and $W_\eta = 0$ means η false.

For our atomic formula $\alpha \equiv V_\zeta > 0$, we begin with observing that the strict inequality $V_\zeta > 0$ is equivalent to $(\exists S_\alpha)(V_\zeta S_\alpha^2 = 1)$, while its negation $V_\zeta \le 0$ is equivalent to $(\exists T_\alpha)(V_\zeta + T_\alpha^2 = 0)$. Thus, by introducing the two auxiliary variables S_α and T_α, we obtain equations instead of inequalities.

In order to have W_α represent the truth value of the subformula α as announced above, we would thus like to add to Γ the subformula

$$(V_\zeta S_\alpha^2 = 1 \wedge W_\alpha = 1) \vee (V_\zeta + T_\alpha^2 = 0 \wedge W_\alpha = 0). \tag{4.1}$$

Sadly, we cannot do that, since we are allowed to add only conjunctions of equations, not disjunctions. But we observe that a conjunction of equations $p = 0 \wedge q = 0$ is equivalent to $p^2 + q^2 = 0$, while a disjunction $p = 0 \vee q = 0$ can be replaced by $pq = 0$. Thus, our desired subformula (4.1) can be added in the following disguise:

$$\big((V_\zeta S_\alpha^2 - 1)^2 + (W_\alpha - 1)^2\big)\big((V_\zeta + T_\alpha^2)^2 + W_\alpha^2\big) = 0$$

Complicated as it may look, this equation still has a constant length, even if we expand the polynomial into the standard form. This will be the case for every equation we add to Γ.

The remaining two atomic formulas, β and δ, are handled analogously and we leave this to the reader. Fortunately, the case of atomic formulas is the most complicated one, and dealing with Boolean connectives is simpler.

The disjunction $\gamma \equiv \alpha \vee \beta$ is, by induction, equivalent to $W_\alpha = 1 \vee W_\beta = 1$, and the equation $W_\gamma = W_\alpha + W_\beta - W_\alpha W_\beta$ will do. This, however, is a bit of a hack, relying on our particular representation of the truth values. A more systematic approach is to first represent the disjunction by $(W_\gamma = 1 \wedge (W_\alpha = 1 \vee W_\beta = 1)) \vee (W_\gamma = 0 \wedge W_\alpha = 0 \wedge W_\beta = 0)$, and then to convert this formula into a single equation using the two tricks, $p^2 + q^2$ and pq, as above.

The negation in the subformula ε is represented by $W_\varepsilon = 1 - W_\delta$, and for the conjunction in $\Phi \equiv \gamma \wedge \varepsilon$ we can simply use $W_\Phi = W_\gamma W_\varepsilon$. Finally, we add the equation $W_\Phi = 1$ to Γ, and this yields the announced conjunction of equations $p_1 = 0 \wedge \cdots \wedge p_m = 0$ that is satisfiable iff the original formula Φ is.

Since Φ has no more than L subformulas, we have $m \le L$. Each p_i has length $O(1)$, and hence Γ has length $O(L)$. Finally, to get the desired instance of FEASIBLE, we replace Γ with $p_1^2 + \cdots + p_m^2 = 0$, which at most doubles the formula length. □

Exercise 4.3.3 In the above proof, we have not dealt with a subformula of the form $\alpha \Leftrightarrow \beta$ (equivalence). Find a suitable equation for replacing such a subformula.

Strict inequalities. Intuitively, the nature of STRICT-INEQ seems to be different from INEQ. Indeed, if $\Phi(X_1, \dots, X_n)$ is a conjunction of strict inequalities, then the subset of $\mathbb{R}^n$ defined by it is *open*. Thus, for example, if it is nonempty, then it always contains a point with rational coordinates; this, of course, is not the case for equations, since, e.g., $X^2 - 2 = 0$ has only irrational solutions.

Yet, as discovered in [SŠ11], INEQ and STRICT-INEQ are equivalent as computational problems. For the proof we need a reasonably difficult result of real algebraic geometry, of independent interest, which we will take for granted.

Theorem 4.3.4 *Let $S \subseteq \mathbb{R}^n$ be a semialgebraic set defined by a quantifier-free formula of length L. If $S \neq \emptyset$, then S intersects the ball of radius $R = 2^{2^{CL\log L}}$ centered at 0, were C is a suitable absolute constant. If S is also bounded, then it is contained in that ball.*

A more refined bound for R is $2^{\tau(\Delta+1)^{O(n)}}$, where n is the number of variables in the formula, Δ is the maximum degree of the polynomials in it, and τ is the maximum number of bits of a coefficient in the polynomials. The ball of the stated radius even intersects every connected component of S and contains every bounded connected component. The bound in the theorem is obtained by substituting the trivial estimates $\tau \leq L$, $\Delta \leq L$, $n \leq L$ into the refined bound.

This kind of result goes back to [Gj88, Lemma 9] (which deals with a special semialgebraic set, namely, the zero set of a single polynomial), and the result as above about a ball intersecting all connected components is [BPR96, Theorem 4.1.1] (also see [BPR03, Theorem 13.14]). A statement directly implying the part with the ball containing all bounded components is [BV07, Proposition 6.1].

Proposition 4.3.5 STRICT-INEQ *is $\exists\mathbb{R}$-complete. Thus, solving a system of strict polynomial inequalities over $\mathbb{R}$ is, up to a polynomial-time reduction, as hard as solving arbitrary system of polynomial inequalities.*

Proof We will reduce FEASIBLE to STRICT-INEQ. Let the input formula for FEASIBLE ask for solvability of $p(X_1, \ldots, X_n) = 0$ and let its length be L.

By Theorem 4.3.4, if $p(X_1, \ldots, X_n) = 0$ is solvable, then it has a solution $(x_1, \ldots, x_n) \in \mathbb{R}^n$ with $\sum_{i=1}^n x_i^2 < R^2$, where $R = 2^{2^k}$ and $k = CL\log L$.

We construct an instance of STRICT-INEQ of the form

$$(\exists X_1 \ldots X_n Y_1 \ldots Y_k Z_1 \ldots Z_\ell)\Phi,$$

where $\ell = C_1 L(\log L)^2$, with a sufficiently large constant C_1, and Φ is the quantifier-free formula

$$\begin{aligned}
&Y_1 > 0 \wedge \cdots \wedge Y_k > 0\\
&\wedge Y_1 < 4 \wedge Y_2 < Y_1^2 \wedge \cdots \wedge Y_k < Y_{k-1}^2 \wedge X_1^2 + \cdots + X_n^2 < Y_k^2\\
&\wedge Z_1 > 4 \wedge Z_2 > Z_1^2 \wedge \cdots \wedge Z_\ell > Z_{\ell-1}^2\\
&\wedge Z_\ell^2 p(X_1, \ldots, X_n)^2 < 1.
\end{aligned}$$

The first two lines say that that $\sum_{i=1}^{n} X_i^2 < R^2$, and the last two lines mean that $|p(X_1, \ldots, X_n)| < \delta := 2^{-2^\ell}$. The length of the formula is clearly bounded by a polynomial in L, even if we convert all polynomials into the standard form.

We need to check that Φ is solvable iff $p(X_1, \ldots, X_n) = 0$ is. First, if $p(X_1, \ldots, X_n) = 0$ is solvable, then there is a solution $(x_1, \ldots, x_n)$ with $\sum_{i=1}^{n} x_i^2 < R^2$, and it can be extended to a solution of Φ.

Conversely, let us suppose that p has no zero. Let us consider the semialgebraic set $S \subseteq \mathbb{R}^{n+k+1}$ given by the following formula $\Xi = \Xi(X_1, \ldots, X_n, Y_1, \ldots, Y_k, Z)$:

$$\begin{aligned}
&Y_1 > 0 \wedge \cdots \wedge Y_k > 0 \\
&\wedge Y_1 < 4 \wedge Y_2 < Y_1^2 \wedge \cdots \wedge Y_k < Y_{k-1}^2 \wedge X_1^2 + \cdots + X_n^2 < Y_k^2 \\
&\wedge Z^2 p(X_1, \ldots, X_n)^2 = 1.
\end{aligned}$$

The length of Ξ is $K = O(k)$. Since we assume that p has no zeros, S is a bounded set, and hence by Theorem 4.3.4, it is contained in the ball of radius $2^{2^{CK \log K}} \le 2^{2^\ell} = 1/\delta$, and in particular, $|Z| \le 1/\delta$ in every solution. This implies that for every $(x_1, \ldots, x_n) \in \mathbb{R}^n$ with $\sum_{i=1}^{n} x_i^2 < R^2$, we have $|p(x_1, \ldots, x_n)| \ge \delta$, and hence Φ has no solution. □

Exercise 4.3.6 Find a bivariate polynomial $p(X, Y)$ with integer coefficients with $p(x, y) > 0$ for all $(x, y) \in \mathbb{R}^2$ and with $\inf_{\mathbb{R}^2} p(x, y) = 0$.

Remark: size of segment representations and Oleinik–Petrovskiĭ–Milnor–Thom. By reasoning very similar to the proof of Theorem 4.3.4, it can also be shown that if S is a nonempty semialgebraic set defined by a formula of length L that is a conjunction of *strict* inequalities, then S contains a rational point whose coordinates are fractions of integers with $2^{O(L \log L)}$ bits (a more refined bound in the notation below Theorem 4.3.4 is $\tau \Delta^{O(n)}$; see [BPR03, Theorem 13.15]).

Using the refined bound on the ETR formula expressing RECOG(SEG) given at the beginning of this chapter, we get that every segment graph on n vertices has a segment representation in which the endpoint coordinates have $2^{O(n)}$ digits (there are $4n$ variables, the polynomials are at most quadratic, and $\tau = O(1)$). Hence Theorem 4.1.1 is tight up to a multiplicative constant in the exponent.

Theorem 4.3.4 is also related to a result of real algebraic geometry that proved truly fundamental for discrete geometry, theoretical computer science, and other fields: the Oleinik–Petrovskiĭ–Milnor–Thom theorem (also often called the Milnor–Thom theorem or Warren's theorem in the literature). This is a result that bounds the maximum number of connected components of semialgebraic sets. In a modern version, it can be stated as follows:

Theorem 4.3.7 *Let $p_i(X_1, \ldots, X_n)$ be polynomials of degree at most Δ, $i = 1, 2 \ldots, m$, and for every sign vector $\sigma \in \{-1, 0, +1\}^m$ let $S_\sigma \subseteq \mathbb{R}^n$ be defined as*

$$\Big\{(x_1, \ldots, x_n) \in \mathbb{R}^n : \bigwedge_{i=1}^{m} \operatorname{sgn} p_i(x_1, \ldots, x_n) = \sigma_i\Big\}.$$

Then for $m \geq n \geq 2$,

$$\sum_{\sigma \in \{-1,0,+1\}^m} \#S_\sigma \leq \left(\frac{50\Delta m}{n}\right)^n,$$

where $\#S$ denotes the number of connected components of S.

Exercise 4.3.8 Apply the theorem just stated to show that there are at most $2^{O(n \log n)}$ nonisomorphic segment graphs on n vertices.

Applications of the Oleinik–Petrovskiǐ–Milnor–Thom theorem abound in the literature, from simple ones as in the exercise to sophisticated uses, and it is useful to be aware of this kind of result.

4.4 Stretchability and the Mnëv Universality Theorem

Here is the main goal of this section.

Theorem 4.4.1 *The problem* RECOG(SEG), *recognizing segment graphs, is* ∃ℝ-*complete.*

As a by-product of the proof, we will also get a proof of a weaker version of Theorem 4.1.1 (exponentially many digits needed for a SEG representation), with $2^{\Omega(\sqrt{n})}$ digits instead of $2^{\Omega(n)}$.

A sample of other geometric ∃ℝ-complete problems. Before starting our development, we list several other ∃ℝ-complete problems; many more such problems can be found in the literature (see, e.g., [SŠ11, Sch10][3]).

- Recognition of intersection graphs of *unit disks* in the plane [KM12].
- Recognition of CONV, intersection graphs of convex sets in $\mathbb{R}^2$ [Sch10].
- Determining the *rectilinear crossing number* $\overline{\mathrm{cr}}(G)$ of a graph [Bie91], that is, the minimum possible number of edge crossings in a drawing of G in the plane, with edges drawn as straight segments.
- (The *Steinitz problem*) Given a partially ordered set, determining whether it is isomorphic to the set of all faces of a convex polytope ordered by inclusion (this partially ordered set is called the *face lattice* of the polytope); see [BLS+99, Corollary 9.5.11].

4.4.1 From Segment Graphs to Line Arrangements

STRETCHABILITY. Let us consider a set L of n lines in the plane. The **arrangement** of L is the partition of the plane into convex subsets induced by L. Four of

[3] Schaefer et al. [SCM25] is aiming at a comprehesive list.

such subsets are marked in the next picture: a *vertex*, which is an intersection of two lines, two *edges*, which are pieces of the lines delimited by the vertices, and a *region*, which is one of the pieces obtained after cutting the plane along the lines.

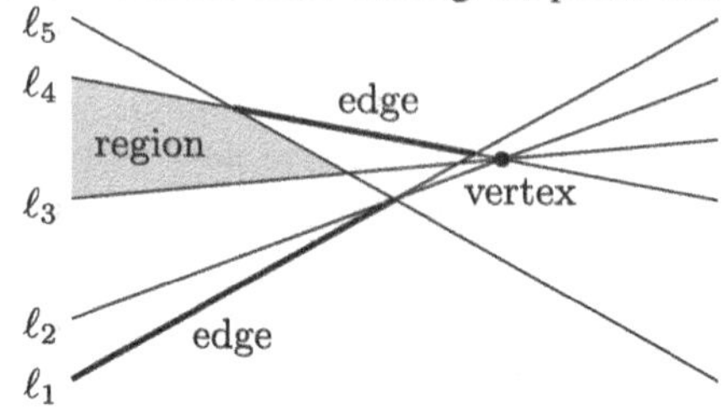

We want to define a *combinatorial description* of such a line arrangement, the idea being that two arrangements with the same description look combinatorially the same. Such descriptions are systematically studied in the theory of *oriented matroids* (see, e.g., [BLS+99]). There are several reasonable definitions of a combinatorial description, all of them essentially equivalent. We will use one that is simple and convenient for our present purposes, although not among the most conceptual ones.

For simplicity, let us assume that none of the lines of L is vertical and every two intersect. Then we number the lines $\ell_1, \ldots, \ell_n$ in the order of decreasing slopes, as in the above picture, and for each $i = 1, 2, \ldots, n$, we write down the numbers of the lines intersecting ℓ_i as we go from left to right. From the picture we thus get the following five lists:

$$(\genfrac{}{}{0pt}{}{2}{5}, 3, 4), (\genfrac{}{}{0pt}{}{1}{5}, \genfrac{}{}{0pt}{}{3}{4}), (5, 1, \genfrac{}{}{0pt}{}{2}{4}), (5, 1, \genfrac{}{}{0pt}{}{2}{3})(4, 3, \genfrac{}{}{0pt}{}{1}{2}).$$

Several numbers in a column mean that the corresponding intersections coincide. These n lists constitute (our way of) the combinatorial description of the arrangement of L.

The decision problem **STRETCHABILITY** can now be defined as follows: given n lists of integers, with some of the entries arranged in columns as above, decide whether they constitute a combinatorial description of an arrangement of n lines.

There are some obvious consistency conditions on the lists: for example, the ith list should contain each element of $[n]\setminus\{i\}$ exactly once, and for $i < j < k$, if ℓ_i intersects ℓ_k before ℓ_j, then ℓ_j must intersect ℓ_i before ℓ_k, etc. It is not hard to formulate conditions on the lists so that they provide a combinatorial description of an arrangement of **pseudolines**. Here a set of n curves is called a set of (affine) pseudolines if every curve intersects every vertical line exactly once, and every two curves cross exactly once.

The real problem in STRETCHABILITY is in recognizing whether a combinatorial description of an arrangement of pseudolines also fits to an arrangement of lines, i.e., if the pseudolines can be "stretched"—this is where the name comes from. The following is a famous example of a non-stretchable arrangement:

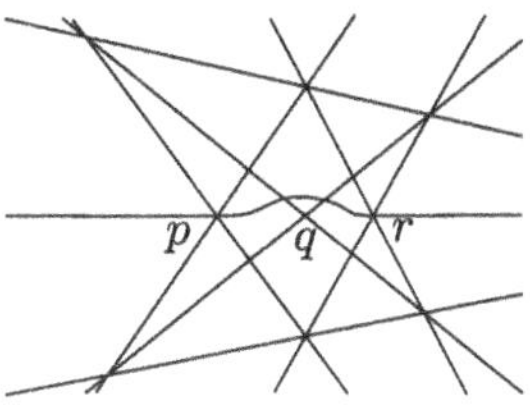

In this picture, all of the pseudolines are straight except for one (and only a part of the arrangement not including all of the intersections is shown, for space reasons). The non-stretchability relies on an ancient theorem of Pappus, which asserts that if eight straight lines intersect as indicated, then the line passing through p and r also has to contain q.

An arrangement of lines is called **simple** (or, sometimes, *uniform*) if no three lines have a point in common. The decision problem **SIMPLE STRETCHABILITY**, a special case of STRETCHABILITY, asks whether given lists form the combinatorial description of a simple arrangement of lines.

In due time, we will see that STRICT-INEQ reduces to SIMPLE STRETCHABILITY, and thus SIMPLE STRETCHABILITY is ∃ℝ-complete. But here we deal with the following result from [KM94], the first step in the proof of Theorem 4.4.1:

Proposition 4.4.2 SIMPLE STRETCHABILITY *reduces to* RECOG(SEG)*; i.e., recognizing segment graphs is no easier than testing simple stretchability.*

Sketch of proof. First we explain a construction of a segment representation of a suitable graph from a given simple line arrangement. We begin with adding a vertical segment v intersecting all the lines and lying left of all intersections in the arrangement. We also add two other segments that together with v enclose all of the intersections in a triangle T. Then we shorten all of the lines to segments ending outside T:

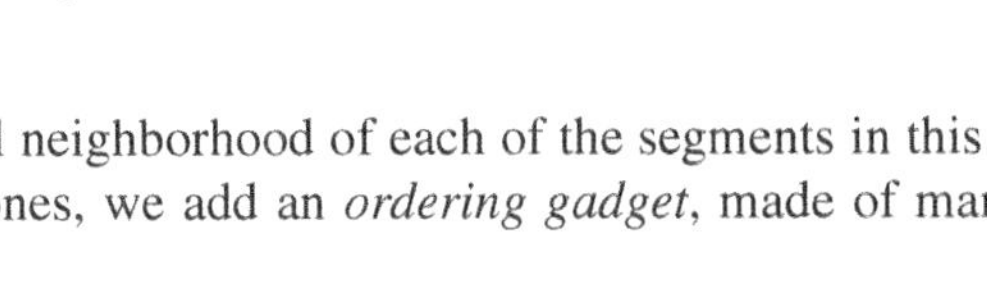

Next, in a small neighborhood of each of the segments in this picture, including the newly added ones, we add an *ordering gadget*, made of many segments, as is indicated below:

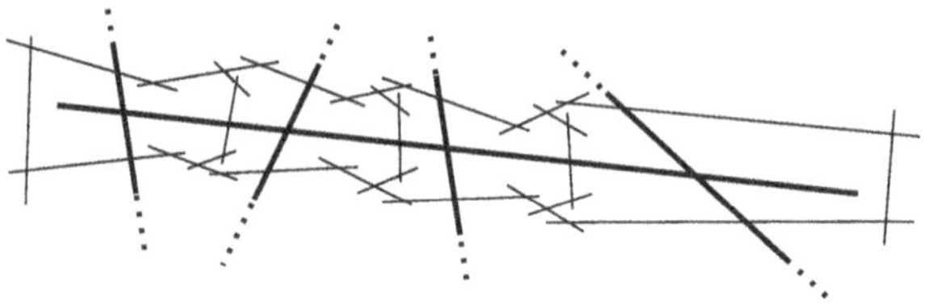

The original segments are drawn thick. Let G be the intersection graph of the resulting set of segments.

We have presented the construction geometrically, but it can obviously be "combinatorialized" so that, given a combinatorial description, which may or may not correspond to an actual line arrangement, it produces a graph G. Clearly, if the description does correspond to a line arrangement, then G has a segment representation, and it remains to prove the converse—if G has a segment representation, then there exists an arrangement with the given description.

The key step in the proof is in showing that the ordering gadget indeed forces the appropriate linear ordering of the intersections along each of the original segments, up to a reversal. This is quite intuitive, and for a detailed proof we refer to [Sch10].

Once we know this, we can discard the segments of the ordering gadgets. We make an affine transformation of the plane so that the segment v is vertical and the triangle T is to the right of it. Then all intersections of the remaining segments must be inside T, and these segments can be extended to full lines. Then either the resulting line arrangement or its upside-down mirror reflection conform to the given combinatorial description. □

4.4.2 From Line Arrangements to Point Configurations

Our main goal now is showing $\exists\mathbb{R}$-completeness of SIMPLE STRETCHABILITY. First we pass from line arrangements to point configurations; these two settings are equivalent, but point configurations appear more convenient for the subsequent development. The passage relies on *line-point duality*, a basic concept of projective geometry, and the material in this section may be rather standard/easy for many readers.

For combinatorially describing a configuration of points in the plane, we use the notion of *order type* (which, under the name of a *chirotope*, is also one of the possible axiomatizations of oriented matroids; see [BLS+99]).

The order type is defined for a *sequence* $(p_1, p_2, \ldots, p_n)$ of points in $\mathbb{R}^2$ (or in $\mathbb{R}^d$, but we will not need that). For an ordered triple (p, q, r) of points, we define the *sign* depending on the direction, left/straight/right, in which we turn when going from p to q and then to r:

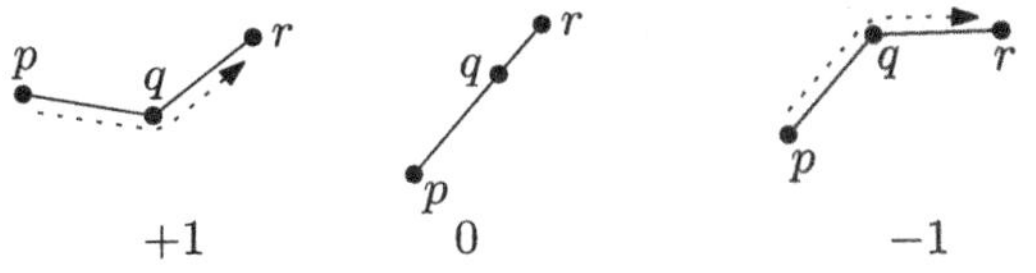

The sign can also be defined algebraically, as the sign of the determinant of the matrix with rows $q - p$ and $r - p$. We define a *combinatorial order type* as a mapping $\binom{[n]}{3} \to \{-1, 0, +1\}$, and the **order type** of a sequence $(p_1, p_2, \ldots, p_n)$ is the mapping assigning to every triple $i < j < k$ the sign of (p_i, p_j, p_k).

The **ORDER-TYPE REALIZABILITY** problem has a combinatorial order type as the input, and asks if it is the order type of an actual point sequence. We also have the **SIMPLE ORDER-TYPE REALIZABILITY** variant, where all the signs are ± 1's, or in other words, the considered point set is in general position, with no collinear triples.

Lemma 4.4.3 SIMPLE ORDER-TYPE REALIZABILITY *reduces to* SIMPLE STRETCHABILITY.

Proof As announced above, the proof is a simple use of a line-point duality. In view of our way of combinatorially describing line arrangements, we use the following version of duality: To a point $p = (a, b) \in \mathbb{R}^2$, we assign the (nonvertical) line $\mathcal{D}(p)$ with equation $y = ax - b$. Conversely, to every nonvertical line ℓ, which has a unique equation of the form $y = ax - b$, we assign the point $\mathcal{D}(\ell) = (a, b)$. The basic property of $\mathcal{D}$, quite easy to check from the definition, is that a point p lies below, on, or above a nonvertical line ℓ iff the dual point $\mathcal{D}(\ell)$ lies below, on, or above the dual line $\mathcal{D}(p)$. In particular, $\mathcal{D}$ preserves line-point incidences and thus maps the intersection of two lines to the line spanned by the two dual points, and vice versa.

Let τ be a simple combinatorial order type. Supposing that there is a point sequence $(p_1, \ldots, p_n)$ realizing it, we may assume that the point p_n lies on the negative y-axis very far below all the other points ("at $-\infty$" for all practical purposes). This can be achieved by a suitable *projective transformation* of the plane; for readers not familiar with this concept we provide an express introduction below.

With this assumption, it is easy to see that we can read off the ordering of the x-coordinates of p_1 through p_{n-1} from the signs of the triples involving p_n. Thus, if we consider the dual lines $\mathcal{D}(p_1), \ldots, \mathcal{D}(p_{n-1})$, we know the ordering of their slopes.

We can also reconstruct the combinatorial description of the line arrangement from the order type. For example, if $i < j < k$ and we want to know which of the lines ℓ_i, ℓ_j intersects ℓ_k first, it suffices to know whether the intersection $\ell_i \cap \ell_j$ lies above or below ℓ_k. This in turn is equivalent to the dual point $\mathcal{D}(\ell_k)$ lying above or below the line through $\mathcal{D}(\ell_i)$ and $\mathcal{D}(\ell_j)$, by the properties of the duality mentioned above, and this last piece of information is contained in the order type. A similar reasoning works for other orderings of the indices i, j, k.

Conversely, the order type of $(p_1, \ldots, p_n)$ can be reconstructed from the combinatorial description of the arrangement of the dual lines $\mathcal{D}(p_1), \ldots, \mathcal{D}(p_{n-1})$.

Summarizing, from the combinatorial order type τ we can construct, in polynomial time of course, a combinatorial description of an arrangement of $n-1$ lines that is stretchable iff τ is realizable. □

Projective transformations. Above and in the sequel we need some properties of projective transformations. For readers not familiar with this topic we provide several introductory sentences, referring, e.g., to Richter-Gebert [RG11] for a solid introduction to projective geometry.

A projective transformation maps the plane onto itself, or more precisely, it maps the plane minus an exceptional line onto the plane minus another exceptional line. Here is a quite intuitive geometric way of thinking of projective transformations. We fix two planes π and σ in $\mathbb{R}^3$ and identify each of them with $\mathbb{R}^2$ by choosing a system of Cartesian coordinates in it. We also choose a point $o \notin \pi \cup \sigma$, and the projective transformation is obtained by projecting π from o into σ.

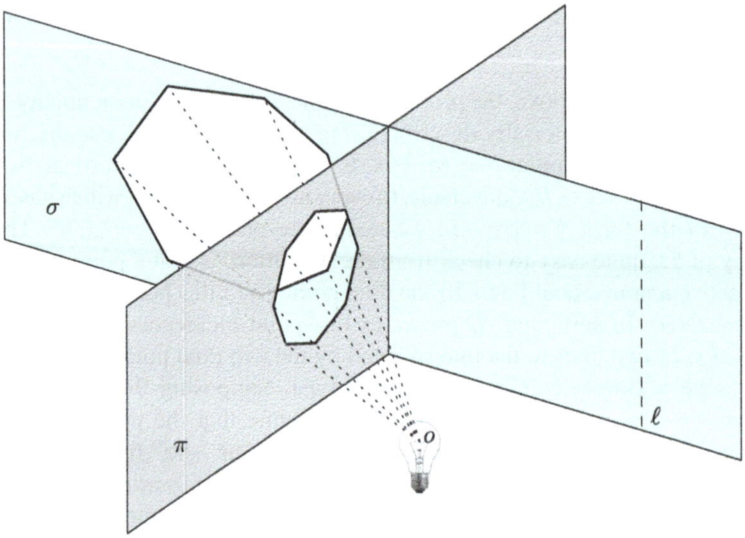

The exceptional line in σ is labeled by ℓ in the picture; it is the intersection of σ with the plane through o parallel to π. In projective geometry, one completes the plane by a line at infinity, to the *projective plane*, and projective transformations map the projective planes bijectively onto itself. In particular, the exceptional line ℓ is mapped onto the line at infinity.

A projective transform maps straight lines to straight lines and preserves line-point incidences. It also preserves the order type, provided that all of the points lie on the same side of the exceptional line.

4.4.3 General Position and Constructible Configurations

So far we are carrying along the simplicity/general position requirement, which came naturally from the setting of segment graphs—there we cannot force three segments to meet at a single point. For reducing the solvability of strict polynomial inequalities, STRICT-INEQ, to a realizability problem for point configurations, the general position requirement would be very inconvenient. Here we replace it with a considerably weaker requirement, called constructibility, by means of an ingenious trick, first used by Las Vergnas (see [BLS+99, Proposition 8.6.3]).

We say that a point sequence $(p_1, p_2, \ldots, p_n)$ is **constructible** if, possibly after renumbering the points suitably, the following hold:

- No three among $p_1, \ldots, p_4$ are collinear (this is usually expressed by saying that $p_1, \ldots, p_4$ form a *projective basis*).
- Each p_i, $i > 4$, lies on at most two of the lines spanned by $p_1, \ldots, p_{i-1}$.

In particular, every sequence in general position is constructible.

The notion of constructibility makes sense for a combinatorial order type, realizable or not. In the corresponding algorithmic problem **CONSTRUCTIBLE ORDER-TYPE REALIZABILITY**, the input is a combinatorial order type for which the points are already ordered as in the definition of constructibility (so that we need not worry about finding the right ordering in polynomial time).

Intuitively, a constructible point configuration is one that can be constructed from the initial four points using only a ruler, i.e., by passing lines through pairs of already constructed points and placing each new point either arbitrarily, or arbitrarily on an already constructed line, or to the intersection of two already constructed lines. Such a ruler construction, though, only takes into account which triples of points should be collinear, and not the signs of triples. Thus, a constructible combinatorial order type may still be unrealizable.

Proposition 4.4.4 CONSTRUCTIBLE ORDER-TYPE REALIZABILITY *is reducible to* SIMPLE ORDER-TYPE REALIZABILITY.

The construction: first version. To prove the proposition, given a constructible combinatorial order type τ, we want to construct, in polynomial time, a simple combinatorial order type ξ such that ξ is realizable iff τ is.

In order to make the proof more accessible, we first explain a somewhat simplified version of the construction in terms of specific point sequences, i.e., assuming that we are given a sequence $\boldsymbol{p}$ realizing the given order type τ.

Thus, let $\boldsymbol{p} = (p_1, \ldots, p_n)$ be a constructible sequence with order type τ, with a numbering as in the definition of constructibility. We inductively construct sequences $\boldsymbol{q}^{(n)} = \boldsymbol{p}, \boldsymbol{q}^{(n-1)}, \ldots, \boldsymbol{q}^{(4)}$. The first t points of $\boldsymbol{q}^{(t)}$ are p_1 through p_t. The final product of this construction is $\boldsymbol{q} = \boldsymbol{q}^{(4)}$, which is a sequence of fewer than $3n$ points in general position.

For obtaining $\boldsymbol{q}^{(t-1)}$ from $\boldsymbol{q}^{(t)}$, we distinguish three cases, depending on the number of the lines spanned by $p_1, \ldots, p_{t-1}$ that pass through p_t (there are at most two by constructibility):

(2) First suppose that p_t lies on two such lines; let they be the lines $p_i p_j$ and $p_k p_\ell$. They divide the plane into four sectors. The sequence $\boldsymbol{q}^{(t-1)}$ is obtained from $\boldsymbol{q}^{(t)}$ by replacing p_t with four points $p_{t,1}, \ldots, p_{t,4}$, one in each of the sectors, lying very close to p_t:

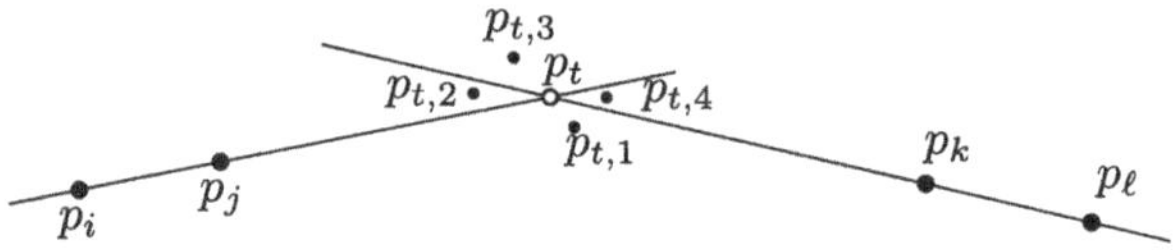

(1) If p_t lies on a single line $p_i p_j$, we replace it with three points $p_{t,1}, p_{t,2}, p_{t,3}$:

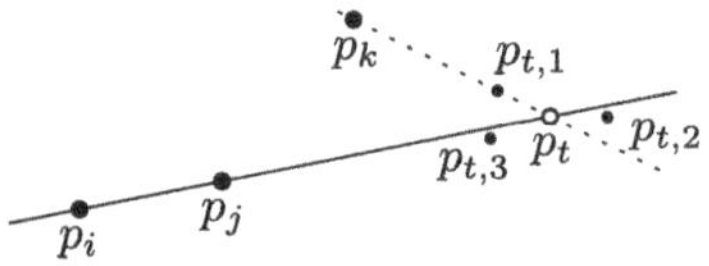

We first choose another point p_k, $k < t$, not lying on $p_i p_j$, we place $p_{t,1}$ on the line $p_t p_k$, and then $p_{t,2}$ and $p_{t,3}$ are on the opposite side of $p_i p_j$ and on different sides of $p_t p_k$. This makes sure that $p_t \in \operatorname{conv}(p_{t,1}, p_{t,2}, p_{t,3})$.

(0) If p_t does not lie on any line spanned by p_1 through p_{t-1}, then we simply set $\boldsymbol{q}^{(t-1)} = \boldsymbol{q}^{(t)}$.

In both of the cases (2) and (1), we place the new points so that they do not lie on any line spanned by the points of $\boldsymbol{q}^{(t)}$ minus p_t. We also place them so close to p_t that every line spanned by the points of $\boldsymbol{q}^{(t)}$ and not passing through p_t has them on the same side as p_t.

By induction, the only collinear triples in $\boldsymbol{q}^{(t)}$ are those spanned by $p_1, \ldots, p_t$, and in particular, $\boldsymbol{q} = \boldsymbol{q}^{(4)}$ is in general position. Here is the main property of the construction.

Lemma 4.4.5 *Given any sequence $\tilde{\boldsymbol{q}}$ realizing the order type ξ of $\boldsymbol{q}$, one can construct a sequence $\tilde{\boldsymbol{p}}$ realizing the order type τ of the original sequence $\boldsymbol{p}$.*

Proof It suffices to check that if $\tilde{\boldsymbol{q}}^{(t-1)}$ has the same order type as $\boldsymbol{q}^{(t-1)}$, we can obtain $\tilde{\boldsymbol{q}}^{(t)}$ with the same order type as $\boldsymbol{q}^{(t)}$. Let us assume that $\boldsymbol{q}^{(t-1)}$ was constructed from $\boldsymbol{q}^{(t)}$ according to case (2), since case (1) is analogous and case (0) trivial.

It is clear how the desired $\tilde{\boldsymbol{q}}^{(t)}$ should be obtained from $\tilde{\boldsymbol{q}}^{(t-1)}$: by deleting the points $\tilde{p}_{t,1}, \ldots, \tilde{p}_{t,4}$ and placing $\tilde{p}_t$ to the intersection of the two lines $\tilde{p}_i \tilde{p}_j$ and $\tilde{p}_k \tilde{p}_\ell$. What needs to be checked is that the resulting $\tilde{\boldsymbol{q}}^{(t)}$ has the order type of $\boldsymbol{q}^{(t)}$.

We need to consider only the signs of the triples involving $\tilde{p}_t$. Let $\tilde{a}$ and $\tilde{b}$ be the other two points in such a triple, and let a and b be the corresponding points in $\boldsymbol{q}^{(t)}$.

The points $\tilde{p}_{t,1}, \ldots, \tilde{p}_{t,4}$ form a convex quadrilateral containing the intersection of the lines $\tilde{p}_i\tilde{p}_j$ and $\tilde{p}_k\tilde{p}_\ell$, since this information is specified by the order type of $\boldsymbol{q}^{(t-1)}$.

Thus, if the line $\tilde{\ell}$ spanned by $\tilde{a}$ and $\tilde{b}$ avoids the quadrilateral, then $\tilde{p}_t$, lying inside the quadrilateral, is on the same side of $\tilde{\ell}$ as $\tilde{p}_{t,1}, \ldots, \tilde{p}_{t,4}$, and so the sign of the triple $(\tilde{a}, \tilde{b}, \tilde{p}_t)$ is the same as for the corresponding triple (a, b, p_t) in $\boldsymbol{q}^{(t)}$.

If $\tilde{\ell}$ does intersect the quadrilateral, then the corresponding line for $\boldsymbol{q}^{(t-1)}$ also intersects the corresponding quadrilateral, and since $p_{t,1}, \ldots, p_{t,4}$ were placed sufficiently close to p_t, the points a, b, p_t must be collinear. But the only collinear triples in $\boldsymbol{q}^{(t)}$ involving p_t lie on the two lines p_ip_j and p_kp_ℓ defining p_t. Hence $\tilde{a}$ and $\tilde{b}$ lie on the corresponding line for $\tilde{\boldsymbol{q}}^{(t)}$ and form a collinear triple with $\tilde{p}_t$ as well. □

Lexicographic extensions. It may seem that Proposition 4.4.4 is already proved, but there is still a problem we need to address. Namely, we need a construction phrased solely in terms of order types; that is, we want to construct the order type $\tau^{(t)}$ of $\boldsymbol{q}^{(t)}$ directly from the order type of $\tau^{(t-1)}$ of $\boldsymbol{q}^{(t-1)}$, without relying on a particular realization of $\tau^{(t-1)}$. Indeed, the construction must also make sense for non-realizable combinatorial order types.

The construction as presented above is not yet suitable for this purpose: when choosing the new points $p_{t,1}, p_{t,2}, \ldots$, we have not specified the order type fully, since the position of a line spanned by two new points among the old points is not determined, and similarly for lines spanned by a new point and an old point. We thus need to be more specific, and for the definition, we use the following general notion.

Let $\boldsymbol{x} = (x_1, \ldots, x_m)$ be an arbitrary point sequence in $\mathbb{R}^2$, let x_i, x_j, x_k be three non-collinear points of $\boldsymbol{x}$, and let $\rho, \sigma \in \{-1, +1\}$ be signs. For every $\varepsilon > 0$, we consider the point $x = x(\varepsilon) = x_i + \rho\varepsilon(x_j - x_i) + \sigma\varepsilon^2(x_k - x_i)$. That is, we move x_i a bit towards x_j (if $\rho = +1$) or away from it, and an even much smaller bit towards x_k or away from it. Let us form a new sequence of $m + 1$ points by inserting $x(\varepsilon)$ into $\boldsymbol{x}$ after x_i. It is not hard to see that for all sufficiently small $\varepsilon > 0$, the order type of the new sequence is the same, and crucially, it can be figured out from the order type of $\boldsymbol{x}$. We call this new sequence a **lexicographic extension** of $\boldsymbol{x}$, and we write the new point x as $[x_i, x_j^\rho, x_k^\sigma]$. Analogously we define $[x_i, x_j^\rho]$, with just one move.

Proof of Proposition 4.4.4. Given a constructible combinatorial order type τ, we produce the simple combinatorial order type ξ essentially according to the construction above, only we specify the way of adding the new points in terms of lexicographic extensions.

Namely, the four new points in case (2) of the construction above are obtained by four successive lexicographic extensions of the current sequence, as $p_{t,1} := [p_t, p_i^+, p_k^+]$, $p_{t,2} := [p_t, p_i^+, p_k^-]$, $p_{t,3} := [p_t, p_i^-, p_k^-]$, and $p_{t,4} := [p_t, p_i^-, p_k^+]$. Thus, a somewhat more realistic illustration to case (2) is this:

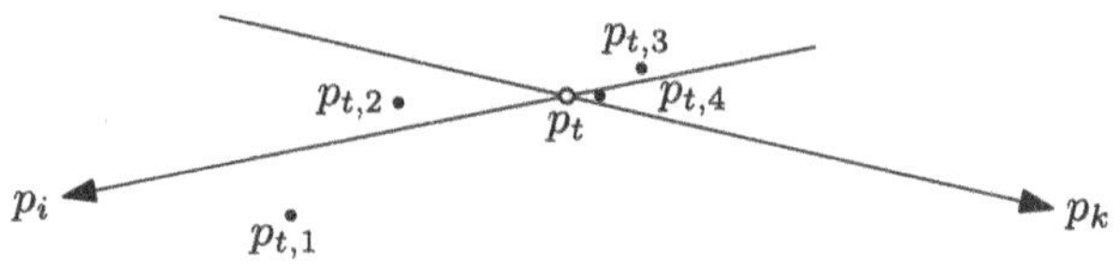

No illustration can be quite realistic, since the ε's should actually be very small and decrease very fast with the successive lexicographic extensions.

Similarly, in case (1), where p_t lies on the line $p_i p_j$ and p_k is another point not lying on that line, we set $p_{t,1} := [p_t, p_k^+]$, $p_{t,2} := [p_t, p_i^+, p_k^-]$, and $p_{t,3} := [p_t, p_i^-, p_k^-]$. Now the construction is fully specified in terms of order types.

If τ is realizable, then, clearly, ξ is realizable (we just perform the construction geometrically). If ξ is realizable, then τ is realizable by Lemma 4.4.5. □

4.4.4 *The Key Part: Modeling* STRICT-INEQ *by Point Configurations*

Here is the most demanding part in our chain of reductions.

Theorem 4.4.6 STRICT-INEQ *reduces to* CONSTRUCTIVE ORDER-TYPE REALIZABILITY.

This result, in a somewhat different context, was first achieved in a breakthrough by Mnëv (pronounce, approximately, "Mnyoff") [Mne89]. A simplified argument was then given by Shor [Sho91], and our presentation below is mostly based on Richter-Gebert's clean treatment [RG95]. The same proof method also works for reducing INEQ to ORDER-TYPE REALIZABILITY.

In STRICT-INEQ, we are given a conjunction of strict polynomial inequalities. The first idea is simple: simulate the evaluation of the polynomials by geometric constructions. If we represent real numbers by points on the x-axis, sums and products can be constructed easily:

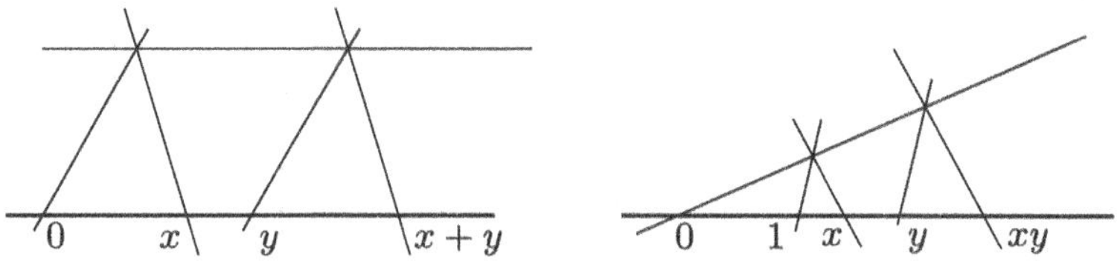

(The parallel-looking lines should be really parallel.)

The first caveat is that these constructions use parallel lines, which cannot be forced by order type. Moreover, given a realization of a given order type, we can always apply a projective transformation and get another realization, so we should better use constructions invariant under projective transformations, and our way of representing real numbers should be invariant as well.

Real numbers as cross-ratios. Both of these issues can be remedied. First, for representing quantities we use cross-ratios, rather than lengths. We recall that if a, b, c, d are points on a line, then their **cross-ratio** (in this order) is the quantity

$$(a, b; c, d) := \frac{|a, c| \cdot |b, d|}{|a, d| \cdot |b, c|},$$

where $|a, b|$ denotes the *oriented* Euclidean distance from a to b, which is positive if a precedes b on the line and negative otherwise. This assumes that an orientation of the line has been chosen, but the cross-ratio does not depend on it. More significantly, the cross-ratio is invariant under projective transformations.

We thus fix a line ℓ on which all quantities will be represented, and choose three points labeled $\mathbf{0}$, $\mathbf{1}$, and $\boldsymbol{\infty}$ on ℓ; we say that we have chosen a **projective scale** on ℓ (we use boldface symbols for the points of a projective scale in order to distinguish them from the usual meaning of 0, 1, ∞). Then a fourth point a on ℓ represents the real number $(a, \mathbf{1}; \mathbf{0}, \boldsymbol{\infty})$.

If ℓ is the x-axis, $\mathbf{0} = (0, 0)$, $\mathbf{1} = (1, 0)$, $a = (x, 0)$, and $\boldsymbol{\infty}$ is at infinity (that is, we take the appropriate limit in the cross-ratio), then $(a, \mathbf{1}; \mathbf{0}, \boldsymbol{\infty}) = x$, which explains the notation.

The sum and product constructions above can be projectivized as well: what used to be parallel lines become lines intersecting on a distinguished line ℓ' (which substitutes the line at infinity). We obtain the following **von Staudt constructions**:

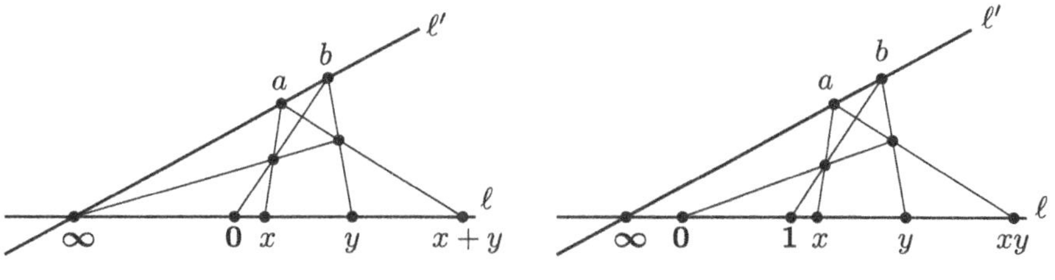

The points x, y, $x + y$ and xy are labeled by the real values they represent. The points $\boldsymbol{\infty}$, $\mathbf{0}$, $\mathbf{1}$, x, and y are given, a and b are chosen arbitrarily on the line ℓ', and the remaining lines and points are constructed as indicated. The correctness of these constructions follows from the correctness of the constructions above with parallel lines and from the invariance of the cross-ratio under projective transformations.

Simulating a polynomial inequality. Let us consider a system of strict polynomial inequalities, which we would like to model by an order type realizability problem.

First, in order to have control over the ordering of the points on the line ℓ, we want to work only with numbers greater than 1, and so we substitute each variable X_j with the expression $X'_j - X''_j$, where X'_j and X''_j are new variables. Then, whenever the resulting system has a solution, it also has one in which all variables exceed 1. From now on, we assume that our system has this property, but we return to calling the variables $X_1, \ldots, X_n$.

We also transform each strict inequality $p_i > 0$ in our system to $p_i^+ > p_i^-$, where both of p_i^+ and p_i^- are polynomials with nonnegative integer coefficients. Here it is useful to have the polynomials in the standard form, for otherwise, it would be difficult to do this splitting.

After these preparations, we can calculate only with numbers greater than 1. Let us consider one of the inequalities, for example $2X + 1 > XY^2$. We place a point x representing the value of X on ℓ, and using a series of von Staudt constructions, we successively construct points v_2, v_{2X}, and v_{2X+1}, each representing the value of the corresponding subexpression. In a similar way, we place y representing the value of Y and construct v_{Y^2} and v_{XY^2}. Then we impose the inequality $2X + 1 > XY^2$ by requiring v_{2X+1} and v_{XY^2} to have the appropriate order along ℓ. (Here it is useful to note that all the considered number-representing points on ℓ may be required to lie on the same side of $\mathbf{1}$, since we can always move $\boldsymbol{\infty}$ sufficiently far from $\mathbf{0}$.) This ordering can be enforced by the order type of the point configuration, by fixing one point not lying on ℓ and considering the signs of triples consisting of this point plus two points on ℓ.

This looks all very nice, until one realizes that we are still far from solving our problem. What we can already do is this: Given an instance of STRICT-INEQ, we can produce in polynomial time a collection of points, some of them lying on ℓ and representing real numbers, others auxiliary, coming from the von Staudt constructions. Certain triples of these points are required to be collinear, which makes the von Staudt constructions possible, and the sign is prescribed for some triples, which enforces the inequalities. Realizability of such a point configuration is equivalent to satisfiability of the given instance of STRICT-INEQ.

But the catch is that we are far from knowing the order type of the configuration. For specifying it completely, we would need to know, for example, what is the order of v_{2X} and v_{Y^2} on ℓ, and all other relations of this kind. Moreover, unless we are careful, the various von Staudt constructions can be intermixed with one another in an uncontrollable fashion. This problem is serious, and overcoming it was Mnëv's main achievement.

Partial order-type realizability. In order not to lose optimism, we point out that we have already achieved something: we have reduced STRICT-INEQ to a decision problem that we may call PARTIAL ORDER-TYPE REALIZABILITY, in which the combinatorial order type is given only partially, by specifying the signs for only some of the triples.

Actually, our partial order type is constructible in a suitable sense, since the von Staudt constructions produce constructible configurations, and since the inequalities we are modeling are strict. (Indeed, constructibility would fail if we required equalities, such as $2X + 1 = XY^2$, since then the point $v_{2X+1} = v_{XY^2}$ would have to lie on three lines spanned by previously constructed points.)

This reduction is already sufficient to see that CONSTRUCTIBLE ORDER-TYPE REALIZABILITY cannot belong to the class NP unless STRICT-INEQ, and hence all problems in $\exists\mathbb{R}$, do. Indeed, assuming that CONSTRUCTIBLE ORDER-TYPE REALIZABILITY is in NP and given an instance of STRICT-INEQ, we set up an instance of CONSTRUCTIBLE PARTIAL ORDER-TYPE REALIZABILITY as above, and then we nondeterministically guess the missing signs so that we get a fully specified constructible combinatorial order type. Then we apply the the nondeterministic polynomial-time decision algorithm for CONSTRUCTIBLE ORDER-TYPE REALIZABILITY whose existence we assume, and this yield a nondeterministic polynomial-time algorithm for STRICT-INEQ. By the reductions done earlier, we also get that RECOG(SEG) and the other problems considered along the way are not in NP unless $\exists\mathbb{R} \subseteq$ NP. However, we do not get $\exists\mathbb{R}$-completeness in this way.

Segment representations with large coordinates: proof of the weaker version of Theorem 4.1.1. By the tools developed so far, we can also obtain segment graphs requiring coordinates with $2^{\Omega(\sqrt{n})}$ digits.

Instead of modeling inequalities by a point configuration as above, we simulate repeated squaring by the von Staudt constructions, obtaining points $v_1, v_2, \ldots, v_k$ on ℓ, with v_i representing the number 2^{2^i}. Since there are no unknowns, the construction can actually be executed with some concrete points. We obtain a specific constructible configuration of $O(k)$ points in which some four collinear points have cross-ratio 2^{2^k}. Moreover, *every* realization of the order type of this configuration has such a fourtuple.

Then we go through the reductions made earlier, obtaining first a simple order type, then a description of a line arrangement, and finally a segment graph with $n = O(k^2)$ vertices. From a segment representation of this graph we can get a realization of the order type we started with—we just follow the proofs of correctness of the reductions. If the segment representation has integer endpoint coordinates with at most M digits, then the lines in the corresponding arrangement have equations with $O(M)$-digit coefficients, and the dual simple point configuration has coordinates with $O(M)$ digits, which we may again assume to be integers.

The passage from the simple point configuration to the constructible one, as in the proof of Lemma 4.4.5, is more subtle, since here the points of the constructible configuration cannot be assumed to have integer coordinates. Instead, we look at the ratio of the largest distance to the smallest distance determined by the points, and we observe that the largest distance may only decrease and the smallest only increase in the process. This is because we always delete some points and add a point in their convex hull. Consequently, the distance ratio in the resulting point configuration also has $O(M)$ digits, and so do all cross-ratios. Therefore, $M \geq 2^{\Omega(k)} = 2^{\Omega(\sqrt{n})}$. □

Separating the variables and constructions: proof of Theorem 4.4.6. After this detour, we return to our main task, reducing STRICT-INEQ to CONSTRUCTIBLE ORDER-TYPE REALIZABILITY. We still need to modify the construction presented earlier so that the order type of the resulting configuration is determined in full, *without knowing the values of the variables*.

It turns out that the crucial task is fixing the order of the points on ℓ; separating the von Staudt constructions is then easier. The idea is to use not one, but many projective scales on ℓ; they all have the ∞ point in common, but the $\mathbf{0}$'s and $\mathbf{1}$'s are different:

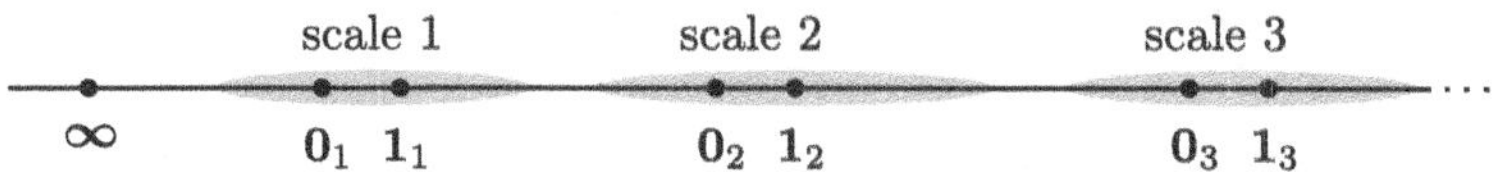

We assume, as we may, that the interval available to each scale is sufficiently long so that all numbers that need to be represented in that scale fit there.

Each of the scales $(\infty, \mathbf{0}_i, \mathbf{1}_i)$ has distinct purpose, and we represent only a small number of quantities (constants or variables) in it, for which we know the ordering. For example, each operation of addition will have its own scale, and so will each multiplication, as well as each comparison of two values. We will also introduce gadgets, similar to the von Staudt constructions, that link the various scales, i.e., make sure that a variable represented in two different scales has the same value in both.

Here by variables we mean both the original variables $X_1, \ldots, X_n$ from the formula and variables V_ξ, where ξ runs through all subexpressions to be evaluated. Moreover, and this is the final trick, we also introduce variables $V_{-\xi}$ and $V_{1/\xi}$ for every subexpression ξ, as well as V_{-X_i} and V_{1/X_i} for every variable X_i.

Let us see, for example, how we should process the inequality $XY + Z > Y$.

1. We place $\mathbf{0}_1$ and $\mathbf{1}_1$ defining scale 1. Then we place a point x representing X to the right of $\mathbf{1}_1$ (since, as we recall, all of the variables X_i and subexpressions ξ have values in $(1, \infty)$). Still in scale 1, we construct a point $1/x$ representing $V_{1/X} = 1/X$, by a suitable inversion gadget—see Fig. 4.1.
2. We place $\mathbf{0}_2$ for scale 2, and obtain $\mathbf{1}_2$ by transferring $\mathbf{1}_1$ from scale 1 using a linking gadget. Then in scale 2, we place y representing Y, we transfer $V_{1/X}$ there from scale 1 by a linking gadget, and by an inverted multiplication gadget we produce a point representing $V_{XY} = XY$. The point is that we do not care which of X and Y is larger, since they never appear in the same scale; in scale 2, we have only $1/X$, Y, and XY, for which the ordering is obvious in view of $X, Y > 1$.
3. We initialize scale 3 by placing $\mathbf{0}_3$ and obtaining $\mathbf{1}_3$ by a linking gadget from scale 1 (or 2, both work). We place a point representing Z and construct $-z$ representing $V_{-Z} = -Z$.
4. We similarly initialize scale 4, and by two linking gadgets, we transfer V_{XY} and V_{-Z} there. By a negated addition gadget we construct a point representing V_{XY+Z} there.
5. We initialize scale 5, we transfer V_{XY+Z} and Y to it, and we enforce $V_{XY+Z} > Y$ there.

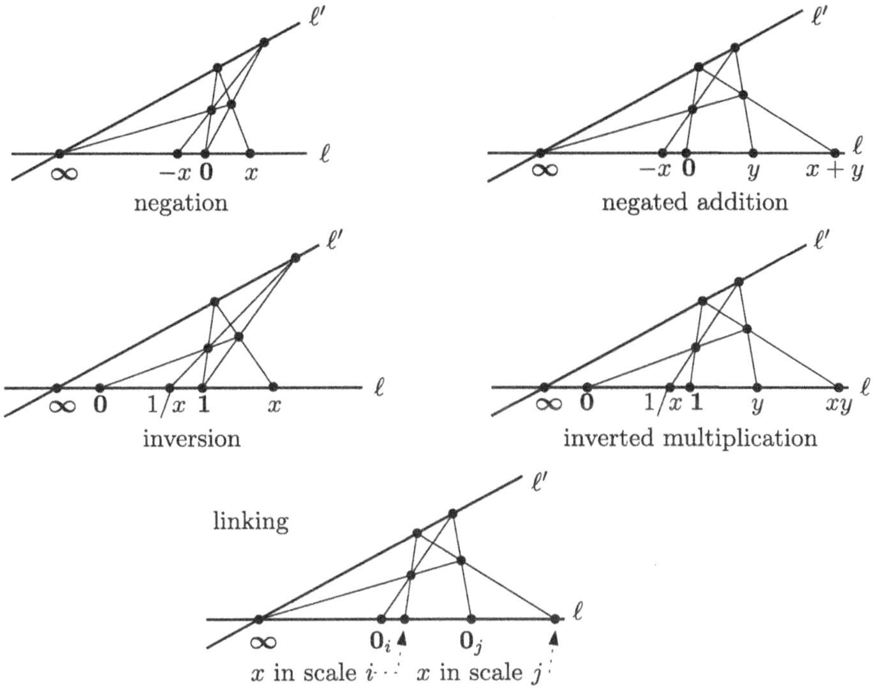

Fig. 4.1 The construction gadgets

In this way, the ordering of the points on ℓ can be fixed, and it remains to untangle the various gadgets. Each gadget has four auxiliary points; let us call them a_i, b_i, c_i, d_i for the ith gadget used in the construction.

The points a_i and b_i lie on the line ℓ' through ∞, which is shared by all of the gadgets. Their position on ℓ' can be chosen freely, and in order to separate the gadgets, we place a_i at distance D_i from ∞ and b_i at distance ε_i from a_i, where the sequence $D_1 \ll D_2 \ll \cdots$ increases extremely fast, and $\varepsilon_1 \gg \varepsilon_2 \gg \cdots$ decreases even faster.

The points c_i and d_i are then determined by the "input values" of the gadget, and if a_i and b_i are sufficiently close, c_i and d_i lie in an arbitrarily small neighborhood of a_i and b_i.

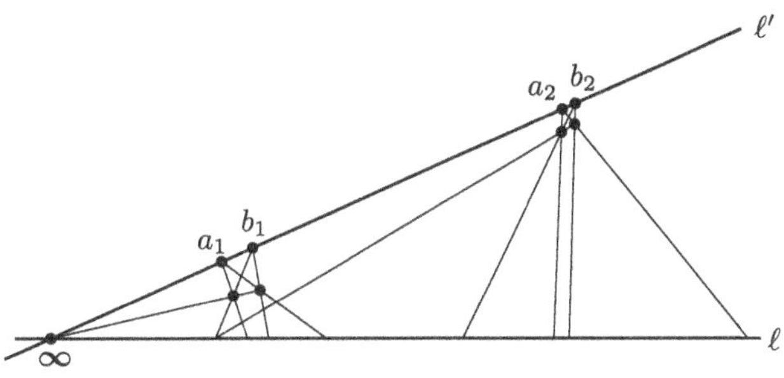

Since we know the order of the points on ℓ, as well as the points where the six lines spanned by each fourtuple $a_i, \ldots, d_i$ intersect ℓ, we know the sign of every triple involving points on ℓ and auxiliary points from a single gadget. By similar considerations, we can determine the sign of all of the remaining types of triples (on ℓ + gadget + another gadget, two same gadget + another gadget, three different gadgets).

In this way, we arrive at a fully specified and constructible combinatorial order type, whose realizability is equivalent to the solvability of the given instance of STRICT-INEQ. This concludes the proof of Theorem 4.4.6. □

The Mnëv universality theorem. Mnëv's original result did not deal with computational complexity—it was actually about a topological question. To formulate it, let us consider a sequence $(p_1, p_2, \ldots, p_n)$ of n points in the plane. This sequence is specified by a list of the point coordinates, i.e., $2n$ real numbers, and so we can regard it as a single point in $\mathbb{R}^{2n}$. For a combinatorial order type τ of n points, we define the **realization space** $\mathcal{R}(\tau) \subseteq \mathbb{R}^{2n}$ as the set of the points corresponding to realizations of τ.

In 1956, Ringel asked, in the equivalent language of line arrangements, whether the realization space has to be path-connected; in other words, whether one realization of τ can always be continuously deformed into any other, while keeping the order type τ along the way. (Strictly speaking, we need to be somewhat careful about mirror reflection, which may trivially disconnect the realization space as defined above into two components—but one usually factors out affine transformations, by fixing the position of three affinely independent points.)

Mnëv's universality theorem shows that the answer to Ringel's question is no in the strongest possible sense: the realization space can be topologically as complicated as one may wish. It may have the "shape" of any prescribed semialgebraic set, or of any prescribed finite simplicial complex; the appropriate term from topology for "having the same shape" here is being *homotopy equivalent*. Mnëv's statement uses the notion of *stable equivalence*, for which we refer to [RG95].

Similar universality theorems also hold for the realization spaces of other kinds of objects from $\exists\mathbb{R}$-complete problems. Essentially, $\exists\mathbb{R}$-completeness and topological universality theorems are just two ways of expressing a great intrinsic complexity of a given class of objects. The proofs are also similar, although in topology one has to watch out for other aspects than in computational complexity.

4.5 Quantifier Elimination According to Muchnik

Here we prove that the first-order theory of $\mathbb{R}$, and thus, in particular, the existential theory of $\mathbb{R}$, are *decidable*. We will actually obtain a stronger result, referred to as **quantifier elimination**.

Theorem 4.5.1 (Tarski [Tar51]) *There is an algorithm accepting as an input a formula Ψ of the first-order theory of the reals, which may contain quantifiers; in general it also contains free variables $Y_1, \ldots, Y_n$, which we write as $\Psi = \Psi(\boldsymbol{Y})$, with $\boldsymbol{Y} = (Y_1, \ldots, Y_n)$. The algorithm outputs a* quantifier-free *formula $\Phi = \Phi(\boldsymbol{Y})$ that is equivalent to Ψ; that is, for every choice of $\boldsymbol{y} \in \mathbb{R}^n$ we have $\Psi(\boldsymbol{y}) \equiv \Phi(\boldsymbol{y})$.*

Geometrically, this result tells us that every subset of $\mathbb{R}^n$ that can be described in the first-order theory of $\mathbb{R}$ is semialgebraic, i.e., can be specified by a quantifier-free formula. Even more geometrically, the essence of the quantifier-elimination result is that the projection of a semialgebraic set onto a coordinate subspace is again semialgebraic.

The running time bounds for the algorithm presented below are actually poor, much worse than for the best known algorithms, and we will not care about them. However, the algorithm is relatively simple and it exhibits some of the ideas also appearing in more sophisticated algorithms.

The main ideas of this algorithm are due to Muchnik (unpublished); our presentation is mostly based on [MO02], also drawing inspiration from the blog [Bha12].

We assume that the given formula Ψ is in prenex form, with all quantifiers on the outside. We eliminate the quantifiers one by one starting from inside. A universal quantifier is converted into an existential one using $(\forall X)\Phi \equiv \neg(\exists X)\neg\Phi$. Thus, it suffices to describe a procedure for eliminating the single existential quantifier from a formula Ψ of the form

$$(\exists X) F(A_1, \ldots, A_m),$$

where F is a Boolean formula, each A_i is an atomic predicate of the form $p_i(X, \boldsymbol{Y})$ rel 0 for some polynomial p_i with integer coefficients, and $\boldsymbol{Y} = (Y_1, \ldots, Y_n)$ is the vector of the free variables of Ψ.

4.5.1 The Univariate Case

In order to explain the method, it is instructive to start with the case $n = 0$, where Ψ is a sentence (no free variables) and we deal with univariate polynomials $p_1(X), \ldots, p_m(X)$.

For $x \in \mathbb{R}$, let the **sign vector** of $(p_1, \ldots, p_m)$ at x be

$$\big(\operatorname{sgn} p_1(x), \operatorname{sgn} p_2(x), \ldots, \operatorname{sgn} p_m(x)\big) \in \{-1, 0, 1\}^m.$$

To decide the validity of the sentence Ψ, we want to know whether there is an $x \in \mathbb{R}$ at which the sign vector attains one of the values allowed by the formula. For example, for the formula

$$(\exists X)\big(p_1(X) > 0 \wedge (p_2(X) > 0 \vee p_3(X) = 0)\big),$$

the allowed sign vectors are $(1, 1, *)$ and $(1, *, 0)$, where $*$ means "arbitrary value".

The algorithm actually computes *all* possible sign vectors of $(p_1, \ldots, p_m)$; this is a common feature of practically all known algorithms for quantifier elimination or for deciding ETR. Thus, the only way the Boolean formula F enters the computation is in checking if any of the resulting sign vectors is allowed by the formula.

The sign table. Let us consider an example with three polynomials $p_1(X) = 4 - X^2$, $p_2(X) = X^3 - 2X^2 - X + 2$, and $p_3(X) = -X^3 + 5X^2 - 6X$. Their graphs are plotted in the next picture,

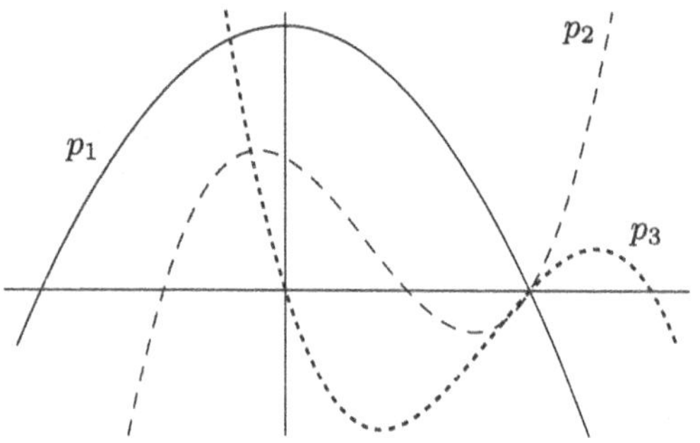

p_1	−	0	+	+	+	+	+	+	+	0	−	−	−
p_2	−	−	−	0	+	+	+	0	−	0	+	+	+
p_3	+	+	+	+	+	0	−	−	−	0	+	0	−

and below the picture we have the **sign table** of the ordered triple (p_1, p_2, p_3). To make the sign table, we partition the real line into maximal subintervals so that each p_i has a constant sign on each of the subintervals. Thus, open subintervals alternate with one-point subintervals; we call the latter ones the **boundaries**. Each subinterval I is then labeled by the sign vector of the p_i at an (arbitrary) point $x \in I$, and the sign table is the sequence of these sign vectors ordered from left to right. We stress that the sign table is *only* the sequence of sign vectors, and it does not include the numerical values of the boundaries (which are roots of the p_i).

We want to compute the sign table, which is of course sufficient to decide the given sentence Ψ, and for reasons which will become apparent later, we want to do so *without* computing the roots of our polynomials (although the sign table provides the number of roots of each p_i and their relative positions).

The algorithm proceeds incrementally. Having computed the sign table for $(p_1, p_2, \ldots, p_{k-1})$, we want to extend it to the sign table for $(p_1, p_2, \ldots, p_k)$. Conceptually, we divide this into two steps:

(i) We compute the sign of p_k at the roots of $p_1, \ldots, p_{k-1}$, i.e., at the boundaries of the old sign table.
(ii) We locate the roots of p_k among the old boundaries, and we add new boundaries and sign vectors accordingly.

Both of the steps should be far from obvious for now. But let us assume for a moment that Step (i) has been accomplished somehow. Thus, for example, we have the sign table for (p_1, p_2, p_3) as in the example above, and we also have the signs of another polynomial p_4 at the boundaries:

		a		b		c							
p_1	$-$	0	$+$	$+$	$+$	$+$	$+$	$+$	$+$	0	$-$	$-$	$-$
p_2	$-$	$-$	$-$	0	$+$	$+$	$+$	0	$-$	0	$+$	$+$	$+$
p_3	$+$	$+$	$+$	$+$	$+$	0	$-$	$-$	$-$	0	$+$	0	$-$
p_4		$-$		$+$		$+$		0		$-$		$-$	

What should be the columns of the new sign table, after adding p_4, between the two boundary columns a and b of the old table? Since $p_4(a) < 0$ and $p_4(b) > 0$, there must be a root of p_4 in the interval (a, b), so a reasonable guess at the new columns is

	a				b
p_1	0	$+$	$+$	$+$	$+$
p_2	$-$	$-$	$-$	$-$	0
p_3	$+$	$+$	$+$	$+$	$+$
p_4	$-$	$-$	0	$+$	$+$

This need not be correct, though, since p_4 may have *several* roots in (a, b). But if it does, then it must have a local minimum there, and hence the derivative p_4' has a root in (a, b).

So if we could make sure that all roots of p_4' are among the boundaries of the old sign table, our guess at the new columns as above would be correct. Under the same assumption, for the next subinterval (b, c), from $p_4(b) > 0$ and $p_4(c) > 0$ we would get that $p_4 > 0$ on (b, c). Similarly we could infer all of the new sign table. (There is a detail that we have skipped—we also need the signs of p_4 "at $-\infty$" and "at $+\infty$," i.e., for all sufficiently small and sufficiently large x, for otherwise, we would not be able to detect roots of p_4 to the left or to the right of all old boundaries.)

This reasoning works in general, for adding a new polynomial p_k, but how do we force all roots of p_k' to be among the old boundaries? The answer is that, at the beginning of the algorithm, we *extend* our collection of polynomials by adding their derivatives, second derivatives, etc. Then it suffices to process the polynomials from smaller degrees to larger ones, and the above condition on p_k' will always be met!

Very good, so how do we perform Step (i), determining the sign of p_k at the old boundaries? Suppose that we want to determine $\operatorname{sgn} p_k(\alpha)$, where α is a root of p_i, $i < k$. The trick is to consider the **remainder** $r := p_k \bmod p_i$; i.e., r is the unique polynomial with $\deg r < \deg p_i$ and $p_k = qp_i + r$ for some polynomial q. (We note that if p_k and p_i have integer coefficients, r may still have rational but non-integral coefficients.)

We have

$$r(\alpha) = p_k(\alpha) - q(\alpha)p_i(\alpha) = p_k(\alpha)$$

since $p_i(\alpha) = 0$, and so, in particular, sgn $p_k(\alpha) = \text{sgn}\, r(\alpha)$. Thus, if r is among the already processed polynomials $p_1, \ldots, p_{k-1}$, we know its sign at the old boundaries, and so we can infer the sign of p_k (without computation). This suggests that, similar to the case with the derivatives, at the beginning of the algorithm we also extend our collection of polynomials by closing it under taking remainders.

Computing the closure. Let $\mathcal{P} = \{p_1, \ldots, p_m\}$ be the initial collection of polynomials, those in the given formula. Let us define $\overline{\mathcal{P}}$ as the closure of $\mathcal{P}$ under taking the derivative and remainders, i.e., an inclusion-minimal set of polynomials containing $\mathcal{P}$ and closed under these two operations.

Exercise 4.5.2 (a) Describe an algorithm to compute $\overline{\mathcal{P}}$ given $\mathcal{P}$, and prove its finiteness (assuming that subroutines for computing the derivative and remainder are available).

(b) What lower and upper bounds can you get for $|\overline{\mathcal{P}}|$ in terms of m and Δ, the maximum degree of the p_i?

Summary of the univariate sign-table algorithm. The input to the algorithm is a set $\mathcal{P}$ of univariate polynomials with integer coefficients, and the output is a sign table for a certain superset of $\mathcal{P}$.

1. We compute the closure $\overline{\mathcal{P}}$ of $\mathcal{P}$ under derivatives and remainders as in Exercise 4.5.2. We number the polynomials in $\overline{\mathcal{P}}$ as $p_1, p_2, \ldots, p_M$ in such a way that deg $p_i \leq$ deg p_j for $i < j$. (This numbering is different from the original numbering of the input polynomials in $\mathcal{P}$; what used to be p_3 may now be p_{1234}.)
2. Let the sign table T_0 consist of a single empty sign vector. For $k = 1, 2, \ldots, M$, compute T_k, the sign table for $(p_1, \ldots, p_k)$, from T_{k-1} according to the next step, and then output T_M.
3. If deg $p_k = 0$, i.e., p_k is a constant, then determine its sign and add the appropriate row to T_{k-1}. Otherwise, for deg $p_k \geq 1$, do the two steps as above:
   (i) Determine the sign of p_k at $-\infty$ and at $+\infty$, by inspecting the leading coefficient and the degree. Then for every boundary a of T_{k-1}, find p_i, $i < k$, with $p_i(a) = 0$, and determine sgn $p_k(a) := \text{sgn}\, r(a)$, where $r := p_k \bmod p_i$. Here sgn $r(a)$ can be found in the sign table T_{k-1}, since $r = p_j$ for some $j < k$.
   (ii) For every interval (a, b), where a, b are consecutive boundaries of T_{k-1} (or $a = -\infty$ or $b = \infty$), inspect the signs of $p_k(a)$ and $p_k(b)$, and determine the behavior of p_k on (a, b): If (sgn $p_k(a)$)(sgn $p_k(b)$) $= -1$, then p_k has a single root in (a, b) and we insert a new column into the sign table as described in the example above. In all other cases, sgn p_k is constant and nonzero on (a, b), and it coincides with the nonzero sign in {sgn $p_k(a)$, sgn $p_k(b)$} (at least one of these must be nonzero—why?).

4.5.2 The Multivariate Case

Now we consider the general case, where we want to eliminate the existential quantifier from $(\exists X)\Phi_0(X, \boldsymbol{Y})$, where $\boldsymbol{Y} = (Y_1, \ldots, Y_n)$ and Φ_0 is quantifier-free. Let us think of the polynomials $p_i(X, \boldsymbol{Y})$ appearing in Φ_0 as polynomials in X whose coefficients are polynomials in $Y_1, \ldots, Y_n$.

The idea, a very computer-science one, is to run the univariate sign-table algorithm on the polynomials $p_1(X, \boldsymbol{Y}), \ldots, p_m(X, \boldsymbol{Y})$. Of course, that algorithm was derived for polynomials with integer (or rational) coefficients, while here the coefficients are polynomials. We thus need to look at the operations performed in the algorithm in more detail.

One thing the univariate algorithm does is computing various expressions in the coefficients using the four basic operations $+, -, *, /$ (no square roots etc.). That we can do with polynomials as well, obtaining rational functions in the Y_j, of the form $u(\boldsymbol{Y})/v(\boldsymbol{Y})$, where $u(\boldsymbol{Y})$ and $v(\boldsymbol{Y})$ are polynomials.[4]

The other kind of operation in the algorithm is *sign testing*: for some already computed expression E, a rational function in our case, the algorithm wants to know the sign of E. When we try to work with rational functions instead of rational numbers, we do not know the answer, since it generally depends on the values of the Y_j.[5]

Here comes the key idea of the multivariate case: we just give *all three* possible answers to the algorithm. Then, instead of following just one computation, we go into three different branches of computation (this may delight fans of the many-worlds interpretation of quantum mechanics).

Such a branching is made at every sign test, so we get a ternary tree, in which every possible computation of the algorithm corresponds to one of the root-leaf paths, and each node is labeled by a rational function $E = E(\boldsymbol{Y})$ from the corresponding sign test. Here is an (artificial and very small) example of such a tree:

[4] Actually, the only place where division appears is the computation of remainders. As in [MO02] and elsewhere in algorithms dealing with multivariate polynomials, one can avoid division completely by replacing the remainder by the **pseudoremainder**: the pseudoremainder of polynomials $a(X) = \sum_{i=0}^{d} a_i X^i$ and $b(X) = \sum_{i=0}^{e} b_i X^i$, $d \geq e$, is the unique polynomial $r(X)$ such that $b_e^{d-e+1} a(x) = q(X)b(X) + r(X)$.

[5] We should remark that sign testing may also occur in perhaps somewhat unexpected parts of the algorithm. For example, when computing the closure $\overline{\mathcal{P}}$, the algorithm repeatedly computes the remainder for two polynomials computed earlier, and for that, it needs to know the degrees of these polynomials. If the coefficients of the polynomials are rational functions of $\boldsymbol{Y}$, then the leading coefficient, as well as some others, may vanish for some values of $\boldsymbol{Y}$, and so the algorithm must test which is the highest power of X whose coefficient is actually nonzero.

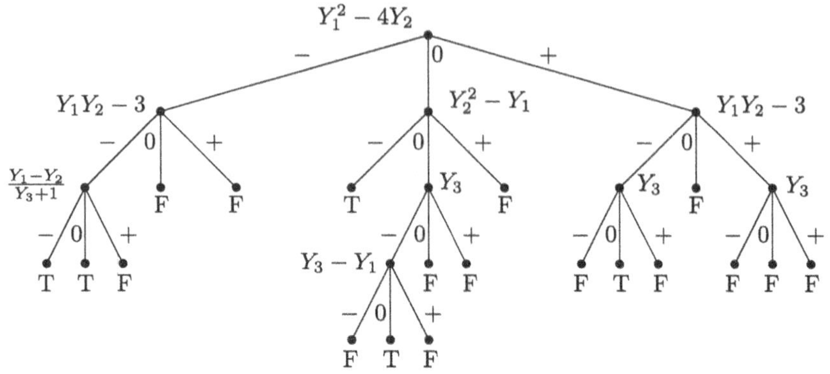

For a real but still simple example see Exercise 4.5.3 below.

At the end of each computation, the univariate algorithm outputs TRUE or FALSE, depending on whether the computed sign table contains one of the sign vectors allowed by the formula Φ_0. We label the leaves of the tree by these output values.

Now we recall our actual goal: we want a quantifier-free formula $\Phi(\boldsymbol{Y})$ equivalent to $(\exists X)\Phi_0(X, \boldsymbol{Y})$. We construct $\Phi(\boldsymbol{Y})$ as a formula describing the conditions under which the univariate algorithm reaches a leaf labeled TRUE. This is easy to do: We make $\Phi(\boldsymbol{Y})$ as a disjunction of subformulas, each corresponding to one TRUE leaf. The subformula corresponding to a given leaf ℓ should say that all the rational functions along the path to ℓ have the signs given by the chosen path. For example, the path to the leftmost leaf in the picture above yields the subformula

$$Y_1^2 - 4Y_2 < 0 \wedge Y_1Y_2 - 3 < 0 \wedge \frac{Y_1 - Y_2}{Y_3 + 1} < 0.$$

This is not yet a formula in the first-order theory of the reals, since it contains division. But we can replace sign testing for rational functions by sign testing for polynomials, by testing the numerator and the denominator separately.

This finishes the presentation of the quantifier-elimination algorithm and the proof of Theorem 4.5.1.

Exercise 4.5.3 Apply the algorithm to the formula $\Psi(A, B, C) \equiv (\exists X)AX^2 + BX + C = 0$, and see how the familiar discriminant materializes in front of your eyes.

4.5.3 On the Complexity of Quantifier-Elimination Algorithms

Let the considered formula Ψ contain m polynomials of degree at most Δ each and with the number of bits in each coefficient bounded by τ. Further let it have

ω blocks of alternating quantifiers, with n_i variables in the ith block, and let n_0 be the number of free variables. Let us set $N := \prod_{i=0}^{\omega}(n_i + 1)$, and let C be a suitable constant. According to [BPR96, BPR03], there is an algorithm that performs quantifier elimination for Ψ using $m^N(\Delta + 1)^{C^{\omega}N}$ arithmetic operations with at most $((\Delta + 1)^{C^{\omega}N}\tau)$-bit integers.

For deciding an existential formula (i.e., for $n_0 = 0$ and $\omega = 1$), this yields an algorithm with $m^{n+1}(\Delta + 1)^{O(n)}$ arithmetic operations with integers having at most $(\Delta + 1)^{O(n)}\tau$ bits.

Why quantifier elimination is doubly exponential. The following exercise provides an example of a quantified formula of length $O(n)$ for which any quantifier elimination has to produce a quantifier-free formula of length doubly exponential in n. It is a simplified variant of an example by Davenport and Heintz [DH88].

Exercise 4.5.4 Let the formula $\Psi_n(X, Y)$ be defined inductively: $\Psi_0(X, Y)$ is $Y = 4X(1 - X)$, and $\Psi_n(X, Y)$ is

$$(\exists Z)(\forall U\ V)\big((U = X \wedge V = Z) \vee (U = Z \wedge V = Y)\big) \Rightarrow \Psi_{n-1}(U, V)$$

(we have used the implication $\Rightarrow$, which strictly speaking wasn't mentioned among the Boolean connectives appearing in the considered formulas, but $A \Rightarrow B$ can be replaced by $\neg A \vee B$).

(a) The formula $\Psi_0(X, Y)$ defines the *logistic map* $f(x) = 4x(1 - x)$. What map is defined by $\Psi_n(X, Y)$?

(b) Show that the formula $\tilde{\Psi}_n(X)$, defined as $\Psi_n(X, \frac{1}{2})$ (wait: fractions were not allowed in formulas of the first-order theory of the reals, so how do we write this properly?), defines a semialgebraic subset of $\mathbb{R}$ whose definition by a quantifier-free formula requires formula length $2^{2^{\Omega(n)}}$. You may want to use Exercise 4.2.3.

Original acknowledgments (2014). I am especially grateful to Marcus Schaefer for answering numerous questions, reading a draft of my exposition, and providing many useful comments. I would also like to thank to Rado Fulek, Vincent Kusters, Jan Kynčl, Jürgen Richter-Gebert, Zuzana Safernová, and Daniel Štefankovič for various contributions, such as proofreading, corrections, answering questions, etc. It was a pleasure to teach courses including the material presented here together with Pavel Valtr in Prague and with Michael Hoffmann and Emo Welzl in Zurich.

Further acknowledgements (2025). Because of the untimely death of Jirka Matoušek in 2015, the present paper has previously never appeared in print. The editors are grateful to Marcus Schaefer, who first suggested that this important work should be published, and carefully read and annotated the manuscript. We are also indebted to Helena Nyklová for implementing the changes and allowing us to publish this paper.

References

[Bha12] Arnab Bhattacharyya. Something you should know about: Quantifier elimination (Part II). Blog, https://cstheory.blogoverflow.com/2012/02/something-you-should-know-about-quantifier-elimination-part-ii, 2012.

[Bie91] Daniel Bienstock. Some provably hard crossing number problems. *Discrete & Computational Geometry*, 6:443–459, 1991.

[BLS+99] Anders Björner, Michel Las Vergnas, Bernd Sturmfels, Neil White, and Günter M. Ziegler. *Oriented Matroids (Second Edition)*. Cambridge University Press, Cambridge, 1999.

[BPR96] Saugata Basu, Richard Pollack, and Marie-Françoise Roy. On the combinatorial and algebraic complexity of quantifier elimination. *J. ACM*, 43(6):1002–1045, 1996.

[BPR03] Saugata Basu, Richard Pollack, and Marie-Françoise Roy. *Algorithms in real algebraic geometry*. Algorithms and Computation in Mathematics 10. Springer, Berlin, 2003. 2nd edition, Springer, Heidelberg, 2006.

[BV07] Saugata Basu and Nicolai N. Vorobjov. On the number of homotopy types of fibres of a definable map. *J. Lond. Math. Soc., II. Ser.*, 76(3):757–776, 2007.

[Can88] John Canny. Some algebraic and geometric computations in PSPACE. In *Proc. 20th Annu. ACM Sympos. Theory Comput.*, pages 460–467, 1988.

[CG09] Jérémie Chalopin and Daniel Gonçalves. Every planar graph is the intersection graph of segments in the plane. In *Proc. 41st Annual ACM Symposium on Theory of Computing (STOC)*, pages 631–638, New York, NY, USA, 2009. ACM.

[DH88] James H. Davenport and Joos Heintz. Real quantifier elimination is doubly exponential. *J. Symbolic Comput.*, 5(1-2):29–35, 1988.

[Gj88] Dima Yu. Grigor'ev and Nikolai N. Vorobjov jun. Solving systems of polynomial inequalities in subexponential time. *J. Symb. Comput.*, 5(1–2):37–64, 1988.

[GPS90] Jacob E. Goodman, Richard Pollack, and Bernd Sturmfels. The intrinsic spread of a configuration in $\mathbf{r}^d$. *J. Amer. Math. Soc.*, 3:639–651, 1990.

[KM94] Jan Kratochvíl and Jiří Matoušek. Intersection graphs of segments. *J. Combin. Theory Ser. B*, 62(2):289–315, 1994.

[KM12] Ross J. Kang and Tobias Müller. Sphere and dot product representations of graphs. *Discrete & Computational Geometry*, 47(3):548–568, 2012.

[MM13] Colin McDiarmid and Tobias Müller. Integer realizations of disk and segment graphs. *J. Comb. Theory, Ser. B*, 103(1):114–143, 2013.

[Mne89] Nikolai E. Mnev. The universality theorems on the classification problem of configuration varieties and convex polytopes varieties. In O. Ya. Viro, editor, *Topology and Geometry - Rohlin Seminar*, volume 1346 of *Lecture Notes Math.*, pages 527–544. Springer-Verlag, 1989.

[MO02] Christian Michaux and Adem Ozturk. Quantifier elimination following muchnik. Preprint Series 10, Univ. de Mons-Hainaut, 2002.

[PKK+14] Arkadiusz Pawlik, Jakub Kozik, Tomasz Krawczyk, Michał Lasoń, Piotr Micek, William T. Trotter, and Bartosz Walczak. Triangle-free intersection graphs of line segments with large chromatic number. *J. Comb. Theory, Ser. B*, 105:6–10, 2014. Preprint, arXiv:1209.1595, 2012.

[RG95] Jürgen Richter-Gebert. Mnëv's universality theorem revisited. *Sémin. Lothar. Comb.*, B34h:1–15, 1995. electronic.

[RG11] Jürgen Richter-Gebert. *Perspectives on projective geometry: a guided tour through real and complex geometry*. Springer Verlag, Berlin Heidelberg, 2011.

[Sch10] Marcus Schaefer. Complexity of some geometric and topological problems. In David Eppstein et al., editor, *Graph drawing. 17th international symposium, GD 2009*, volume 5849 of *Lecture Notes in Computer Science*, pages 334–344. Springer, Berlin, 2010.

[SCM25] Marcus Schaefer, Jean Cardinal, and Tillmann Miltzow. The existential theory of the reals as a complexity class: A compendium. In J. Pach and G. Tóth, editors, *Courses in Discrete and Computational Geometry*, Bolyai Soc. Math. Studies, volume 31, Springer, 2026. arXiv:2407.18006.

[Sho91] Peter W. Shor. Stretchability of pseudolines is np-hard. In P. Gritzman and B. Sturmfels, editors, *Applied Geometry and Discrete Mathematics: The Victor Klee Festschrift*, volume 4 of *DIMACS Series in Discrete Mathematics and Theoretical Computer Science*, pages 531–554. AMS Press, 1991.

[SŠ11] Marcus Schaefer and Daniel Štefankovič. Fixed points, nash equilibria, and the existential theory of the reals. Manuscript, DePaul University, Chicago, 2011.

[Tar51] Alfred Tarski. *A decision method for elementary algebra and geometry*. Univ. of California Press, Berkeley, CA, 1951.

Chapter 5
The Existential Theory of the Reals as a Complexity Class: A Compendium

Marcus Schaefer, Jean Cardinal, and Tillmann Miltzow

We survey the complexity class $\exists\mathbb{R}$, which captures the complexity of deciding the existential theory of the reals. The class $\exists\mathbb{R}$ has roots in two different traditions, one based on the Blum-Shub-Smale model of real computation, and the other following work by Mnëv and Shor on the universality of realization spaces of oriented matroids. Over the years the number of problems for which $\exists\mathbb{R}$ rather than NP has turned out to be the proper way of measuring their complexity has grown, particularly in the fields of computational geometry, graph drawing, game theory, and some areas in logic and algebra. $\exists\mathbb{R}$ has also started appearing in the context of machine learning, Markov decision processes, and probabilistic reasoning. We have aimed at collecting a comprehensive compendium of problems complete and hard for $\exists\mathbb{R}$, as well as a long list of open problems. This compendium is the third part of our survey; a tour through the compendium and the areas it touches on makes up the second part. The first part introduces the reader to the existential theory of the reals as a complexity class, discussing its history, motivation and prospects as well as some technical aspects.

The first author was the primary contributor, while the second and third authors contributed equally and are listed alphabetically.

T. Miltzow: Generously supported by the Netherlands Organisation for Scientific Research (NWO) under project no. VI.Vidi.213.150.

M. Schaefer (✉)
DePaul University, Chicago, IL, USA
e-mail: mschaefe@depaul.edu

J. Cardinal
Université libre de Bruxelles (ULB), Brussels, Belgium
e-mail: jean.cardinal@ulb.be

T. Miltzow
Utrecht University, Utrecht, The Netherlands
e-mail: t.miltzow@uu.nl

J. Pach and G. Tóth (eds.), *Courses in Discrete and Computational Geometry*,
Bolyai Society Mathematical Studies 31,
https://doi.org/10.1007/978-3-032-10503-5_5

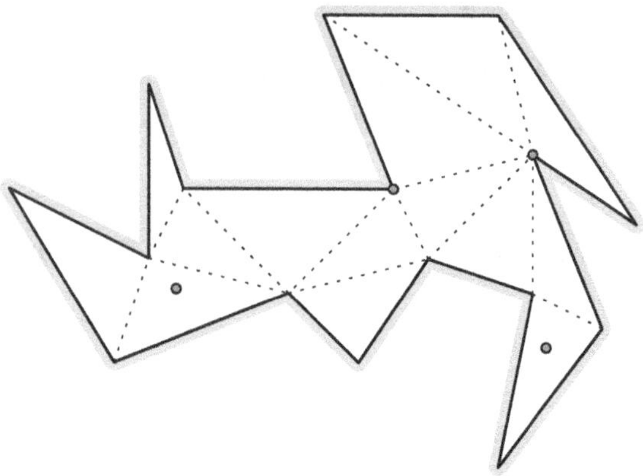

Fig. 5.1 A gallery with 16 vertices that can be guarded with $4 < \lfloor 16/3 \rfloor$ guards

5.1 Introduction

How many guards are necessary to guard an art gallery—a simple polygon on n vertices—if every location inside the gallery must be seen by at least one of the guards? Chvátal [Chv75] famously proved that at most $n/3$ guards are needed and that this bound is tight in general.

Figure 5.1 shows that for specific layouts fewer than $n/3$ guards may be sufficient to guard a gallery, and we may ask how hard it is to find the smallest number of guards. It was shown early on that the problem is NP-hard [O'R87], and variants in which guards are stationed at vertices could be shown to lie in NP, but membership of the general problem remained elusive. The main difficulty lies in determining the coordinates of the guards, which could (potentially) be arbitrary real numbers. How can we handle real coordinates within NP?

A closer look shows that the art gallery problem is not alone, many computational problems rely on real numbers. For some of these problems, researchers asked whether they lie in NP, including the rectilinear crossing number [BMP05, Sect. 9.4] and the collinearity problem [Aar11], but no positive answers were forthcoming. We now know why these questions were so hard to answer: all of these problems are $\exists\mathbb{R}$-complete, they are equivalent, up to a polynomial-time reduction, to deciding the existential theory of the reals (L1). For some problems, this had been known for a while, for the rectilinear crossing number (CG50) from a paper by Bienstock [Bie91] and for collinearity logic (L9) from a paper by Davis et al. [DGC99]; in comparison, the complexity of the art gallery problem (CG16) was not resolved until more than twenty years later in the papers by Abrahamsen et al. [AAM18, AAM22].

Since these questions were first asked, the existential theory of the reals has become more established and better known. It has its own Wikipedia page [Wik24] and has even found shelter in the complexity zoo [Aar24]. The time now seems right to attempt a survey of what we know about this complexity class. There is some material in that direction already, including the second author's computational geometry column [Car15], Bilò and Mavronicolas catalogues of results on Nash equilibria [BM16, BM21], and Matoušek's introduction to the technical framework underlying $\exists\mathbb{R}$ [Mat14], but no comprehensive coverage has been published so far. (The first author's unpublished "The real logic of drawing graphs", first announced in [Sch10], was never completed, but has served as a quarry for this compendium.)

Our survey consists of three parts: The first part introduces the existential theory of the reals covering basic definitions, motivation, history, prospects as well as some tools and techniques. The second part takes the reader on a brief tour of the compendium, followed by the compendium itself in the third part; this third part lists, to the best of our knowledge, (nearly) all problems that are known to be $\exists\mathbb{R}$-complete or $\exists\mathbb{R}$-hard. All parts are indexed and cross-referenced which we hope will help the reader discover interesting old and new problems. For the convenience of the reader, we supply a link to our bibliography.

https://tinyurl.com/ER-bibliography

We have aimed at giving an easily accessible overview of current knowledge about $\exists\mathbb{R}$, to help researchers find relevant literature, and to serve as an entry point for further study. We hope this compendium will lead to many new and exciting discoveries in the computational world of the theory of the reals. We would love to hear from readers if they prove new results that belong in the compendium.

5.2 Background and Overview

5.2.1 A Brief Introduction to the Existential Theory of the Reals

5.2.1.1 What is the Existential Theory of the Reals?

The easiest way to define the existential theory of the reals as a complexity class is through one of its complete problems; there are several possible candidates including feasibility of polynomials (A1) and stretchability of pseudoline arrangements (CG9), two of the original complete problems, but before both of these, "the existential theory of the reals" existed as a formal language.

The theory of the reals is built on arithmetical terms, that is, constants 0, 1, as well as variables, combined with operations $+, -, \cdot$ (and parentheses), for example $x \cdot y - (1 + 1) \cdot (1 + 1 + 1)$; an atomic formula compares two such terms using $<, \leq, =, \geq, >$; quantifier-free formulas are then obtained by combining atomic formulas using Boolean operators $\wedge, \vee, \neg$; for example,

$$\neg(x < z) \vee x < y \wedge y < z \wedge x \cdot z = y \cdot y.$$

Finally, we can quantify free (unquantified) variables in a formula; we can always assume that the resulting formula is in *prenex* form, that is, all quantifiers occur at the beginning of the formula, for example $\exists y, z : x \cdot x \cdot x = y + 1 + 1 \wedge x \cdot x = 1 + 1 - z \wedge x \geq 0 \wedge y \geq 0 \wedge z \geq 0$, which over the reals is equivalent to $x \in [\sqrt[3]{2}, \sqrt{2}]$. If all variables in a formula are quantified, it is called a *sentence*, e.g.

$$\forall x, z \exists y : \neg(x < z) \vee x < y \wedge y < z \wedge x \cdot z = y \cdot y$$

is a sentence; interpreted over the reals, so all variables range over real numbers, it expresses that every pair of real numbers $x < z$ has a real geometric mean y. The sentence is false, unless we also assume that $x > 0$. Examples of true sentences include $\exists x, y : x \cdot x = 1 + 1 \wedge y \cdot y = 1 + 1 + 1 \wedge x + y > 3$ and $\forall x, y : \neg(x \cdot x = 1 + 1) \vee \neg(y \cdot y = 1 + 1 + 1) \vee x + y > 3$, both of these sentences express that $\sqrt{2} + \sqrt{3} > 3$. The quantifier-free part of a quantified formula is known as the *matrix* of the formula.

The language ETR, the existential theory of the reals (L1), is defined as the set of all existentially quantified sentences which are true if interpreted over the real numbers. For example, $\exists x, y : x \cdot x = y \in$ ETR, while $\forall x \exists y : y \cdot y < x \cdot x \notin$ ETR.

From ETR it is a short step to defining the existential theory of the reals as a *complexity class* by defining it as the downward closure of ETR under polynomial-time many-one reductions. We denote this class by $\exists\mathbb{R}$ (read "exists R" or "ER"). Intuitively, $\exists\mathbb{R}$ contains all decision problems that can be expressed efficiently in ETR. This is analogous to how NP can be defined from the Boolean satisfiability problem (as the existential theory of GF[2]). Let us emphasize that even though the sentences in ETR express true facts about the real numbers, the sentences themselves are discrete objects (strings over a finite alphabet) and the decision question asks for a binary yes/no decision; real numbers do not occur as either inputs or outputs, though, implicitly, they may occur as witnesses, e.g. $\exists x : x \cdot x = 1 + 1 \in$ ETR is witnessed by $x = \sqrt{2}$.

Hardness and completeness are defined as usual for complexity classes: A problem is $\exists\mathbb{R}$-*hard*, if every problem in $\exists\mathbb{R}$ polynomial-time many-one reduces to it. It is $\exists\mathbb{R}$-*complete* if it lies in $\exists\mathbb{R}$ and is $\exists\mathbb{R}$-hard.

To show that a problem lies in $\exists\mathbb{R}$, it is then sufficient to express it in the language of the existential theory of the reals; for that it is useful to know that the definition of ETR can be relaxed to include integer, rational or even algebraic constants, and integer exponents without changing the computational power of the class. To show $\exists\mathbb{R}$-hardness of a problem, we need to prove that ETR (or some other problem already established as $\exists\mathbb{R}$-hard) reduces to it. We cover proving $\exists\mathbb{R}$-completeness in more detail in Sect. 5.2.1.4.

Let us present some sample computational problems which are complete for $\exists\mathbb{R}$, see Fig. 5.2 for illustrations.

Feasibility (A1). In the feasibility problem we are given a family of polynomials $(f_i)_{i \in [k]} : \mathbb{R}^n \to \mathbb{R}^k$ with integer coefficients and we ask whether they have a common zero, that is an $x \in \mathbb{R}^n$ such that $f_i(x) = 0$ for all $i \in [k]$. Given the polynomials, it is easy to express the problem in the existential theory of the reals. Showing hardness is trickier, since ETR allows strict inequalities and Boolean

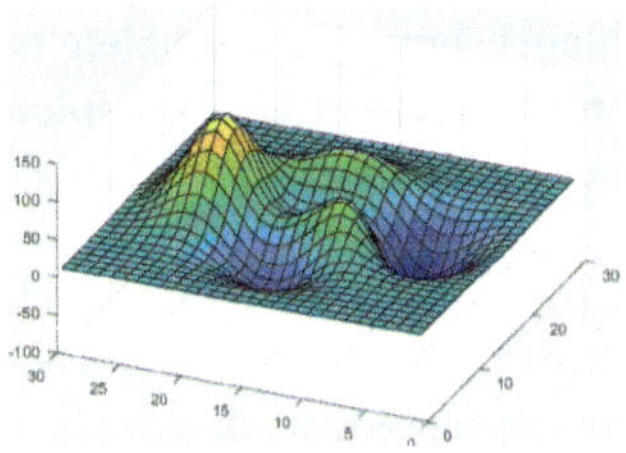

			χ
p_1	p_2	p_3	1
p_1	p_2	p_4	1
p_1	p_3	p_4	-1
p_2	p_3	p_4	1

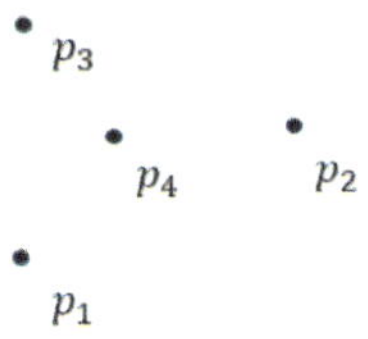

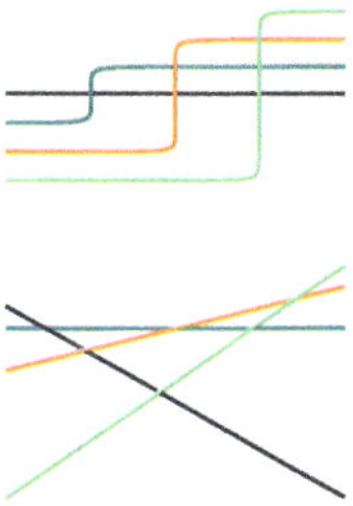

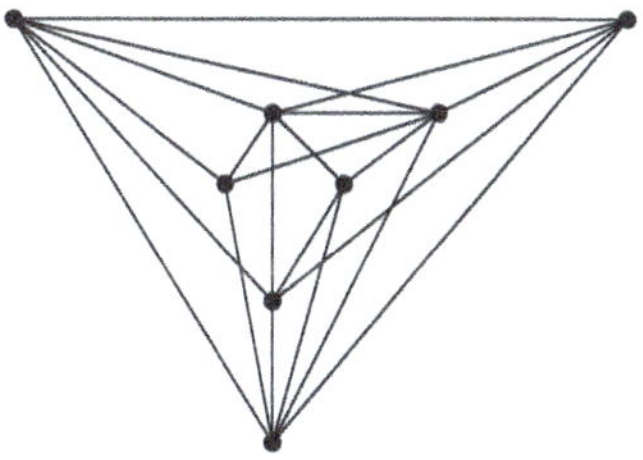

Fig. 5.2 *Top left*: Feasibility problem in 3d; are there $(x, y) \in \mathbb{R}^2$ such that $f(x, y) = 0$? *Top right*: Order-type realizability problem, with the chirotope specified as a table and a realizing point set in the plane next to it. *Bottom left*: A pseudoline arrangement (top) and an isomorphic line arrangement (bottom). *Bottom right*: A straight-line drawing of the complete 4-partite graph $K_{2,2,2,2}$ with 8 crossings, which is minimal for straight-line drawings. If the drawing does not have to be straight-line, six crossings are sufficient

operators, which we need to express using feasibility. This can be done (and was first done in [BSS89]). For example, the (true) sentence $\exists x, y : x < y \wedge x^2 = 2$ can be expressed as $f_1(x, y, z) := (y - x)z^2 - 1$ and $f_2(x, y, z) := 2 - x^2$ having a common zero $(x, y, z) \in \mathbb{R}^3$. Note that the term feasibility is often applied to the case $k = 1$ specifically.

Order Type Realizability (CG1). For three points p, q and r in the plane we distinguish three relative positions: r lies above (1), on (0) or below (-1) the line through p and q, directed from p to q. An *order type* (or *chirotope*) is a function $\chi : [n]^3 \to \{-1, 0, 1\}$ for some n which specifies the relative positions of all triples of n points. An order type captures, purely combinatorially, the relative positions of a set of points. It is *realizable* if there are points $p_1, \ldots, p_n$ in the plane such that the relative position of p_i, p_j, p_k corresponds to the specification $\chi(i, j, k)$ for all triples $(i, j, k) \in [n]^3$. Membership in the existential theory of the reals is easy to show, since the relative position of three points p,q,r can be expressed arithmetically as the sign of a determinant:

$$\operatorname{sgn} \begin{vmatrix} p_x & p_y & 1 \\ q_x & q_y & 1 \\ r_x & r_y & 1 \end{vmatrix}.$$

Order type realizability was one of the original problems shown $\exists\mathbb{R}$-complete by Mnëv [Mnë88, Sho91]. Mnëv also showed that the problem remains $\exists\mathbb{R}$-complete for *uniform* or *simple* order types, in which no three points are collinear, so $\chi : [n]^3 \to \{-1, 1\}$.

Stretchability (CG9). A *pseudoline arrangement* is a family of x-monotone curves (i.e. intersecting each vertical line exactly once) in the plane so that each pair crosses exactly once. We can describe pseudoline arrangements discretely as wiring diagrams, see Fig. 5.2. The stretchability problem asks whether a pseudoline arrangement is *stretchable*, that is, if there is a homeomorphism of the plane that turns the pseudoline arrangement into a (straight) line arrangement. Expressing the problem in ETR is again easy, but hardness is difficult; it is a byproduct of Mnëv's universality theorem [Mnë88, Sho91, Mat14] (See Sect. 5.2.1.3). Via projective duality, the problem is equivalent to order type realizability in the plane (CG1). Uniform order types correspond to simple pseudoline arrangements, in which not three pseudolines intersect in a point; it follows, from the corresponding result on order types, that *simple stretchability* (CG9), the stretchability of simple pseudoline arrangements, is also $\exists\mathbb{R}$-complete [Mnë88].

Rectilinear Crossing Number (CG50). Given a graph and a number k, is there a straight-line drawing of the graph with at most k crossings? Membership is not difficult, but cumbersome, worked details can be found in [Dea02, Sch10]. Mirroring what happened with NP, many papers now skip $\exists\mathbb{R}$-membership proofs. Bienstock [Bie91] showed how to reduce stretchability, which, as we just saw, is $\exists\mathbb{R}$-hard, to the rectilinear crossing number problem, showing that it, in turn, is $\exists\mathbb{R}$-hard.

There is another way of defining $\exists\mathbb{R}$, which corresponds more closely to how (some) complexity classes have been defined traditionally, and that is through a machine model. For example, NP was first defined as the class of problems solvable by non-deterministic polynomial-time Turing machines, and *then* satisfiability was shown NP-complete. The NP-completeness of satisfiability, now known as the Cook-Levin theorem, laid the foundation of computational complexity. There is a Cook-Levin theorem for $\exists\mathbb{R}$ as well, if we define $\exists\mathbb{R}$ in the Blum-Shub-Smale (BSS) model of computation, where it is known as NPR or $\mathsf{BP}(\mathsf{NP}^0_{\mathbb{R}})$. Roughly speaking, in the BSS model of real computation we work with register machines that can read, store, and write real numbers in a single cell or register (with full precision), and perform exact arithmetic and comparisons on these numbers; registers can be preloaded with real constants. As inputs, BSS machines take tuples of real numbers (of arbitrary arity), so a BSS machine accepts a set of tuples of real numbers. In the BSS model, we can define polynomial time and non-deterministic polynomial time (real numbers have unit length and operations on real numbers take unit time). This allows a definition of problems that can be decided in polynomial time, $\mathsf{P}_{\mathbb{R}}$, and non-deterministic polynomial time, $\mathsf{NP}_{\mathbb{R}}$. If we restrict the non-deterministic machines so they are *constant-free*, that is, only constants 0 and 1 are allowed in the program code, and we only consider Boolean input tuples (all entries are 0 or 1) we obtain the complexity class $\mathsf{BP}(\mathsf{NP}^0_{\mathbb{R}})$. The original paper by Blum et al. [BSS89] already showed that the feasibility problem (with real coefficients) is complete for $\mathsf{NP}_{\mathbb{R}}$, this is the Cook-Levin theorem in the BSS model since it expresses a machine computation as a (real) satisfiability problem; the real Cook-Levin theorem easily implies that the feasibility problem we introduced earlier (with integer coefficients), is complete for $\mathsf{BP}(\mathsf{NP}^0_{\mathbb{R}})$, and this implies $\exists\mathbb{R} = \mathsf{BP}(\mathsf{NP}^0_{\mathbb{R}}) = \mathsf{NPR}$, so the BSS-model is a machine model for $\exists\mathbb{R}$.

As we will see in Sect. 5.2.3 on the history of the existential theory of the reals, these two alternative definitions arose independently.

We have defined ETR as the existentially quantified subset of true sentences in the theory of the reals. In analogy with the polynomial-time hierarchy building on NP, we can similarly define a real polynomial hierarchy. The $Q_1 \cdots Q_k$*-fragment* of the theory of the reals is defined by the true sentences which start with k alternating blocks of quantifiers of the types $(Q_1, \dots, Q_k)$, where $Q_i \in \{\exists, \forall\}$. The corresponding complexity classes are called $\Sigma_k\mathbb{R}$ and $\Pi_k\mathbb{R}$, depending on whether $Q_1 = \exists$ or $Q_1 = \forall$. The most common classes, apart from $\exists\mathbb{R}$, are $\forall\mathbb{R} = \Pi_1\mathbb{R}$, at the first level, and $\exists\forall\mathbb{R} = \Sigma_2\mathbb{R}$ and $\forall\exists\mathbb{R} = \Pi_2\mathbb{R}$ at the second level [SŠ23, Jun23b]. These classes can also be defined in the BSS-model by considering alternating machines (machines with existential and universal states) or using an oracle model [Min24].

Only a few problems have been shown complete for higher levels of the hierarchy (reflecting the situation for the polynomial-time hierarchy), we collect them in Sect. 5.4.8.

5.2.1.2 Where is the Existential Theory of the Reals?

We do not know much about the relationship of $\exists\mathbb{R}$ to traditional complexity classes, but we do know a little. Tarski [Tar48] was the first to show that the problem is decidable; his algorithm is not elementary; that is, its running time is not bounded by a finite tower of exponentials. Shor [Sho91] proved that $\mathsf{NP} \subseteq \exists\mathbb{R}$ by reducing satisfiability to stretchability; using ETR as the complete problem leads to an easier proof, since satisfiability can be expressed directly in ETR; for example, $(p \vee \overline{q} \vee r) \wedge (\overline{p} \vee \overline{r})$ can be expressed as $\exists p, q, r : (p = 1 \vee q = 0 \vee r = 1) \wedge (p = 0 \vee r = 0)$ in ETR.

The second result we know is due to Canny [Can88b, Can88a], who in his thesis showed that $\exists\mathbb{R} \subseteq \mathsf{PSPACE}$ using his new notion of roadmaps (also see Renegar [Ren88]). PSPACE is the class of problems decidable by (traditional) Turing machines in polynomial space. Complete problems for PSPACE include quantified Boolean formulas, and many game-related problems, such as Sokoban.

And this is all we know about $\exists\mathbb{R}$ with respect to traditional complexity classes. The status of $\exists\mathbb{R}$ with respect to PH, for instance, is wide open, and there are results suggesting that improving the PSPACE upper bound on $\exists\mathbb{R}$ will be hard using Canny's method [GKP13]. The possibility that $\mathsf{NP} = \exists\mathbb{R}$ cannot yet be ruled out, but is usually considered unlikely. Natural polynomial-sized witnesses of positive instances of $\exists\mathbb{R}$-complete problems remain elusive. $\exists\mathbb{R}$ also contains possibly simpler problems that are not known to lie in NP, we will review some of these next.

Inside $\exists\mathbb{R}$. $\exists\mathbb{R}$ contains several problems which are unlikely to be $\exists\mathbb{R}$-complete but are often relevant to $\exists\mathbb{R}$-complete problems. We discuss three of these problems here: the sum-of-square-roots problem, PosSLP and semidefinite/convex programming.

The *sum-of-square-roots problem* (SSQR) asks whether $\sum_{i\in[n]} a_i^{1/2} \geq k$ for positive integers $a_1, \ldots, a_n$ and k. Clearly, SSQR $\in \exists\mathbb{R}$, but this innocent-looking special case of $\exists\mathbb{R}$ has proved difficult to handle. There is a highly non-trivial upper bound on SSQR, via PosSLP which we discuss next, at the third level of the counting hierarchy (which is expected to be lower than PSPACE, but way above NP), see [ABKPM09]. This is a serious first obstacle in proving NP $= \exists\mathbb{R}$, since we cannot even prove good upper bounds on SSQR, which belongs to $\exists\mathbb{R} \cap \forall\mathbb{R}$. The SSQR-problem is powerful enough to solve some interesting problems in computational geometry: the complexity class $\mathsf{P}^{\mathrm{SSQR}}$ contains problems such as 3-connected planar graph realizability for triangulations [CDR07, Theorem 1], as well as minimum-link path problems [KLPS17].

Also in $\exists\mathbb{R}$ and not known to lie in NP, is the *Euclidean traveling salesperson problem* (ETSP), in which we are given n cities in the plane and ask for a shortest tour through all cities (in the Euclidean metric). It is often asked whether this problem is NP-complete; the obstacle to placing ETSP in NP is the computation of the length of the route, which seems to require the solution of an SSQR-problem. The best upper bound on the Euclidean traveling salesperson is $\mathsf{NP}^{\mathrm{SSQR}}$. There are other problems that fall into this class, most prominently minimum weight triangulation, minimum dilation graphs, and the Steiner Tree problem [PS02, Sect. 10.4]; all of these problems are non-trivially NP-hard, see [Pap77, MR08, GGJ77, GKK+10].

If the n cities the Euclidean traveling salesperson must visit lie on an integer grid of size polynomial in n, then the problem lies in a slightly smaller class, $\mathsf{NP}^{\mathrm{USSR}}$, where USSR is the variant of SSQR in which the a_i are given in unary (and this applies to grid versions of other problems in $\mathsf{NP}^{\mathrm{SSQR}}$ as well). Balaji and Datta [BD23] recently announced that USSR $\in$ P/poly, but this still does not imply that this grid version of the Euclidean salesperson lies in NP. It is quite possible that USSR and SSQR are polynomial-time solvable, and many researchers believe so, but their exact complexity remains open.

Let us turn to PosSLP, short for positive straight line program. A *straight line program (SLP)* is a finite sequence of instructions of the form $a_0 = 1$, and $a_k = a_i \circ a_j$, where $i, j < k$ and $\circ \in \{\cdot, +, -\}$. An SLP computes a sequence of numbers $(a_k)_{k=0,\ldots,n}$. PosSLP asks whether $a_n > 0$, that is, whether the final number computed by the SLP is positive. (For a related problem, see (A-Open1)). The value of a_n may require exponentially many bits, since SLPs can perform repeated squaring, so PosSLP does not even obviously lie in PSPACE, but it does, since PosSLP lies in $\exists\mathbb{R}$.

The best current upper bound on PosSLP is that, like SSQR, it lies in the third level of the counting hierarchy [ABKPM09]. One of the most striking results is that $\mathsf{P}^{\mathrm{PosSLP}}$, polynomial time on the word RAM with access to a PosSLP oracle, is as powerful as polynomial time on the real RAM on integer inputs [ABKPM09, Proposition 1.1]: the real RAM can run the SLP for PosSLP, since it can perform computations on reals, and therefore integers, in constant time; and we can simulate any real RAM algorithm on the word RAM using the PosSLP oracle whenever the program branches. This also implies that SSQR can be solved with oracle access to PosSLP, as the real RAM can solve SSQR.

In other words, PosSLP explains the gap between the word RAM and the real RAM. The third author conjectures that PosSLP also explains the gap between NP and $\exists\mathbb{R}$:

Conjecture

$$\mathsf{NP}^{\mathrm{PosSLP}} = \exists\mathbb{R}.$$

The currently best lower bound on PosSLP is by Bürgisser and Jindal [BJ24]. PosSLP can be solved using semidefinite programming [ABKPM09], which is a special case of convex programming, and both of these problems belong to $\exists\mathbb{R}$. Thus the conjecture implies that convex programming is not polynomial-time solvable unless $\mathsf{NP} = \exists\mathbb{R}$.

Semidefinite programming also belongs to $\forall\mathbb{R}$ [SŠ23], so it lies in $\exists\mathbb{R} \cap \forall\mathbb{R}$, making it unlikely to be $\exists\mathbb{R}$-complete, unless $\forall\mathbb{R} = \exists\mathbb{R}$, which would be quite surprising. This also implies that $\exists\mathbb{R}$-complete problems, such as the rectilinear crossing number, cannot be solved using semidefinite programming, unless $\exists\mathbb{R} = \forall\mathbb{R}$.

Are there any other problems in $\exists\mathbb{R} \cap \forall\mathbb{R}$? Yes, Junginger [Jun23b] argues that polynomial identity testing lies in in $\exists\mathbb{R} \cap \forall\mathbb{R}$; the problem also lies in coRP, the complement of randomized polynomial time, but is not known to belong to P.

Above $\exists\mathbb{R}$. If we look upwards from $\exists\mathbb{R}$, towards PSPACE, we are at the start of a hierarchy, the real polynomial hierarchy. All the finite levels of this hierarchy, such as $\exists\forall\mathbb{R}$ and $\forall\exists\mathbb{R}$, see (L17), still lie in PSPACE which follows from work of Renegar [Ren92a], also see [BPR06, Remark 13.10]. Unbounded quantification of the full theory of the reals takes us outside of PSPACE, to EXPSPACE, exponential space [BOKR86]; in practice, one of the fastest methods is still Collins' Cylindrical Algebraic Decomposition, which runs in double-exponential time [Col75]. In this survey we restrict ourselves to the finite levels of the hierarchy, but in the BSS-model there is structural research on the whole hierarchy, for example, an analogue of Toda's theorem [BZ10]. One can also consider looking at mixing real and discrete quantifiers; in the BSS-model discrete quantifiers were studied under the name "digital nondeterminism", leading to the class $\mathsf{DNP}_{\mathbb{R}}$, the restriction of $\mathsf{NP}_{\mathbb{R}}$ to quantifiers over $\{0, 1\}$, see the survey by Meer and Michaux [MM97] and Table 5.1. The Vapnik-Červonenkis dimension of a family of semialgebraic sets (A-Open5) is a natural problem that can be captured by mixed quantification [SŠ23, Sect. 4].

Table 5.1 Complexity classes corresponding to discrete versus real inputs, and discrete (also called digital) versus real nondeterminism

		Nondeterminism	
		Discrete	Real
Input	Discrete	NP	$\exists\mathbb{R}$
	Real	$\mathsf{DNP}_{\mathbb{R}}$	$\mathsf{NP}_{\mathbb{R}}$

Another recent development is the introduction, by van der Zander et al. [vdZBL23] of the complexity class succ∃ℝ, a succinct version of ∃ℝ, which is allowed to encode arithmetic formulas using circuits. The class succ∃ℝ is the real analogue of NEXP, just like ∃ℝ is the real analogue of NP[BDLvdZ24]; it has turned out to be useful for studying problems in probabilistic reasoning, see [vdZBL23, DvdZBL24, BDLvdZ24, IIM24].

Where is ∃ℝin Practice? Many mathematical software systems such as Mathematica and Maple contain solvers for not just the existential but for the full theory of the reals. These systems, as well as special-purpose solvers such as Z3 [Z3], Redlog [Redb], and CVC5 [CVC] are all based on Collins' Cylindrical Algebraic Decomposition [PJ09, Kre20]; in spite of its double-exponential running time it appears to be more successful for practical instance sizes compared to the theoretically superior methods following Canny and Renegar. We should mention though that the last systematic comparison study, which is due to Hong [Hon91], dates to the nineties, and much has happened since then. More recent references like [PJ09] and [Kre20, Sect. 1] seem to confirm that the practical status quo has not changed though; for a competition (affiliated with the SAT-25 conference), see https://smt-comp.github.io/2025/.

5.2.1.3 Universalities

∃ℝ-completeness theory is closely related to an algebraic geometric notion usually referred to as *universality*. Informally, universality results state that spaces of solutions of simple geometric problems over the reals can behave "arbitrarily bad". The occurrence of irrational numbers, and more generally numbers of arbitrary algebraic degree, or numbers requiring exponential precision, can sometimes be a hint towards universality, and ∃ℝ-hardness. In fact, algebraic universality can be seen as the strongest geometric statement about the intrinsic complexity of a realizability problem involving real numbers.

Irrationality. The famous Perles configuration, described by Micha Perles in the 1960s, is composed of nine points and nine lines, with prescribed incidences, any realization of which in the Euclidean planes requires at least one irrational coordinate, see Fig. 5.3. A much earlier example of such "nonrational" configurations with 11 points was given by MacLane in 1936 [Mac36], using von Staudt's "algebra of throws". Perles also constructed an 8-dimensional combinatorial polytope that could be realized with vertex coordinates in $\mathbb{Q}[\sqrt{5}]$, but not with rational coordinates only [Grü03, Zie08]. Much later, Brehm constructed three-dimensional polyhedral complexes, every realization of which requires irrational coordinates (see Ziegler [Zie08]).

These three examples can be understood as early indications of the much more general universality property for point configurations, polytopes, and polyhedral complexes. In general, ∃ℝ-complete problems often require solutions of arbitrary high algebraic degree. Note, however, that this is not always the case; realizations of simple line arrangements, for instance, can always be slightly perturbed, but deciding the existence of one still remains ∃ℝ-hard. This leads us to the next topic.

Precision. Pursuing the idea that solutions of simple geometric realizability problems may "behave badly", let us note that ETR formulas of size $O(n)$ can encode doubly exponential numbers $x_n = 2^{2^n}$ as follows

$$x_0 = 2, \; x_i = x_{i-1}^2, \; \text{for all } i = 1, 2, \ldots, n.$$

For ∃ℝ-complete geometric realizability problems, this implies that the number of bits, hence the precision, required to encode the coordinates of a realization, if one exists, can be exponentially large.

Early papers related to ∃ℝ-complete problems emphasized this precision issue. For instance Goodman et al. [GPS89, GPS90] showed two things: order types (combinatorial descriptions of point sets giving the orientation of every triple) can always be realized on an integer grid of double exponential size, and there are order types that need what they named "exponential storage". The upper bound follows from the work of Grigoriev and Vorobjov [GV88] (the "Ball Theorem", see Sect. 5.2.2.1). Bienstock [Bie91] and later Kratochvíl and Matoušek [KM94] prove double exponential precision lower bounds for realizations with minimum crossing numbers and segment intersection graphs, respectively. Similarly, Richter-Gebert and Ziegler [RGZ95] explicitly state double-exponential lower bounds on coordinates for realizations of polytopes, and McDiarmid and Müller [MM10, MM13] actually cast this as the main result for unit disk graphs and disk graphs, while ∃ℝ-hardness is only mentioned in passing.

Like irrationality, the "exponential storage" phenomenon is a consequence of the much more general property of algebraic universality.

Algebraic Universality. Algebraic universality is the culmination of those partial results on the "bad behavior" of polytopes and configurations of points and lines, and it is a property of the so-called *solution*, or *realization spaces* of a satisfiability problem over the reals. In the stretchability problem, for instance, we are given the combinatorial type of an arrangement of n pseudolines, consisting of the order, from left to right, of the intersection points of each pseudoline with the other $n-1$ pseudolines. The corresponding realization space then is the set of all possible

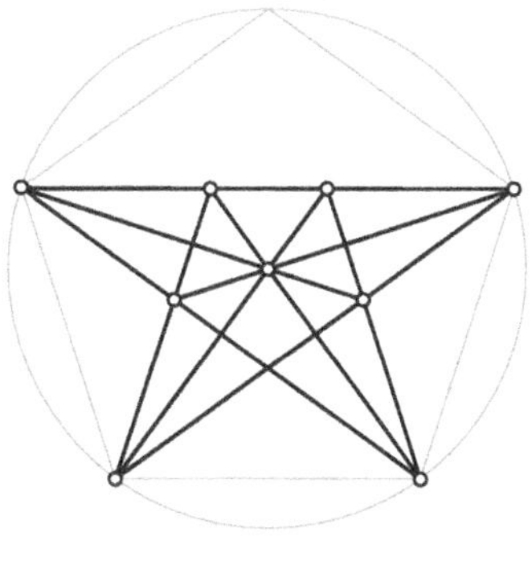

Fig. 5.3 The Perles configuration

coordinates $(a_1, b_1, a_2, b_2, \ldots, a_n, b_n) \in \mathbb{R}^{2n}$ such that the n lines defined by the equations $y = a_i x + b_i$ realize the same arrangement.

A *semialgebraic set* is the set of solutions in $\mathbb{R}^n$ of a Boolean combination of finitely many polynomial equalities and inequalities (both strict and nonstrict). More precisely, given collections $\{f_i\}_{i\in[r]}, \{g_i\}_{i\in[s]}, \{h_i\}_{i\in[t]}$ of polynomials with integer coefficients in $\mathbb{Z}[x_1, x_2, \ldots, x_n]$, the set

$$\begin{aligned}\{x \in \mathbb{R}^n : \ & f_i(x) = 0 \text{ for all } i \in [r], \\ & g_i(x) < 0 \text{ for all } i \in [s], \\ & h_i(x) \leq 0 \text{ for all } i \in [t]\},\end{aligned}$$

is a *basic semialgebraic set.* It is *closed* if it involves only non-strict inequalities (hence if $r = s = 0$), *open* if it involves only strict inequalities (hence if $r = t = 0$), and *primary* if it involves only equalities and strict inequalities (hence if $t = 0$). Semialgebraic sets are defined as finite unions, or equivalently finite Boolean combinations, of basic semialgebraic sets. One can prove that projections, interiors and closures of semialgebraic sets remain semialgebraic (using quantifier elimination [Tar48]). Note that in algebraic geometry, the coefficients of the polynomials are usually allowed to be reals, but for the purposes of this survey coefficients are always assumed to be integers.

The concept of *stable equivalence* between semialgebraic sets combines rational changes of coordinates with so-called *stable* projections that preserve the homotopy type and the algebraic number type. Different definitions of stable equivalence have been used in various contexts. We refer to Richter-Gebert [RG96, Sect. 2.5] for a strong version of stable equivalence, and to Boege [Boe22b] and Verkama [Ver23, Sect. 4.1] for a recent discussion. With this terminology we can now state Mnëv's universality theorem [Mnë88, RG99].

Theorem 5.2.1 (Mnëv) *Given any primary basic semialgebraic set S, there exists a pseudoline arrangement whose realization space is* stably equivalent *to S.*

Maybe the simplest application of this theorem is to a semialgebraic set S consisting of two points. Mnëv's Theorem implies that there exists two arrangements of n lines in the plane which have the same combinatorial type (orders of intersections) but which cannot be continuously transformed into each other without changing the combinatorial type in between. So Mnëv gives a strong negative answer to Ringel's isotopy conjecture [Rin55, RG99].

This phenomenon of "disconnected realization spaces" had been observed before for polytopes [BGdO90]. Using *Gale diagrams*, Mnëv also gave a universality result for polytopes: Any semialgebraic set is stably equivalent to the realization space of a d-dimensional polytope with $d + 4$ vertices. This gave rise to a series of results refining and generalizing Mnëv's proof [RGZ95, RG96, RG99, PT14, APT15, AP17]. Richter-Gebert, in particular, dedicated a complete monograph on the universality of realization spaces of 4-dimensional polytopes [RG96], implying that the problem of deciding whether a given lattice is realizable as the face lattice of a 4-dimensional polytope is ∃ℝ-complete (CG69). The case $d = 4$ is best possible: Steinitz Theorem gives a simple characterization of face lattices of three-dimensional polytopes that is decidable in polynomial time.

Besides the implications in computational complexity, it is worth noting that Mnëv's result spawned a large body of work on the so-called "Murphy's law in algebraic geometry", initiated in particular by Vakil [Vak06], and stating that unless there is some a priori reason otherwise, the deformation space of a given object can be "as bad as possible".

While algebraic universality and ∃ℝ-hardness are closely related, they do not always imply each other: stable equivalences do not have to be effective (e.g. Datta [Dat03]), and ∃ℝ-hardness reductions rarely maintain stable equivalence. In many cases, stable equivalences can be made effective though; on the other hand, ∃ℝ-hardness results typically do not automatically yield algebraic universality results, but the proofs often allow the conclusion of weaker universality properties, e.g. Bienstock's proof of ∃ℝ-hardness result for the rectilinear crossing number shows that every semialgebraic set occurs as the projection (*not* stable) of the realization space of a rectilinear crossing number problem. Our compendium indicates if universality results are known.

5.2.1.4 Proving ∃ℝ-Completeness

How can we tell whether a problem is a candidate for ∃ℝ-hardness? There are no hard-and-fast rules, but there are some tell-tale signs. First of all, there should be solutions to the problem that use real parameters. NP-hardness, without NP-membership can be a next clue: why is the problem not known to lie in NP, are real numbers the issue? To strengthen a suspicion of ∃ℝ-hardness, one can look for evidence of algebraic universality: can one construct instances that require irrational solutions or rational solutions of exponential bitlength? None of these clues are sufficient, of course, for example, the Euclidean traveling salesperson problem is NP-hard and is not known to lie in NP, but it is unlikely to be ∃ℝ-hard, since it belongs to $\mathsf{NP}^{\mathsf{SSQR}}$; and representing a plane graph as a disk contact graph may require irrational (albeit algebraic) radii, but testing planarity can be decided in linear time. Topological universality can also be misleading, there are problems in NP that exhibit topological universality [SW23, BEMM+22, ST22].

∃ℝ-Membership

Since ETR is a very expressive language, it is typically not difficult, though sometimes cumbersome, to establish membership in ∃ℝ, and we can even use the result of ten Cate, Kolaitis, and Othman [tCKO13, Theorem 4.11] that ETR is closed under NP-many-one reductions, for example, see (MDP4) and (ML3).

Compare this situation to NP. We can show membership by encoding the problem as a satisfiability problem, yet it is often easier to show that the problem can be accepted by a Turing machine in non-deterministic polynomial-time. Or we can even work in the word RAM model, which allows random access to an unlimited number of registers storing integers on which we can perform arithmetical operations.

Erickson et al. [EvdHM20] showed that we have the same option for $\exists\mathbb{R}$: if we replace the word RAM by a real RAM, which allows us to guess real, not just discrete, numbers and has unlimited storage for real numbers, we obtain $\exists\mathbb{R}$. The real RAM model can simplify $\exists\mathbb{R}$-membership arguments significantly, as describing an algorithm is much easier than writing down an ETR formula, and researchers can make use of the huge number of real RAM algorithms already developed in computational geometry and other areas.

Efficient membership proofs may also lead to algorithmic solutions of problems, e.g. Dean [Dea02] showed how to express the rectilinear crossing number problem as an optimization problem with linear constraints. While solvers for these problems may not yet be as fast as solvers for NP, where a linear programming formulation for the (topological) crossing number problem has been implemented and can compute crossing numbers of reasonably large graphs on the web [CW16], this may very well change in the future.

$\exists\mathbb{R}$-Hardness

In this section, we discuss two primary techniques to show $\exists\mathbb{R}$-hardness, one based on algebra, and the other on geometry. The first technique, which we refer to as algebraic encoding, involves encoding real variables, addition and multiplication directly. The second technique, known as stretchability encoding, is based on encoding the stretchability problem. These two techniques, with variations, cover most of the reductions we have encountered in compiling this compendium.

Algebraic Encoding. Algebraic encoding work by encoding algebraic constraints on variables, such as $x = y + z$ and $x = yz$; so in a first step, we need to encode real variables. How to do so is problem-specific but often straightforward. For instance, in the art gallery problem we can encode variables by using the coordinates of a guard's position, see Fig. 5.4 for an illustration.

Similarly, to show stretchability $\exists\mathbb{R}$-hard, we can work with the parameters of the lines, or the coordinates of the intersections.

It is often useful to know that the variables we need to encode can be assumed to be bounded, e.g. ETR- INV (L5) allows us to assume that all variables belong to the interval $[1/2, 2]$. Bounded variables can be encoded in the coordinates of a geometric object with limited range, such as a guard in an art gallery. One can even reduce the range to much smaller intervals, such as $[0, n^{-200}]$, where n is the size of the input.

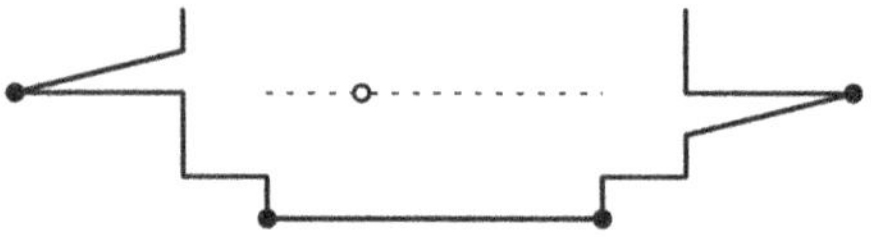

Fig. 5.4 The white guard needs to be placed anywhere on the dotted line segment in order to see all the marked vertices. We can interpret the position of the guard as a value in a real-valued interval

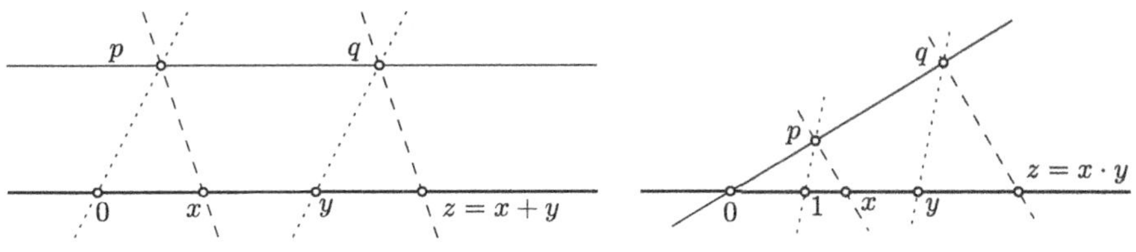

Fig. 5.5 Simplified von Staudt addition gadget, *left*, and multiplication gadget, *right*. The original gadgets are projective, not Euclidean

This may seem counter-intuitive as it leaves very little wiggle room for the variables, but because we can encode variables with very high precision, it is sufficient.

In the second step, we encode addition, of the form $x = y + z$. This is typically simple as it is a linear relation between three variables. For geometric problems, addition can often be encoded by forcing two geometric objects to lie next to each other, e.g. two pieces in the packing problem, with the combined object encoding the sum.

Remark 5.2.2 (*Addition with von Staudt*) Many proofs work with variants of the von Staudt gadgets originally used by Mnëv [Mnë88] to show that order type (CG1), and thereby stretchability (CG9), is universal. Figure 5.5 shows a simplified von Staudt addition gadget on the left. Variables are encoded as length of segments (starting at 0). Assuming the drawing of the gadget satisfies that $0, x, y$ and z are collinear and $pq \parallel 0x$, $0p \parallel yq$ and $px \parallel qz$, where $\parallel$ denotes two segments being parallel, then $z = x + y$.

In some rare cases, though, the encoding of addition is trickier, as, for example, in the case of the boundary-guarding variant of the art gallery problem; addition here is the main obstacle as there is no direct interaction of three guards at all, so it is not obvious how to encode a ternary relation.

In the third step, we encode multiplication, i.e., $x = y \cdot z$.

Remark 5.2.3 (*Multiplication with von Staudt*) Figure 5.5 shows how multiplication can be encoded using von Staudt gadgets (again simplified, removing the projective setting): as long as $\{0, p, q\}$ and $\{0, 1, x, y, z\}$ are each collinear and distinct, and $p1 \parallel qy$ and $px \parallel qz$, we have $z = x \cdot y$. It follows that geometric problems that can express parallelism and collinearity are $\exists\mathbb{R}$-hard. By interpreting the von Staudt gadgets in the projective plane (as von Staudt and Mnëv did), both collinearity and parallelism can be expressed using order types.

Encoding multiplication directly can be tricky, but it can be simplified by simulating multiplication through inversion, as is done in ETR- INV (L5). To see how to do this, note that

$$(y + z)^2 - (y - z)^2 = 4x$$

is equivalent to

$$x = y \cdot z.$$

This equivalence indicates that multiplication can be simulated by squaring, i.e., $x = y^2$. Consider

$$\frac{1}{\frac{1}{x-1} - \frac{1}{x}} + x = x^2.$$

We can conclude that inversion, i.e. $x \cdot y = 1$, is sufficient to encode squaring. Inversion is often much easier to encode compared to multiplication as it involves only two variables in a symmetric way. The inversion constraint can be replaced by almost any non-linear constraint [MS24] of the form

$$f(x, y) = 0,$$

see (L8). Although this is very powerful, we are only aware of one application, namely showing that geometric packing (CG18), packing convex polygons into a square container, is ∃ℝ-complete [AMS20]. (The insight that we just need one concave and one convex constraint between x and y could also simplify the ∃ℝ-hardness proofs of the art gallery problem and the minimum convex cover problem, but the papers were written before the result.) The idea of the proof is that any non-linear constraint that is sufficiently smooth can be approximated well by a quadratic function, using the Taylor expansion, in a sufficiently small neighborhood.

Finally, we need to connect the variables to the different gadgets. This step, which looks like simple housekeeping, can be surprisingly challenging in many reductions. As Richter-Gebert writes in [RG96]:

> The main difficulty in Mnëv's proof is to organize the construction in a way such that different basic calculations do not interfere and such that the underlying oriented matroid stays invariant for all instances of a geometric computation.

To see what the issue is, recall the von Staudt gadget for addition (or multiplication) in Fig. 5.5. If we want to encode the gadget using order types, we need to know whether $x < y$, $x = y$ or $x > y$. How would we know? And even if we did, how would we know for other, intermediate variables? This problem has been solved using various normal forms; maybe the most elegant solution is due to Richter-Gebert [RG96], this is the approach presented by Matoušek [Mat14].

It can also be helpful to know that the variable-clause incident graphs are of a restricted type, for example planar, as in PLANAR- ETR- INV (L6).

Stretchability Encoding. The stretchability problem (CG9) asks whether a given pseudoline arrangement can be *stretched*, that is, if there is an *equivalent* line arrangement, a line arrangement with the same combinatorial structure (Fig. 5.6).

Reductions from stretchability are particularly popular in the graph drawing and computational geometry community. The segment intersection graph problem (CG26) is an early example: we can represent each pseudoline by a segment and surround each pseudoline-segment with additional segments that enforce the combinatorial structure of the pseudoline arrangement, see Fig. 5.7.

If there is a representation of the graph using segments then the underlying pseudoline arrangement is stretchable, see [KM94, Sch10]. Another, earlier, example due to Bienstock [Bie91] shows how to embed stretchable arrangements in the rectilinear crossing number problem.

5.2.1.5 What is Next for ∃ℝ?

In this section, we want to point at possible future research directions that appear of interest to us. We focus on broad research lines. For specific candidate problems see Sect. 5.4.9.

Breadth. Most of the early ∃ℝ-complete problems arose in graph drawing and algebraic complexity. As the compendium shows, many more areas like probabilistic reasoning, machine learning, polytope theory etc. have been added since then, and it seems reasonable to believe that this trend continues and that more ∃ℝ-complete problems are found in other research areas.

Depth. After a problem has been shown ∃ℝ-complete, we can start asking whether more restricted versions of the problem remain ∃ℝ-complete. This is particularly important for problems that are common starting points for reductions; for NP, this is evidenced by the numerous satisfiability variants that have been shown to be NP-complete. For ∃ℝ, the original ETR (L1) has been refined in many ways, including the many variants of ETR- INV (L5), which have significantly simplified reductions for many problems.

Similarly, the partial version of order type realizability (CG1) was introduced in [Sch21b] to show that many problems remain hard if some parameters are constant. As an example, consider the simultaneous geometric embedding (CG56) problem, which is ∃ℝ-complete [CK15]. We now know that this problem remains ∃ℝ-complete, even if the number of input graphs is bounded (the current best upper

Fig. 5.6 A (stretchable) pseudoline arrangement on the *left*, and an equivalent line arrangement on the *right*

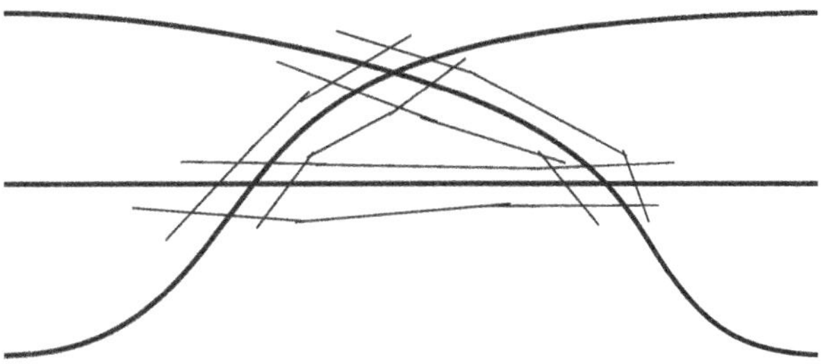

Fig. 5.7 Forcing the combinatorial structure of a pseudoline arrangement using segment intersections

bound is 31 [FKM+24, Sch21b]), or all the input graphs are edge-disjoint [KR23]. There are many examples like this already, and we expect that many more problems with constant natural parameters will be shown $\exists\mathbb{R}$-complete.

Parameterized Theory of $\exists\mathbb{R}$. Many NP-complete problems become considerably easier if some of their parameters are bounded, the point of view taken by the area of parameterized complexity.

For example, we can decide whether a graph on n vertices contains a vertex cover of size at most k in time $O(2^k n^2)$. For fixed k, this gives a quadratic-time algorithm for the vertex cover problem. Even if k is not fixed, but small compared to n, this running time is a tremendous speed-up to the general worst-case complexity of $O(c^n)$. We are not aware of a single $\exists\mathbb{R}$-complete problem with a running time of $f(k)n^{O(1)}$ for some natural parameter k. It would be interesting to see if such a problem exists and what type of parameterization is required. On the other hand, we are aware of some $\exists\mathbb{R}$-complete problems for which the parameterized version is $W[1]$-hard [BM20]. It would be interesting to be able to find the correct parameterized complexity class for those problems.

Structural Complexity. The study of structural properties of complexity classes has matured since NP was first introduced, see for instance the book by Arora and Barak [AB09]. Structural questions have been studied widely in the BSS-model of computation, starting with the book by Blum et al. [BCSS98]; for a more recent survey, see Baartse and Meer [BM13]. Yet, for many classical results from computational complexity, we do not have analogue results for $\exists\mathbb{R}$ or the real hierarchy. We include some examples below; some of these questions have been answered since the original version was posted and we include answers we are aware of:

- Fagin's theorem characterizes NP as the set of properties captured by monadic second-order logic. Can we characterize $\exists\mathbb{R}$ in a similar fashion? An affirmative answer was found recently by Meer and Wurm [MW25, Theorem 3]. A similar theorem for $\mathsf{NP}_{\mathbb{R}}$ has been known, see [MM97, Sect. 7].
- Toda's theorem, that the polynomial-time hierarchy reduces to PP, probabilistic polynomial time, was a breakthrough result in classical complexity theory [AB09, Theorem 17.4]. Basu and Zell [BZ10] showed that the result can be adopted to the BSS-model. Is there a discrete analogue for the real hierarchy?
- The PCP theorem shows that problems in NP have (p)robabilistically (c)hackable (p)roofs using at most $O(\log n)$ random bits and checking at most $O(1)$ bits of the proof. Is there a discrete version of this result? Baartse and Meer show that there is a PCP theorem for $\mathsf{NP}_{\mathbb{R}}$ [BM15].
- $\exists\mathbb{R}$ can be relativized, e.g. using its machine model definition, so we can ask whether there are any oracle results for $\exists\mathbb{R}$, either separations or collapses. Here are a couple of specific questions:
  - Is there an oracle relative to which $\mathsf{NP} = \exists\mathbb{R} \subsetneq \mathsf{PSPACE}$?
  - Is there an oracle relative to which $\mathsf{NP} \subsetneq \exists\mathbb{R}$?
  - Does Canny's result relativize? Or is there an oracle for which $\exists\mathbb{R} \not\subseteq \mathsf{PSPACE}$?

There already are some answers, but they depend on how $\exists\mathbb{R}$ is relativized, that is, the type of oracle access granted to$\exists\mathbb{R}$. If we allow $\exists\mathbb{R}$ to make real queries (real numbers being written in a single cell of the oracle tape), then the second question was answered affirmatively by Meer and Wurm [MW25] who showed that the separation can be obtained by using $\mathbb{Z}$ as the oracle: the reason is that $\exists\mathbb{R}$ relativized to $\mathbb{Z}$ contains undecidable problems, whereas NP relativized to $\mathbb{Z}$ does not; similarly PSPACE relativized to $\mathbb{Z}$ contains only decidable problems, which implies that the answer to the third question is negative: $\exists\mathbb{R} \subseteq$ PSPACE does not relativize. The results suggest that allowing $\exists\mathbb{R}$ to make real queries gives it too much power compared to a traditional complexity classes. What then happens if we restrict the oracle access for $\exists\mathbb{R}$ to the same model that we use for NP and PSPACE, that is, $\exists\mathbb{R}$ can only write binary strings to the oracle tape? This is a difficult question, but Hamm et al. [HMMP25] where able to show that in this binary oracle model $\exists\mathbb{R} \subseteq$ PSPACE relativizes and that there is a binary oracle relative to which $\exists\mathbb{R} \subsetneq$ PSPACE; they can even show that the whole real hierarchy is proper and lies in PSPACE. The second question, whether $\exists\mathbb{R}$ can be separated from NP remains open in the binary model (and appears hard).
For oracle results in the BSS-model, see [Eme94, Gaß09].
- Instead of working with oracles, one could also add predicates to the language of $\exists\mathbb{R}$; one could then ask whether there are any natural predicates that can be added so that the extended version of $\exists\mathbb{R}$ is PSPACE-complete (this question was suggested by a referee).
- Stockmeyer and Meyer [SM73] showed that deciding the truth of quantified Boolean formulas is PSPACE-complete. Is there an analogue for the real hierarchy? That is, is the full theory of the reals EXPSPACE-complete? An affirmative answer was conjectured already by Ben-Or, Kozen, and Reif, who showed membership [BOKR86].
- What are the connections between $\exists\mathbb{R}$ and other complexity classes defined to capture working with continuous solution spaces? Specifically, how does $\exists\mathbb{R}$ relate to CLS [DP11] and FIXP [EY10]? This can be studied by looking at finding a Nash Equilibrium problem that is either $\exists\mathbb{R}$-complete or FIXP-complete, depending on the technical details of the definition.

We expect positive answers to many of these questions, and some of them may not be that difficult. As examples of some structural questions for which we know the answer we present the following.

- If NP $=$ coNP, then the polynomial-time hierarchy collapses to the first level. Minkin [Min24] was able to show that the same is true for the real hierarchy: if $\forall\mathbb{R} = \exists\mathbb{R}$, then the real hierarchy collapses to the first level (and a similar statement is true at higher levels). To make these statements meaningful, we need to work with a model for $\exists\mathbb{R}$ that allows for oracles, as introduced in [Min24].
- Ladner's theorem states that P $\neq$ NP implies that there are problems in NP which are not NP-complete [Lad75]. Meer and Wurm [MW25] managed to show that NP $\neq \exists\mathbb{R}$ implies that there are problems in $\exists\mathbb{R}$ which are not $\exists\mathbb{R}$-complete. Is it sufficient to assume P $\neq \exists\mathbb{R}$?

Topics that have not been explored in the context of ∃ℝ include promise problems, see (L-Open1) for a sample candidate, or (integer-valued) function classes, e.g. a real analogue to FNP. Real-valued functions are probably better investigated within the BSS-framework, see for example [Høg23], which introduces ∃ℝ-search functions.

Counting Complexity. A counting version of ∃ℝ was introduced by Bürgisser and Cucker [BC06] as $\mathsf{BP}(\#\mathsf{P}^0_{\mathbb{R}})$, which we could call #ℝ; they show that counting the number of points in a semialgebraic set is complete for this class, and computing the Euler-Yao characteristic of a semialgebraic set is complete for this class under Turing reductions. The paper leaves many interesting open questions, not least of it being the question of where #ℝ is located with respect to traditional complexity classes.

Better Techniques to Show Hardness. Reductions for ∃ℝ-hardness tend to start with a relatively small set of problems. Maybe the three most prominent starting problems are simple stretchability (CG9), feasibility (A1) and ETR- INV (L5).

Simple stretchability is useful as a geometric problem that has turned out to be very adaptable to different contexts; one advantage it has (particularly over stretchability) is that its solution space is open, which is the case for many geometric problems. ETR- INV is quite versatile and can be adapted to many situations: the range of variables can be bounded, multiplication can be replaced by inversion. Feasibility has served as the starting point for nearly all problems from algebraic geometry and analysis; that may not be that surprising, since it is the oldest and best-studied problem.

These results can be refined even further. Instead of Stretchability, we can now use pseudo-segment stretchability (CG10). This makes the reduction more tedious, but it allows to bound certain parameters by a constant [Sch21a]. Similarly, the inversion in ETR- INV can be replaced by virtually any curved function [MS24]. In analogy with NP, we expect that the core problems will be further refined to become more flexible, easier to apply, and yield stronger results.

Real Hierarchy. There is a small number of papers that study the real polynomial-time hierarchy starting with a paper by Bürgisser and Cucker [BC09] in the BSS-tradition. Some fundamental problems like the Hausdorff distance are complete for the second level of this hierarchy [JKM22]. Studying the second level of the hierarchy is more challenging as many standard techniques that can be used for the first level are not yet known to work for the second level.

Developing techniques for higher levels of the hierarchy and identifying additional natural problems captured by these levels appear of interest to us. There are also structural questions, for example related to the power of exotic quantifiers that have not been answered yet. It is known that the real hierarchy is proper relative to an oracle [HMMP25].

Unifying Algorithmic Techniques. Although ∃ℝ is a relatively new complexity class, the problems that are complete for it have a very long and rich history. Yet, the study of different algorithms for ∃ℝ-complete problems is almost independent.

Consider the $\exists\mathbb{R}$-complete problem of training neural networks for example (ML1). Neural networks with billions of parameters are routinely trained in industry applications; the insights and techniques developed in this application of gradient descent to neural networks seem to have little effect on the way that the graph drawing community is drawing graphs. See [BBP20, ALD+20] for some first work to transfer knowledge from machine learning to graph drawing. For NP-complete problems algorithmic techniques that work well for one problem also work well for another problem. Therefore, those problems are often studied together. Yet the graph drawing, packing, and machine learning communities are almost disjoint.

It may be the case that the problem-specific differences are more important for $\exists\mathbb{R}$-complete problems and thus a more unified approach might just not be very fruitful. Even then it would be nice to capture this in some way.

5.2.2 A Toolbox for $\exists\mathbb{R}$

In this section we review some tools that have shown themselves useful in various aspects of $\exists\mathbb{R}$-completeness, including the ball and gap theorems, which allow us to juggle with equalities and inequalities, quantifier elimination, and exotic quantifiers, which extend the language of the theory.

5.2.2.1 The "Ball Theorem"

The original "ball theorem" is due to Vorobjov and Grigoriev [Vor84, GV88]. The following statement is weaker than the currently best known results, see [BR10, Theorem 3], but sufficient for all of our applications to the existential theory of the reals. A version stated for semialgebraic sets can be found in [SŠ17, Corollary 3.4], and Renegar [Ren92b, Proposition 1.3] stated a version that is useful at higher levels of the hierarchy.

Theorem 5.2.4 (Vorobjov and Grigoriev) *If $f_i \in \mathbb{Z}[x_1, \ldots, x_n]$, $i \in [k]$, is a family of polynomials with coefficients of bitlength at most L and total degree at most d, then every connected component of $\{x \in \mathbb{R}^n : f_i(x) \geq 0$ for all $i \in [k]\}$ intersects the ball of radius $2^{(L+\log k)d^{cn}}$, for some constant $c > 0$ independent of the f_i.*

We want to point out that this theorem relies on the coefficients being integers. If the coefficients are arbitrary real numbers then no upper bound on the radius is known. To see the difficulty in translating such an upper bound to real coefficients consider the linear equation $ax - bx = 1$. The solution is $x = \frac{1}{a-b}$, which can be arbitrarily large.

The innocent-looking Theorem 5.2.4 has some far-reaching consequences. To start with, it can be used to write a decision algorithm for $\exists\mathbb{R}$ that runs in exponential time [GV88].

The ball theorem also implies that solutions to computational problems in $\exists\mathbb{R}$ can be bounded in size. As an example, consider the order type realizability problem (CG1). One way to measure the size of a solution of a geometric problem like order type realizability is by working with *area*; to make this notion non-trivial—a realizable order type can be realized in any arbitrary small region—one typically adds some resolution requirement, say that the points must have at least unit distance. We can then define the *area* of a realization to be the area of a smallest bounding box; the area of a realizable instance of order type then is the smallest area of a realization in which every pair of points has distance at least one. The following result, stated for spread rather than area, can be found in Goodman, Pollack, and Sturmfels [GPS90, Theorem 2].

Theorem 5.2.5 (Goodman, Pollack, and Sturmfels) *A realizable order type on n points can be realized in area $2^{2^{O(n)}}$, and there are order types that require area $2^{2^{\Omega(n)}}$.*

We can sketch a proof of the upper bound: the realizability of an order type on points $(p_i)_{i\in[n]}$ can be expressed as the feasibility of a system of quadratic polynomials $f_i(p) \geq 0$, $i \in [k]$, where $k = O(n)$. We add $O(n^2)$ polynomial conditions of the form $\|p_i - p_j\|^2 - 1 \geq 0$ to enforce the minimum resolution. By Theorem 5.2.4, there is a solution in which all variables have distance at most $R = 2^{2^{O(n)}}$ from the origin; in particular, this applies to the coordinates of the points $(p_i)_{i\in[n]}$, implying the area upper bound.

The main obstacle to upper bounds like this is, occasionally, the encoding of the conditions in ETR. A recent example is right-angle drawability [Bie20, Sch23b]. This issue becomes even more intricate if one aims at optimal bounds. This can require significant additional work, see, for example, the papers by McDiarmid and Müller [MM13] and Kang and Müller [KM12].

Goodman et al. [GPS90] also proved the corresponding lower bound, using a construction that lends itself to adaptation to other settings. Double-exponential lower bounds were reproduced in several early papers for various problems including rectilinear crossing number (CG50) in [Bie91], segment intersection graphs (CG26) in [KM94], and realizability of 4-polytopes (CG69) in [RGZ95]. For most $\exists\mathbb{R}$-hardness reductions, double-exponential precision carries over, so these lower-bound results are nearly automatic, and are rarely stated explicitly these days, unless extra work is required.

Since the order type problem is algebraically universal, realizations may require irrational points [Grü72, Theorem 2.25], so we cannot expect its realizations to be possible on a grid; this applies, for example, to the Perles configuration we saw in Sect. 5.2.1.3. But if we restrict ourselves to point sets in general position, so no three points are collinear, then there always is a solution on a grid.

Theorem 5.2.6 (Goodman, Pollack, and Sturmfels) *If an order type is realizable by n points in general position, then there is a realization on a grid of size $2^{2^{O(n)}} \times 2^{2^{O(n)}}$, and there are order types that require grids of size $2^{2^{\Omega(n)}} \times 2^{2^{\Omega(n)}}$.*

The proof extends the proof of Theorem 5.2.5 with a careful analysis on how far points can be perturbed without changing the order type. Kachiyan and Porkolab [KP97, Corollary 2.6] observed that work by Basu, Pollack and Roy implies the general result that every open (actually, every full-dimensional) semialgebraic set always contains a rational point of at most exponential complexity. This automatically applies to any problem in $\exists\mathbb{R}$ whose solution space is open, such as the simple stretchability (CG9) and the rectilinear crossing number (CG50). In the case of right-angle drawings with one bend, a similar upper bound can be obtained even though the realization space is not open [Sch23b, Theorem 15].

Tarski's decision algorithm for the existential theory of the reals works via quantifier elimination [Tar48], see Sect. 5.2.2.3 for some details on quantifier elimination. Since the quantifier-elimination step is the same for every real-closed field, Tarski's algorithm implies that a formula can be satisfied over $\mathbb{R}$ if and only if it can be satisfied over the algebraic numbers (which are real-closed). It follows that any non-empty semialgebraic set contains an algebraic point. Kachiyan and Porkolab [KP97, Proposition 2.2] strengthened this result by showing that (non-empty) semialgebraic sets always contain algebraic points of at most exponential complexity; that is, there is an exponential bound on the size of an encoding of such a point (algebraic geometers work with Thom encodings, see, for example [BPR06, Definition 2.29]). This conclusion applies to most of the remaining problems, such as stretchability (CG9) and the art gallery problem (CG16).

Theorem 5.2.4 can also be used to show that feasibility (A1) remains $\exists\mathbb{R}$-complete if restricted to a compact domain, say $[-1, 1]^n$. As far as we know, this was first shown by Schaefer [Sch13, Lemma 3.9], and there is no result like this (for compact domains) for $\mathsf{NP}_{\mathbb{R}}$.

Theorem 5.2.7 (Schaefer) *Testing whether a polynomial $f \in \mathbb{Z}[x_1, \ldots, x_n]$ has a root in $[-1, 1]^n$ is $\exists\mathbb{R}$-complete.*

Feasibility over a compact domain plays a role in the $\exists\mathbb{R}$-hardness proofs of several problems: Brouwer Fixed Point (A5), H_2N (A6) and consensus halving (GT21).

For open domains, like $(-1, 1)^n$, the $\exists\mathbb{R}$-hardness proof of feasibility is much simpler, since $z \mapsto \frac{z}{1-z^2}$ is a surjective map from $(-1, 1)$ to $\mathbb{R}$, but there cannot be a continuous surjective map from $[-1, 1]$ to $\mathbb{R}$, since the image of such a map would be compact. An extension of this result to the second level of the hierarchy was obtained by D'Costa, Lefaucheux, Neumann, Ouaknine, and Worrellin [DLN+21, Lemma 10] and Jungeblut et al. [JKM22].

We will look at further consequences of Theorem 5.2.4 in Sect. 5.2.2.2, where we will see how it can be combined with gap theorems to replace equalities with strict inequalities. It can also be used to remove the exotic H-quantifier, see Lemma 5.2.12.

5.2.2.2 Gap Theorems

Gap theorems establish lower bounds on the distance between two algebraic objects that have positive distance. One of the oldest is Cauchy's theorem, which establishes upper and lower bounds on roots of a univariate polynomial [Yap00]. These bounds can be generalized, under certain assumptions, to multivariate polynomials, as was done for example by Canny [Can88b, Can88a]. One can derive the multivariate separation bounds via quantifier elimination from Cauchy's theorem, but there are also direct proofs. The following theorem follows from a strong separation result due to Jeronimo and Perruci [JP10]. A simple derivation can be found in [SŠ17, Corollary 3.6].

Theorem 5.2.8 (Jeronimo, Perruci) *If* $f \in \mathbb{Z}[x_1, \ldots, x_n]$ *is a polynomial with coefficients of bitlength at most L and total degree at most d, for which* $\inf_{x\in\mathbb{R}^n} f(x) > 0$, *then*

$$\inf_{x\in\mathbb{R}^n} f(x) > 2^{-(Ln^d+1)d^{n+1}} d^{-(n+1)d^{n+1}}.$$

With this theorem, one can establish useful separation bounds for semialgebraic sets, e.g. the following bound from [SŠ17, Corollary 3.4]. For a formula φ we write $|\varphi|$ for its bitlength (in a reasonable encoding).

Lemma 5.2.9 (Schaefer, Štefankovič) *Suppose the distance between two semialgebraic sets* $S = \{x \in \mathbb{R}^n : \varphi(x)\}$ *and* $T = \{x \in \mathbb{R}^n : \psi(x)\}$ *is positive. Then that distance is at least* $2^{-2^{L+5}}$ *for* $L := \max(|\varphi| + |\psi|, 5n)$.

Combining this separation with Theorem 5.2.4 one can show that testing whether $f(x) > 0$ for all $x \in \mathbb{R}^n$ for a given polynomial f is $\forall\mathbb{R}$-complete (A3): Given f, if $f(x) = 0$ for some $x \in \mathbb{R}^n$, then by Theorem 5.2.4, there is an $x \in [-R, R]^n$, where $R = 2^{2^{O(L)}}$ and L is the bitlength of f. Consider the sets $S = \{(x, y) \in \mathbb{R}^{n+1} : f(x) = y, x \in [-R, R]^n\}$ and $T = \{(x, 0) : x \in [-R, R]^n\}$. If $f(x) = 0$ for some x, then the distance between S and T is 0, otherwise, the distance is positive (since both sets are compact), and so by Lemma 5.2.9 that distance is at least $2^{-2^{O(L)}}$. Here we use repeated squaring to express R in the definition of S and T, and we can also use it to express a lower bound $\delta > 0$ on the distance between S and T (if $f(x) > 0$) of bitlength at most $O(L)$. It follows that $f(x) = 0$ for some $x \in \mathbb{R}^n$ if and only if $|f(x)| < \delta$ for some $x \in \mathbb{R}^n$. It follows that strict global positivity (A3) is $\forall\mathbb{R}$-complete.

Since this idea allows us to replace equalities with strict inequalities, it also follows that ETR remains $\exists\mathbb{R}$-complete even if equality and negation are not allowed. This result, proved in [SŠ17], is an indication that the definition of $\exists\mathbb{R}$ is quite robust. It implies, as we mentioned earlier, that problems definable without equality, like simple stretchability and segment intersection graphs have the same computational complexity as problems like feasibility and stretchability, which use equality.

A similar approach was used by Bürgisser and Cucker [BC09] to eliminate the exotic H-quantifier, see Lemma 5.2.12 below.

We saw that $f(x) = 0$ can be equivalently replaced by $f(x) < \varepsilon$, where ε is a doubly exponential small constant (depending on f). Another way of reading this is that finding an approximate solution to the feasibility problem, with a double exponential error, is as hard as the feasibility problem itself. This type of statement could be made for nearly every $\exists\mathbb{R}$-complete problem, though it is rarely made explicitly. Examples include drawings that are nearly right-angle drawings (CG58), and approximating the Hausdorff distance (A44) to within a double-exponential error.

5.2.2.3 Quantifier Elimination

Suppose we are given a two-dimensional region $S = \{x \in \mathbb{R}^2 : \varphi(x)\}$ and we want to determine whether S has diameter at most 1. How hard is that? Using the first-order theory of the reals, we can express the diameter condition as

$$(\forall x, y \in \mathbb{R}^2)\ x \in S \wedge y \in S \rightarrow \|x - y\| \leq 1.$$

Let us be a bit more honest and replace $\in S$ with φ, and eliminate the hidden square roots in $\|\cdot\|$. That gives us

$$(\forall x, y \in \mathbb{R}^2)\ \varphi(x) \wedge \varphi(y) \rightarrow \|x - y\|^2 \leq 1. \tag{5.1}$$

We can decide the truth of that sentence by using quantifier elimination. There are very sophisticated quantitative versions of quantifier elimination depending on the number of variables, degree of polynomials, and other parameters. For our purposes, a very simple version, which only eliminates a single quantifier is sufficient. This version follows, for example, from [BPR06, Algorithm 14.5], and that reference presents many more variants. As earlier, we write $|\Phi|$ for the bitlength of formula Φ.

Lemma 5.2.10 *Let $\Phi(x, y)$ be a quantifier-free formula, with $x \in \mathbb{R}$ and $y \in \mathbb{R}^n$. In time $|\Phi|^{O(n)}$ one can construct a formula $\Psi(y)$, of size $|\Phi|^{O(n)}$, such that*

$$(\exists x \in \mathbb{R})\ \Phi(x, y) \leftrightarrow \Psi(y).$$

The lemma allows us to remove one quantifier at a time, and it can be used for both existential and universal quantifiers (using negation). Let us return to the diameter problem. Applying the lemma four times (twice for each point) to (5.1) yields, in polynomial time since n is bounded, a variable-free formula whose truth can then be evaluated in polynomial time. So having diameter at most 1 can be decided in polynomial time for semialgebraic sets in $\mathbb{R}^2$, since there is only a constant number of quantifiers that need to be eliminated. An analogous argument works for every fixed dimension d. If d is part of the input, so not fixed, then the problem becomes $\forall\mathbb{R}$-complete (A36). Eliminating quantifiers no longer works in polynomial time in this case, since we have $\Theta(d)$ many variables. Quantifier elimination still gives us an (exponential-time) decision procedure, of course.

Quantifier elimination can be used to show that a large number of problems are polynomial-time solvable in fixed dimension; this includes problems like convexity (A33), radius (A35), Brouwer fixed point (A5), and many variants of the Nash equilibrium problem, see Sect. 5.3.7.

5.2.2.4 Replacing Exotic Quantifiers

Sometimes problems are not captured well by traditional quantifiers. Let us have a closer look at the dimensionality problem of semialgebraic sets. A semialgebraic set $S \subseteq \mathbb{R}^d$ is *full-dimensional* if it contains an open subset, or, equivalently, an open ball. We can express this formally as

$$(\exists x \in \mathbb{R}^d)(\exists \delta \in \mathbb{R})(\forall y \in \mathbb{R}^d)\ \|y - x\| < \delta \rightarrow y \in S.$$

So testing whether S is full-dimensional lies in $\exists\forall\mathbb{R}$, the second level of the hierarchy. The condition can be expressed more succinctly using $\exists^*$, one of the *exotic quantifiers* introduced by Koiran [Koi99], $(\exists^* x \in \mathbb{R}^d)\ \varphi(x)$ is defined as $(\exists x \in \mathbb{R}^d)(\exists \delta \in \mathbb{R}_{>0})(\forall y \in \mathbb{R}^d)\ \|y - x\| < \delta \rightarrow \varphi(y)$. With that, our characterization of full-dimensionality of S simplifies to

$$(\exists^* x \in \mathbb{R}^d)\ x \in S.$$

Koiran then went on to show that $\exists^*$ can be replaced with $\exists$ by adding some variables, also see [BC09, Sect. 8]:

Lemma 5.2.11 (Koiran) *Given a formula $\Phi(x, y)$ in prenex form with a fixed number of quantifier alternations, one can in polynomial time construct a formula $\Psi(x', y)$ such that*

$$(\exists^* x \in \mathbb{R}^d)\ \Phi(x, y)\ \leftrightarrow (\exists x' \in \mathbb{R}^{d'})\ \Psi(x', y),$$

d' is polynomially bounded in d, and $\Psi(x', y)$ has the same pattern of quantifier alternations as $\Phi(x, y)$.

For the full-dimensionality example we can let Φ be the defining formula of $S = \{x \in \mathbb{R}^d : \varphi(x)\}$, so there are no additional quantifiers or variables. It follows that being full-dimensional lies in $\exists\mathbb{R}$, and it is easily seen to be $\exists\mathbb{R}$-hard, so the real dimension of a semialgebraic set (A31) is $\exists\mathbb{R}$-complete.

The second exotic quantifier, $\forall^*$, is the negation of $\exists^*$, so it intuitively corresponds to saying "for a dense set", and, as the negation of $\exists^*$ can be replaced by $\forall$ using Koiran's lemma.

Koiran's result can significantly simplify proving membership in the existential theory of the reals for properties that can be expressed using $\exists^*$; additional examples include positivity (A3), continuity (A11), and image density (A46), also see [BC09, Table 2].

The third exotic quantifier, H, is a bit different. Whereas $\exists^*$ and $\forall^*$ are concerned with density, H is concerned with limits, and tends to appear when suprema or infima are involved.

Define $(\mathsf{H}\varepsilon)$ as $(\exists\varepsilon' > 0)(\forall\varepsilon \in (0, \varepsilon'))$; H can be read as "for all sufficiently small ε"; it is introduced in Bürgisser and Cucker [BC09]. We saw earlier that having diameter at most 1 lies in $\forall\mathbb{R}$. What about requiring the diameter to be strictly less than 1? It is not sufficient to say

$$(\forall x, y \in \mathbb{R}^d)\ x \in S \wedge y \in S \rightarrow \|x - y\| < 1,$$

since the diameter is the supremum of the distances of points in S. The H-quantifier offers a solution:

$$(\mathsf{H}\varepsilon > 0)(\forall x, y \in \mathbb{R}^d)\ x \in S \wedge y \in S \rightarrow \|x - y\| \leq 1 - \varepsilon,$$

By a result of Bürgisser and Cucker [BC09, Theorem 9.2], a leading H-quantifier can be eliminated; roughly speaking, it works by replacing ε with a sufficiently small constant, which depends on Φ.

Lemma 5.2.12 (Bürgisser and Cucker) *Let $\Phi(\varepsilon)$ be a formula in prenex form. We can construct a formula Ψ in polynomial time such that*

$$(\mathsf{H}\varepsilon > 0)\ \Phi(\varepsilon)$$

is equivalent to Ψ, and Ψ has the same pattern of quantifier alternations as $\Phi(\varepsilon)$.

In particular, if Φ is an existentially or universally quantified formula, then so is Ψ. (This special case was independently discovered by Schaefer and Štefankovič [SŠ17], already used in [Sch13].) From this we can conclude that the diameter problem lies in $\forall\mathbb{R}$.

Several other computational problems have been placed in $\exists\mathbb{R}$ or $\forall\mathbb{R}$ using the elimination of a leading H-quantifier. This includes several properties of semialgebraic sets such as unboundedess (A34), a specific point belonging to the closure of a set (A38), the distance between semialgebraic sets (A41), as well as linkage rigidity (CG67), and a point not being an isolated zero of a polynomial (A7). As an example from a different area we mention the angular resolution of a graph (CG60); angular resolution is defined as the supremum (of some expression); since the supremum can be expressed using a leading H-quantifer, it can be eliminated.

5.2.3 A History of the Existential Theory of the Reals

The existential theory of the reals has roots in several areas of mathematics. In logic, it traces back to the axiomatization of the real numbers around the turn of the 19th century. One of the major early results in that tradition is Tarski's decision procedure for the first-order theory of the reals based on quantifier elimination. Tarski's work implies that the first-order theory of the reals is decidable [Tar48]. This result has been sharpened over the years, with seminal contributions by Collins, Renegar, Canny and many others (for a modern treatment, see [Bas17]), and decision procedures are implemented in modern solvers, such as Microsoft's Z3 and the Redlog system [Redb, Reda] as well as in mathematical software systems such as Mathematica and Maple.

The logic roots branched in various ways, but there are two main branches along which, independently, the existential theory of the reals was discovered as a complexity class. One branch stems from Mnëv's universality theorem and Shor's presentation of it, and the other branch from the Blum-Shub-Smale model of real computation. These two traditions did not intermingle until quite recently.

5.2.3.1 $\exists\mathbb{R}$ in the BSS Tradition

Let us first have a look at the discovery of $\exists\mathbb{R}$ in the BSS-model, where $\exists\mathbb{R}$ is known as $\mathsf{BP}(\mathsf{NP}^0_{\mathbb{R}})$ or sometimes NPR. The BSS-model model of computation was introduced by Blum, Shub, and Smale in [BSS89], followed by the book [BCSS98]. This seminal paper defined the real complexity classes $\mathsf{P}_{\mathbb{R}}$ and $\mathsf{NP}_{\mathbb{R}}$, and asked whether $\mathsf{P}_{\mathbb{R}} \neq \mathsf{NP}_{\mathbb{R}}$, the real equivalent of the $\mathsf{P} \neq \mathsf{NP}$ question. The paper also establishes a foundational result, an "analogue of Cook's theorem for $\mathbb{R}$", namely that the feasibility of a polynomial of degree at most 4 (with real coefficients) is complete for $\mathsf{NP}_{\mathbb{R}}$ [BSS89, Main Theorem, Sect. 6]. This result and the definition of $\mathsf{NP}_{\mathbb{R}}$ are the first two steps towards establishing $\mathsf{BP}(\mathsf{NP}^0_{\mathbb{R}})$ as a complexity class. Two more steps are needed, the real Turing machines need to be restricted in two ways: they need to be constant-free (see below), denoted by the 0 superscript in $\mathsf{NP}^0_{\mathbb{R}}$, and the input needs to be binary, denoted by the BP operator.

The real Turing machines in the BSS-model are allowed real constants, making them quite powerful (every countable set can be encoded in a real constant, and a real Turing-machine can extract that information, so the halting problem, for example, encoded in unary lies in $\mathsf{P}_{\mathbb{R}}$). In the *constant-free* (also called parameter-free) model, the only constants allowed are 0 and 1; this is equivalent to allowing the machine to use rational or even algebraic constants [ABKPM09]. One of the earliest papers we have found studying constant-free machines is by Cucker and Koiran [CK95, Section 7]; they study a complexity class based on existentially quantified formulas (over the reals) in which only integer coefficients, addition (and subtraction) are allowed, no multiplication and division; this leads them to so-called additive machines, which were well-studied at the time, but with the restriction to integer constants (instead or arbitrary real constants). There are other contemporary papers that looked at constant-free machines, including Koiran [Koi97c, Koi97a], but the superscript 0 notation, $\mathsf{NP}^0_{\mathbb{R}}$, first occurs in papers by Fournier and Koiran [FK99] and Chapuis and Koiran [CK99]. An explanation of the notation is given in the paper by Chapuis and Koiran [CK99]: for a real complexity class C, the authors introduce the notation C^k, where the real machines defining C are restricted to have at most k constants.

The earliest papers we can find that consider restricting the inputs of the real Turing machines to binary inputs are due to Koiran [Koi92, Koi94], but neither paper introduces a name for the restriction. Probably the first paper to do so is by Cucker and Koiran [CK95, Definition 1]; it defines and names the *Boolean part* operator

BP explicitly, describing it as capturing the "the computational power of real Turing machines over binary inputs". Starting with this paper, the Boolean part (or binary part) is studied in various papers, including [CM96, CG97, Koi97c, MM97].

Some papers, e.g. the paper by Cucker and Koiran [CK95] work with both Boolean parts and constant-free machines, but they do not combine them to define $\mathsf{BP}(\mathsf{NP}^0_{\mathbb{R}})$. This seems to have first happened in the paper by Bürgisser and Cucker [BC06], in which the authors consider both $\mathsf{BP}(\mathsf{NP}^0_{\mathbb{C}})$ and $\mathsf{BP}(\mathsf{NP}^0_{\mathbb{R}})$, and show that the feasibility (A1) and real dimension problem (A31) (based on Koiran [Koi99]) are complete for $\mathsf{BP}(\mathsf{NP}^0_{\mathbb{R}})$, see their Proposition 8.5.

There are relatively few papers with completeness results in the BSS-tradition, even for the real complexity class $\mathsf{NP}_{\mathbb{R}}$, something commented on in the introduction to Bürgisser and Cucker [BC06]: "these first completeness results were not followed by an avalanche of similar results".

This changed somewhat with a later paper by Bürgisser and Cucker [BC09]. Section 9 of that paper introduces the operator BP^0, which restricts the real Turing machines to be both constant-free and the input to be in binary. The authors then define the "constant-free Boolean part" $\mathsf{BP}^0(C)$ of a complexity class C. (So $\mathsf{BP}^0(\mathsf{NP}_{\mathbb{R}})$ and $\mathsf{BP}(\mathsf{NP}^0_{\mathbb{R}})$ become two different ways to denote the same class.) With this operator, Bürgisser and Cucker define several complexity classes, including $\mathsf{PR} = \mathsf{BP}^0(\mathsf{P}_{\mathbb{R}})$, $\mathsf{NPR} = \mathsf{BP}^0(\mathsf{NP}_{\mathbb{R}})$, the existential theory of the reals, equal to $\exists\mathbb{R}$, and $\mathsf{coNPR} = \mathsf{BP}^0(\mathsf{coNP}_{\mathbb{R}})$, the universal theory of the reals, equal to $\forall\mathbb{R}$, and they look at higher levels of the hierarchy as well. (The real hierarchy in the BSS-model was introduced by Cucker [Cuc93].)

They then argue that several known completeness results for $\mathsf{NP}_{\mathbb{R}}$, $\mathsf{coNP}_{\mathbb{R}}$ and even higher levels transfer to the discrete setting nearly automatically (see their Table 1), namely feasibility (A1), first shown $\mathsf{NP}_{\mathbb{R}}$-complete in the original paper by Blum et al. [BSS89], convexity, by Cucker and Rosselló [CR92], and the real dimension problem, shown $\mathsf{NP}_{\mathbb{R}}$-complete by Koiran [Koi99] (using exotic quantifiers). Also in this category is an early result by Zhang [Zha92] on neural networks (ML1).

Bürgisser and Cucker [BC09] also added several new problems (see Table 2 in their paper), including unboundedness of semialgebraic sets (A34), which is $\exists\mathbb{R}$-complete, and surjectivity of a function (A45), which is complete for $\forall\exists\mathbb{R}$, or $\mathsf{BP}^0(\forall\exists)$ in the Bürgisser-Cucker notation. The main thrust of that paper was not actually the discrete setting, which may explain why it took more than ten years for researchers in the Mnëv/Shor-tradition to find these results; as its title suggested, the paper focused on the power of the exotic quantifiers H, $\forall^*$ and $\exists^*$ introduced by Koiran [Koi99]. For example, $\exists^* x$ means "for an open set of x". Bürgisser and Cucker, building on the work of Koiran, were able to show that in many cases, and in all interesting cases at the first level, the exotic quantifiers $\forall^*$ and $\exists^*$ can be replaced with their non-exotic counterparts, and a fixed number of H-quantifiers can be eliminated, see Sect. 5.2.2.4. This shows that the classes PR, NPR, and coNPR are robust under modifications of the quantifiers. For more recent work extending this see [Jun23b].

A comprehensive survey of (structural) complexity results in the BSS-model was written by Baartse and Meer [BM13].

5.2.3.2 ∃ℝ in The Mnëv/Shor Tradition

Perhaps the first paper to view the existential theory of the reals as a complexity class, without defining it explicitly, is Shor's presentation of Mnëv's universality theorem [Mnë88, Sho91]. Mnëv showed that every semialgebraic set is "equivalent", in a technical sense, to the realization space of a stretchable arrangement of pseudolines (see 5.2.1.3). Mnëv's work was in algebraic geometry, and his notion of equivalence aimed at capturing algebraic similarity. Shor's paper recasts Mnëv's result with a computer science audience in mind; Mnëv's universality theorem is phrased as a reduction, showing how to translate a system of equalities and inequalities into a stretchability problem, where equivalence is now at the decision level: the system of equalities and inequalities is solvable if and only if the pseudoline arrangement constructed by the reduction is stretchable. Shor's paper is also the first to show that NP reduces to the stretchability problem, implying that NP $\subseteq$ ∃ℝ (a result that is immediate if one starts with ETR as the complete problem for ∃ℝ).

Together with Kempe's 19th-century universality theorem on linkages [Kem77], Mnëv's result led to other universality theorems in algebraic geometry, but these often were not concerned with being effective reductions. Kempe's result can be made effective, implying that linkage realizability (CG66) is ∃ℝ-complete [KM02, Abb08, Sch13]), but this is not the case for all universality results, e.g. [Stu87, Dat03].

Shortly after Shor's paper was published, Kratochvíl and Matoušek [KM94] showed that the segment intersection graph problem (CG26) is hard for the existential theory of the reals, reducing from a system of strict inequalities (they cite a manuscript version of Shor's paper dated to 1988). Kratochvíl and Matoušek add two observations (based on work by Goodman et al. [GPS89], which in turn is based on results of Grigoriev and Vorobjov [GV88] in algebraic geometry): first, the segments realizing a particular segment intersection graph may require double exponential precision to specify (in the number of segments), and double exponential precision is sufficient.

A manuscript version of Kratochvíl and Matoušek [KM94] then influenced Bienstock [Bie91] who showed that the rectilinear crossing number (CG50) is as hard as solving a system of strict inequalities, and also derives upper and lower bounds on the required precision of the vertex coordinates.

Other early papers influenced by the work of Shor, Kratochvíl and Matoušek include Tanenbaum, Goodrich, and Scheinerman [TGS95] on the point-halfspace order problem (CG14); Richter-Gebert and Ziegler [RGZ95] on the Steinitz problem; Davis, Gotts, and Cohn [DGC99] on topological inference with convexity (L13); Buss, Frandsen, and Shallit [BFS99] on various matrix problems (see Sect. 5.4.2.2), and Bose, Everett, and Wismath [BEW03] on the arrangement graph problem (CG11). The paper by Buss, Frandsen, and Shallit [BFS99] in particular comes pretty close to treating the existential theory of the reals (and other fields) as a complexity class.

Schaefer [Sch10] defined the existential theory of the reals as we defined it in this survey, as a computational complexity class based on ETR, and introduced the notation ∃ℝ; this paper also mentioned the result (not published until 2017 by

Schaefer and Štefankovič [SŠ17]) that the definition of $\exists\mathbb{R}$ is robust in the sense that if we restrict the language of ETR to not allow equality and negation—so we can only express strict inequalities, then the resulting complexity class $\exists\mathbb{R}_<$ is the same as $\exists\mathbb{R}$. This implies that problems like simple stretchability, rectilinear crossing number and segment intersection graphs, which do not require equality, are as hard as problems like feasibility and stretchability, which do.

In the following years, a majority of the research in the Mnëv/Shor tradition stayed in the areas of graph drawing and computational geometry, but some of the research branched off, looking at problems in analysis [OW14, Shi16a, OW17, SŠ17, SŠ18], algebra [HL13, BK15, Shi16b, BRS17, Shi17b], and other areas [HZ16, Pet16, Pet17].

Inevitably, the Mnëv/Shor tradition also discovered the higher levels of the real polynomial hierarchy; this happened in a paper by Dobbins et al. [DKMR18, DKMR23], who introduced $\exists\forall\mathbb{R}$ and $\forall\exists\mathbb{R}$; several subsequent papers located problems beyond the first level of the hierarchy, including [BH21, DLN+21, JKM23]. Robustness results for all higher levels, similar to the ones known for the first level, were shown by Schaefer and Štefankovič [SŠ23, SŠ25] extending work in [DLN+21, DKMR23].

5.2.3.3 Statistics

Figure 5.8 shows the number of publications each year that (explicitly or implicitly) contain $\exists\mathbb{R}$-hardness results. If a paper was published in multiple forms (conference/journal/preprint), we only counted the first version. Many of these papers contain multiple $\exists\mathbb{R}$-hardness results.

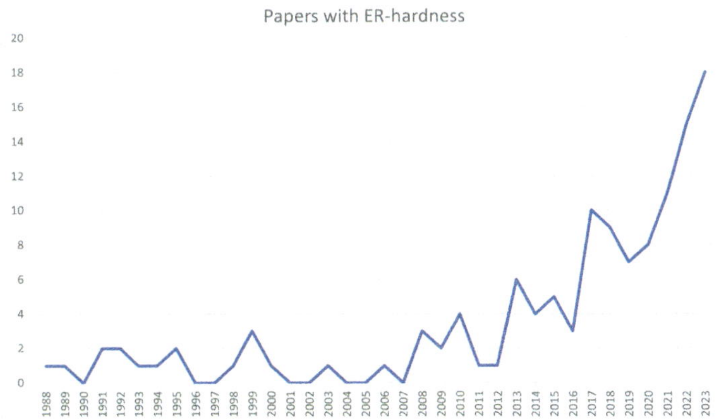

Fig. 5.8 Number of papers with $\exists\mathbb{R}$-hardness results by year

Computational geometry is by far the most active area for $\exists\mathbb{R}$-complete problems, though we have to admit that the classification of problems is somewhat ambiguous.

5.3 A Tour of the Compendium

What can the reader expect from our compendium? We collect $\exists\mathbb{R}$-complete problems, but we go farther by including $\exists\mathbb{R}$-hard problems, as long as they belong to higher levels of the polynomial hierarchy beyond $\exists\mathbb{R}$, that is, problems that can be captured by the theory of the reals with a bounded number of quantifier alternations.

Within the entries, we will point out related universality results if we are aware of them. As we mentioned, there are many different notions of universality, depending on the equivalence relation that is used. We refer to Verkama [Ver23] for a recent discussion on the notion of stable equivalence used by Richter-Gebert [RG96].

We will not discuss problems complete for several other complexity classes that are related to $\exists\mathbb{R}$, but probably different. This includes $\mathsf{NP}^{\mathsf{SSQR}}$, $\mathsf{P}^{\mathsf{SSQR}}$, and $\exists\mathbb{R} \cap \forall\mathbb{R}$—which we already discussed earlier, variants of the BSS computation model, and classes such as FIXP and PPAD, which are important in the context of game theory.

We also make no attempt to include problems complete for other existential theories, such as the existential theory of arbitrary rings and fields introduced by Buss et al. [BFS99].

Remark 5.3.1 (*More Existential Theories*) The existential theory of GF(2) yields NP. Over the rational numbers we obtain the class $\exists\mathbb{Q}$, whose complexity is still unknown (possibly undecidable); the best-known result in this direction is that $\forall\exists\mathbb{Q}$ is undecidable [Poo09]. The existential theory of the natural numbers (or, equivalently, integers), $\exists\mathbb{N}$, is known as Hilbert's Tenth problem, and is famously undecidable by results of Davis, Robinson, and Matiyasevich [Mat06]. Finally, the existential theory of the complex numbers, $\exists\mathbb{C}$, can be interpreted in $\exists\mathbb{R}$ by viewing complex numbers as pairs of real numbers; $\exists\mathbb{C}$ lies in $\mathsf{RP}^{\mathsf{NP}}$ and therefore belongs to the second level of the polynomial hierarchy by a result of Koiran's, assuming the generalized Riemann hypothesis [Koi96]. We have $\mathsf{NP} \subseteq \exists\mathbb{C} \subseteq \exists\mathbb{R} \subseteq \exists\mathbb{Q} \subseteq \exists\mathbb{N}$, in particular, all these theories are NP-hard.

If natural variants of $\exists\mathbb{R}$-hard problems turn out to be complete for other existential theories, we will mention this fact. For example, $\exists\mathbb{Q}$ is relevant in the field of graph drawing, since it naturally occurs when looking at grid variants of graph drawing problems such as planar slope number, visibility graph, and right-angle crossing drawings, or in rational Nash Equilibria [BH22].

The entries in the compendium are mostly self-contained, though some common notions are defined at the beginning of each section. To keep the entries brief and focused on complexity aspects, we do not include definitions of some of the more intricate notions, relying on references in that case. We include cross-references between related entries. Within each category we have also attempted to group similar problems close together, but that has not always been possible.

There are plenty of open problems, both within the entries, and in Sect. 5.4.9 on candidate problems. We have not carefully vetted all papers listed in this compendium, so we cannot guarantee correctness of results. Instead we opted to be inclusive to cover more ground, so we also include preprints; this allows the reader to follow up with the original sources. We will mention issues with papers if we are aware of them.

5.3.1 *Logic*

The existential theory of the reals (L1), as the name suggests, is anchored in logic, as it is defined by the language ETR, a fragment of the theory of the reals. As we have seen, ETR is a powerful, and expressive language, which makes it useful for upper bounds, but inconvenient for reductions; we would prefer restricted versions of this language, and, as it turns our, $\exists\mathbb{R}$ is quite robust when it comes to the exact definition and signature of the underlying logic. Variants include forcing variables to be distinct (L4), or ordered (L2), or allowing them to be approximate (L3). Another family of variants is based on replacing multiplication with inversion, starting with ETR- INV (L5), and including PLANAR- ETR- INV (L6) in which the variable constraint graph is planar, and other nonlinear constraints (L8).

These restrictions are useful as intermediate problem for reductions to show that other problems are $\exists\mathbb{R}$-complete.

A number of problems in this section also involve other formal systems such as the *collinearity logic* (L9) the *$\mathcal{LR}$-calculus* involving left/right orientations of triples of points in the plane (L10), the *region-connection calculus* (L13), or Birkhoff and von Neumann's *quantum logic* [BvN36] (L14), (L15).

5.3.2 *Algebra*

The section on algebra is split into three parts. We deal with properties of polynomials and arithmetic circuits; matrices and tensors; as well as semialgebraic sets.

5.3.2.1 Polynomials and Arithmetic Circuits

To us arithmetic circuits and polynomials are ways to encode functions $f : \mathbb{R}^n \to \mathbb{R}$. Every arithmetic circuit can be expressed as a polynomial and every polynomial can be expressed as an arithmetic circuit. The difference between the two is that certain functions can be expressed very succinctly using arithmetic circuits, while requiring much longer polynomial expressions. The function $f(x, y) = (x + y)^{2^n}$ may serve as an example; in this notation (allowing arbitrary exponents), f can be encoded using roughly $\log n$ bits. Arithmetic circuits can encode this function using using a linear number of bits, as shown in the drawing.

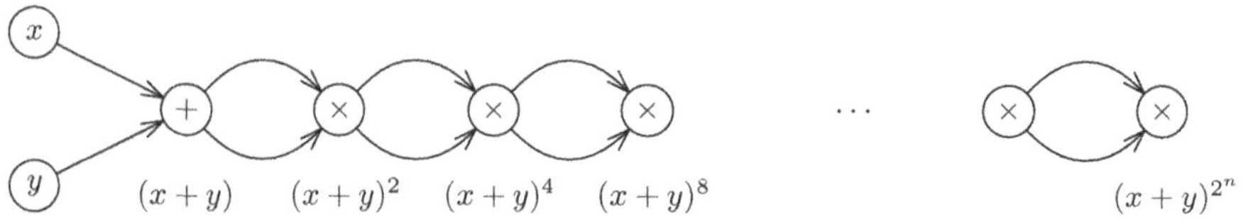

Finally, an explicit polynomial expression without arbitrary exponents will require an exponential number of bits.

$$f = \sum_{i=0}^{N} \binom{N}{i} x^i y^{N-i},$$

where $N = 2^n$. The explicit polynomial representation requires an exponential number of coefficients, many of them having an exponential-size binary representation.

We are interested in properties like, does the function have a zero, can it be positive, is it convex, etc. Table 5.2 contains an overview of the properties of functions whose complexity lies in the real hierarchy.

5.3.2.2 Matrices and Tensors

Given an $n \times n$ matrix M, we can compute a decomposition $M = A \cdot B$, into an $n \times k$ matrix A and a $k \times n$ matrix B quickly, if such a decomposition exists. The underlying reason is a hidden linear structure. The rank is the smallest k for which such a decomposition exists.

Can we also quickly compute concepts closely related to decomposition and rank? The answer seems to be no in most cases. The easiest example is arguably the nonnegative rank of a matrix. Here, we require that both that all entries of the input and the output are nonnegative numbers. A close inspection of the matrix

$$M = \begin{pmatrix} 1 & 1 & 0 & 0 \\ 0 & 1 & 1 & 0 \\ 0 & 0 & 1 & 1 \\ 1 & 0 & 0 & 1 \end{pmatrix}$$

shows that the nonnegative rank can be different from the ordinary rank, as was observed [Tho74] in 1974. It was not until 2017 though that it was noted that a nonnegative decomposition may require irrational numbers [CKM+17]. Shortly afterwards this result was improved to $\exists\mathbb{R}$-completeness of the nonnegative rank problem

Table 5.2 Properties of functions and their complexity. We use the abbreviation AC to indicate that the result is known for arithmetic circuits

Property	Complexity	Source
Brouwer fixed point (A5)	$\exists\mathbb{R}$-complete	Schaefer, Štefankovič [SŠ17]
Continuity (AC) (A11)	$\forall\mathbb{R}$-complete	Bürgisser and Cucker [BC09, Corollary 9.4]
Continuity at a point (AC) (A12)	$\forall\mathbb{R}$-complete	Bürgisser and Cucker [BC09, Corollary 9.4]
Domain density (AC) (A13)	$\forall\mathbb{R}$-complete	Bürgisser and Cucker [BC09, Proposition 5.7, Corollary 9.4]
Feasibility (A1)	$\exists\mathbb{R}$-complete	Folklore
Fixed point (A4)	$\exists\mathbb{R}$-complete	Schaefer and Štefankovič [SŠ17]
(Strict) Global positivity (A3)	$\forall\mathbb{R}$-complete	Ouaknine and Worrell [OW17, Theorem 7]
Image density (AC) (A46)	$\forall\exists\mathbb{R}$-complete	Combining results from Bürgisser and Cucker [BC09, Proposition 5.7, Corollary 9.4] and Junginger [Jun23b, Theorem 4.27]
Injectivity (polynomial) (A8)	$\forall\mathbb{R}$-complete	See entry
Isolated zero (A7)	$\forall\mathbb{R}$-complete	Koiran [Koi00], Schaefer [Sch13]
(Strict) Local positivity (A2)	$\exists\mathbb{R}$-complete	Schaefer and Štefankovič [SŠ17], also Ouaknine and Worrell [OW17, Theorem 7]
Nontrivial homogeneous Zero (A6)	$\exists\mathbb{R}$-complete	Schaefer [Sch13, Corollary 3.10]
Polynomial projection (Linear) (A14)	$\exists\mathbb{R}$-complete	Bläser et al. [BRS17]
Regularity (polynomial) (A9)	$\forall\mathbb{R}$-complete	Cucker and Rosselló [CR92]
Surjectivity (A45)	$\forall\exists\mathbb{R}$-complete	Schaefer and Štefankovič [SŠ23]
Totality (AC) (A10)	$\forall\mathbb{R}$-complete	Bürgisser and Cucker [BC09, Proposition 4.1, Corollary 9.4]

by Shitov [Shi17b]. We see that the seemingly minor requirement of non-negativity changes the computational nature of the matrix rank problem drastically.

Tensors can be viewed (from a computer science perspective at least) as higher-dimensional matrices, so it is not surprising that tensors make an appearance here. Hillar and Lim [HL13] offer a thorough survey of computational problems for tensors and show that a fair number of these problems are NP-hard, many of them may be candidates for $\exists\mathbb{R}$-completeness.

Table 5.3 lists matrix and tensor properties for which $\exists\mathbb{R}$-completeness results are known.

Table 5.3 Properties of matrices and tensors

Property	Complexity	Source
Matrix singularity (A15)	∃ℝ-complete	Buss et al. [BFS99]
Minimum matrix rank (A16)	∃ℝ-complete	Buss et al. [BFS99]
Minimum rank of linear combination of matrices (A17)	∃ℝ-complete	Bläser et al. [BIL+21, Theorem 2.3]
Minimum rank of sign pattern matrix (A18)	∃ℝ-complete	Bhangale and Kopparty [BK15]
Monotone matrix rank (A19)	∃ℝ-complete	Lienkaemper [Lie23]
Nonnegative matrix factorization (A21)	∃ℝ-complete	Shitov [Shi16b]
Symmetric nonnegative matrix trifactorization (A22)	∃ℝ-complete	Shitov [Shi25a]
(Real) PSD rank (A23)	∃ℝ-complete	Shitov [Shi23, Theorem 3.4]
(Real) CPSD rank (A24)	∃ℝ-complete	Shitov [Shi23, Theorem 3.10]
Phaseless rank (A25)	∃ℝ-complete	Shitov [Shi23, Theorem 3.7]
Rank-constrained optimization (A26)	∃ℝ-complete	Bertsimas et al. [BCWP22]
Tensor rank (A27)	∃ℝ-complete	Schaefer and Štefankovič [SŠ18]
Tensor with eigenvalue 0 (A28)	∃ℝ-complete	Schaefer [Sch13, Corollary 3.10]
Orbit closure containment (A29)	∃ℝ-complete	Bläser et al. [BIL+21]

5.3.2.3 Properties of Semialgebraic Sets

The study of properties of semialgebraic sets and their complexity began in the BSS-model, and, in a way, was a motivating influence. Early papers include Cucker and Rosselló [CR92], settling boundedness, convexity, cardinality, and closure; and Koiran [Koi99] on the dimension problem. Later, many more problems were added by Bürgisser and Cucker [BC06, BC09], and there are now additional results [SŠ17, JKM23, SŠ23]. Some problems remain intriguingly open, such as the complexity of deciding whether a semialgebraic set is closed. Table 5.4 summarizes the known results.

Other areas of algebra, such as group theory, are currently absent; Meer and Ziegler [MZ09] asked an interesting question in this context: is there a group whose word-problem is equivalent to $\mathsf{NP}_{\mathbb{R}}$? We can ask the same question for ∃ℝ as well.

5.3.3 *Matroids, Arrangements, and Polytopes*

Mnëv's Universality Theorem [Mnë88, RG99], and its later complexity-theoretic interpretation by Shor [Sho91], is one of the seminal results on which much of the more recent literature on ∃ℝ builds upon.

Table 5.4 Properties of semialgebraic sets and their complexity. Membership for closedness and compactness only applies to basic semialgebraic sets

Property	Complexity	Source
Cardinality (Emptiness) (A30)	∀ℝ-complete	Cucker and Roselló [CR92]
Convexity (A33)	∀ℝ-complete	Cucker and Roselló [CR92]
Closedness (A40)	∀ℝ-complete	Bürgisser and Cucker [BC09, Theorem 6.15, Corollary 9.4]
Closure (A38)	∃ℝ-complete	Cucker and Roselló [CR92]
Compactness (A40)	∀ℝ-complete	Bürgisser and Cucker [BC09, Theorem 6.15, Corollary 9.4]
Density (A37)	∀ℝ-complete	Koiran [Koi99] and Bürgisser and Cucker [BC09, Corollary 9.3], also see Schaefer and Štefankovič [SŠ23]
Diameter (A36)	∀ℝ-complete	Schaefer and Štefankovič [SŠ23]
Dimension (A31)	∃ℝ-complete	Koiran [Koi99], also see Bürgisser and Cucker [BC09, Corollary 9.4]
Distance (A41)	∃ℝ-complete	Schaefer and Štefankovič [SŠ17]
Isolated Point (A39)	∀ℝ-complete	Bürgisser and Cucker [BC09, Corollary 9.4]
Hausdorff distance (A44)	∀∃ℝ-complete	Jungeblut et al. [JKM23], also see Schaefer and Štefankovič [SŠ23]
Hyperplane Local Support (A42)	∃ℝ-complete	Bürgisser and Cucker [BC09], also see Schaefer and Štefankovič [SŠ23]
Local Dimension (A32)	∃ℝ-complete	Bürgisser and Cucker [BC09, Corollaries 6.6 and 9.4]
Radius (A35)	∀ℝ-complete	Schaefer and Štefankovič [SŠ23]
Unboundedness (A34)	∃ℝ-complete	Cucker and Roselló [CR92], also see Bürgisser and Cucker [BC09, Corollary 9.4]

The theorem is usually stated in terms of realization spaces of *oriented matroids*: Any semialgebraic set is *stably equivalent* to the realization space of an oriented matroid. Just like matroids, oriented matroids can be defined in a variety of ways (*cryptomorphisms*). They are intended to be a common abstraction of directed graphs, configurations of vectors, and arrangement of hyperplanes [BLVS+99]. One standard way of defining an oriented matroid of rank d is via a *chirotope*: a map from ordered d-tuples of elements of a ground set to the set $\{0, -1, +1\}$, that satisfy certain relations, coined the *chirotope axioms*, derived from properties of determinants. In the computational geometry literature, rank-three chirotopes, or their isomorphism classes, are often referred to as *order types* [GP83, AAK02]. The oriented matroid of rank three corresponding to a configuration of n points in the plane is defined by the order type that encodes the orientation of each ordered triple of points: clockwise, counterclockwise, or collinear. In general, in dimension d, the order type simply encodes the sign of the determinants defined by the ordered d-tuples of points. However, not all oriented matroids correspond to arrangements of points. The *realization space* of an oriented matroid M of rank d on n elements is the subset of $\mathbb{R}^{dn}$, every point of which corresponds to a set of n points in $\mathbb{R}^d$ realizing M. Deciding whether this set is nonempty, hence whether M is realizable in $\mathbb{R}^d$, is a canonical $\exists\mathbb{R}$-complete problem (CG2). Completeness actually holds for matroids as well (CG3).

The *Topological Representation Theorem* of Folkman and Lawrence [FL78] gives a convenient representation of *any* oriented matroid, not only the representable ones: Oriented matroids of rank d are in bijection with equivalence classes of arrangements of *pseudo-hemispheres* of $\mathbb{S}^{d-1}$, whose boundaries are not geometric great circles, but behave combinatorially so. In rank three, for instance, oriented matroids are one-to-one with *oriented pseudocircle arrangements* on the 2-sphere. Since Mnëv's Universality Theorem holds for fixed rank three, and applying some elementary simplifications, we obtain that the decision problem for the realizability of rank-three oriented matroid is essentially the pseudoline stretchability problem described earlier (CG9). Simple pseudoline arrangements are in fact one-to-one with *uniform acyclic oriented matroids of rank 3*. We refer the interested reader to the standard textbook on oriented matroids [BLVS+99] for more details about the cryptomorphisms of oriented matroids and the definitions of the terms. Detailed and simplified proofs of the Topological Representation Theorem were given by Bokowski et al. [BMS01], and Bokowski et al. [BKMS05]. An alternative treatment of order types corresponding to rank-three uniform oriented matroids was also given by Knuth [Knu92]. The projective duality between sets of points and arrangements of hyperplanes directly yields that realizability of order types as point sets in the plane is also $\exists\mathbb{R}$-complete (CG1).

Combinatorial abstractions of configurations of points or vectors have been defined in various ways in computational geometry. From the previous developments, it was shown that geometric realizability of many of them was $\exists\mathbb{R}$-hard to determine. These include Goodman and Pollack's allowable sequences (CG4) [GP80], radial systems (CG5), directional walks (CG6), and convex geometries (CG7) as defined by Edelman and Jamison [EJ85].

The Universality Theorem also applies to realization spaces of combinatorial types (face lattices) of polytopes [RGZ95, RG96, RG99, PT14, APT15]. As a consequence, the problem of deciding whether a given combinatorial type of polytope is realizable is also a standard $\exists\mathbb{R}$-complete problem (CG68). A major achievement in the theory of $\exists\mathbb{R}$-completeness is the proof, due to Richter-Gebert [RG96], that the problem is hard already in dimension 4 (CG69). The result also holds for simplicial and inscribed polytopes in arbitrary dimension (CG71). *Delaunay triangulations* are projections of inscribed simplicial polytopes, and are therefore also hard to recognize in arbitrary dimension (CG73).

A list of problems related to realizability of matroids and polytopes is given in Table 5.5.

Table 5.5 Combinatorial structures for which realizability over the reals is known to be $\exists\mathbb{R}$-complete. Realizability is $\exists\mathbb{R}$-complete already in the plane for all the structures in the top part of the table

Structure	Source
Order types (CG1)	Mnëv [Mnë88, Sho91]
Oriented matroids (CG2)	Mnëv [Mnë88, Sho91]
Pseudoline arrangements (CG9)	Mnëv [Mnë88, Sho91]
Pseudo-segment arrangements (CG10)	Mnëv [Mnë88, Sho91]
Arrangement graphs (CG11)	Bose et al. [BEW03]
Pseudocircle arrangements (CG12)	Kang and Müller [KM14], Felsner and Scheucher [FS20]
Matroids (CG3)	Explicitly by Kim et al. [KdMM23]
Allowable sequences (CG4)	Hoffmann and Merckx [HM18]
Radial systems (CG5)	Cardinal and Kusters [CK15]
Directional walks (CG6)	Schaefer [Sch21b]
Convex geometries (CG7)	Adaricheva and Wild [AW10], Hoffmann and Merckx [HM18]
Polytopes (CG68)	Mnëv [Mnë88, Sho91]
4-dimensional polytopes (CG69)	Richter-Gebert [RG96]
Simplicial polytopes (CG70)	Mnëv [Mnë88], Adiprasito and Padrol [AP17]
Inscribed polytopes (CG71)	Adiprasito et al. [APT15]
Delaunay triangulations (CG73)	Adiprasito et al. [APT15]

5.3.4 Graph Drawing

Graph drawing forms the theoretical core of information visualization; it studies drawing styles of graphs—such as topological, polyline and straight-line drawings, and parameters that measure a drawing's readability, such as crossing number and rectilinear crossing number. Figure 5.9 illustrates a more recent drawing style, Lombardi drawings in which all vertices must have perfect angular resolution, that is, at each vertex, the angles between consecutive edge ends must be the same, and edges are realized by circular arcs (CG59).

Computational problems in graph drawing are often solvable in polynomial time if they are related to planarity and embeddability, or NP-complete, typically when non-planarity comes into play. As soon as one replaces topological drawings, in which edges are represented as curves, with geometric drawings in which edges are straight-line or polygonal, the complexity often jumps from NP to $\exists\mathbb{R}$. This applies to problems such as crossing number or pseudolinear crossing number versus rectilinear crossing number (CG51), or local crossing number versus rectilinear local crossing number (CG52), weak realizability vs. weak rectilinear realizability (CG54), simultaneous embedding vs. simultaneous geometric embedding (CG56), and thickness versus geometric thickness (CG57).

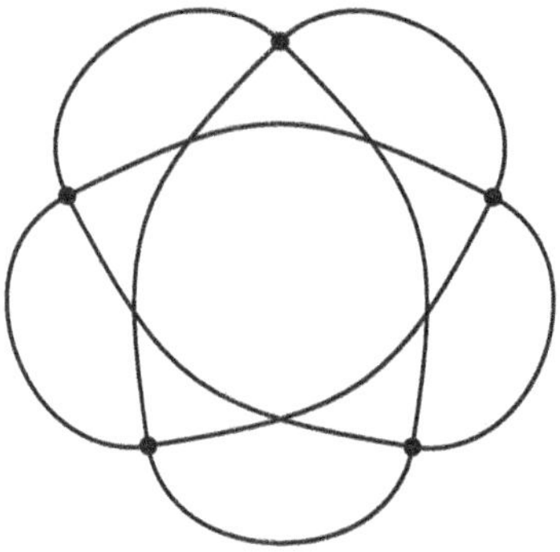

Fig. 5.9 A Lombardi drawing of K_5, each vertex has perfect angular resolution and edges are drawn as circular arcs

$\exists\mathbb{R}$-hardness of straight-line realizations also arises from *metric constraints*: graph realizability (CG63), unit spanning ratio (CG62), from constraints on the *geometric complexity* of the drawing: planar slope number (CG41), line cover number (CG43), segment number (CG44), and triangulation stretchability (CG47), and from *resolution constraints*: angular resolution (CG60), crossing resolution (CG58), and perfect resolution in a Lombardi drawing (with rotation system) (CG59). The last result is one of the few graph drawing results for a drawing style in which edges are neither straight-edges, nor arbitrary curves, but circular arcs.

A tantalizing and fundamental open question in this area is the complexity of the partial drawing extensibility problem (CG-Open7): Is it $\exists\mathbb{R}$-complete to test whether a partial straight-line embedding of a graph can be extended to a straight-line embedding of the full graph? The problem is only known to be $\exists\mathbb{R}$-hard if the realization has to lie within a polygon (CG48).

5.3.5 *Intersection Graphs*

Intersection graphs are graphs representing the intersection relation between objects. We refer to the book by McKee and McMorris for a general theory of such graphs [MM99]. In computational geometry, *geometric* intersection graphs are typically defined as follows: For a family $\mathcal{F}$ of subsets of $\mathbb{R}^d$, a graph $G = (V, E)$ is an $\mathcal{F}$-intersection graph if there exists a map $f : V \to \mathcal{F}$ such that $uv \in E$ if and only if $f(u) \cap f(v) \neq \emptyset$. Perhaps the simplest family to consider is the family of intervals in $\mathbb{R}$, which yields *interval graphs*. *Disk* intersection graphs, defined as intersection graphs of disks in the plane, have been used in the context of wireless communications for modeling the network induced by antennas with bounded transmission ranges. Other well-known families of geometric intersection graphs are *segment intersection graphs*, intersection graphs of straight line segments in the plane, and *string graphs*, intersection graphs of arbitrary curve segments ("strings") in the plane.

The *recognition* problem for a class of graphs is the problem of deciding whether a given graph belongs to the class. When defined as a language, the recognition problem is simply the set of graphs in the class, hence we can refer to the computational complexity of the recognition problem for a class simply as the complexity of the class.

Early investigations of Kratochvíl and Matoušek [KM94] on the complexity of segment intersection graphs, and the finding that they were ∃ℝ-complete, are one of the milestones of ∃ℝ-completeness theory (CG26). Since then, many classes have been proved ∃ℝ-complete. Table 5.6 summarizes the complexities of some of the best-known geometric intersection graphs. ∃ℝ-completeness holds in particular for convex sets in the plane (CG32), unit disks (CG21), and rays in the plane (CG29). Figure 5.10 shows relationship between classes of geometric intersection graphs. More recently, there have also been ∃ℝ-completeness results on intersection graphs in hyperbolic geometry [BBDJ23, BFKS23]. Matoušek [Mat14, Theorem 3.17] discusses the Oleinik–Petrovskii–Milnor–Thom theorem, which allows one to give an $2^{O(n \log n)}$ upper bound on the number of non-isomorphic intersection graphs; see also Sauermann [Sau21] on this topic.

Geometric contact graphs are defined similarly: A graph $G = (V, E)$ is an $\mathcal{F}$-contact graph for a family $\mathcal{F}$ of compact subsets of $\mathbb{R}^d$ if there exists a map $f : V \to \mathcal{F}$ such for any pair $u \neq v$ of vertices the sets $f(u)$ and $f(v)$ are interior-disjoint, and $uv \in E$ if and only if $f(u) \cap f(v) \neq \emptyset$. The Koebe-Andreev-Thurston Theorem states that planar graphs are exactly the contact graphs of disks. However,

Table 5.6 Classes of geometric intersection graphs and their computational complexity

Intersection graphs of	Complexity	Source
Circle chords	P	Gabor, Supowit, and Hsu [GSH89], Spinrad [Spi94]
(Unit) intervals	P	Booth and Lueker [BL76]
Outerstrings	NP-complete	Kratochvíl [Kra91a] (see also Rok and Walczak [RW19])
Strings	NP-complete	Kratochvíl [Kra91a], Schaefer and Sedgwick [SSŠ03]
Convex sets (CG32)	∃ℝ-complete	Schaefer [Sch10]
(Unit) disks (CG21)	∃ℝ-complete	McDiarmid and Müller [MM13]
Ellipses (CG32)	∃ℝ-complete	Schaefer [Sch10]
Outer segments (CG29)	∃ℝ-complete	Cardinal et al. [CFM+18]
k-polylines (CG28)	∃ℝ-complete	Hoffmann et al. [HMWW24]
(Downward) rays (CG29)	∃ℝ-complete	Cardinal et al. [CFM+18]
Segments (CG26)	∃ℝ-complete	Kratochvíl and Matoušek [KM94] (see also [Sch10, Mat14])
3D segments (CG30)	∃ℝ-complete	Evans et al. [ERS+19]
3D lines (CG31)	∃ℝ-complete	Cardinal [Car24]
Unit balls (CG24)	∃ℝ-complete	Kang and Müller [KM12]
Unit segments (CG27)	∃ℝ-complete	Hoffmann et al. [HMWW24]

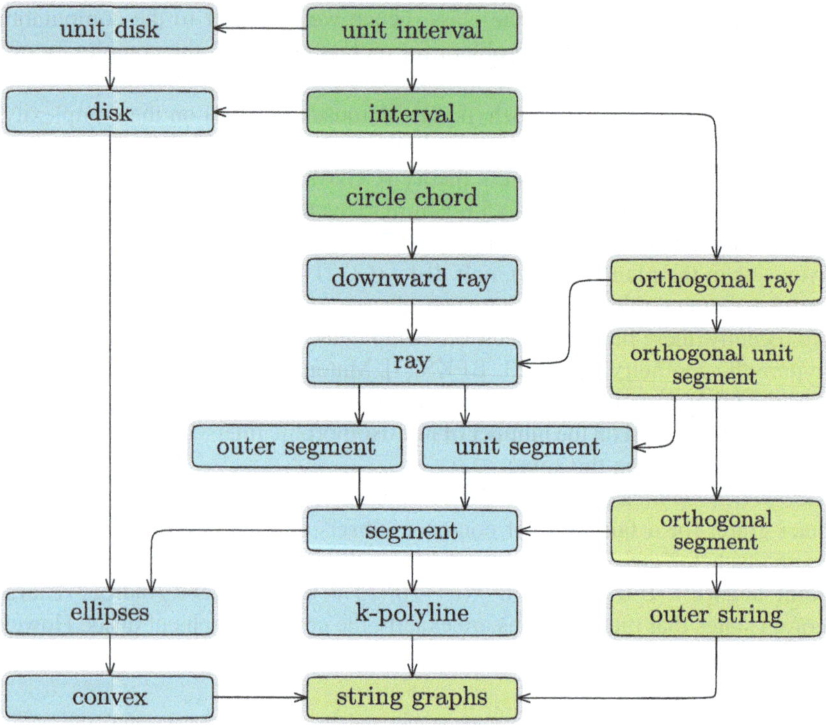

Fig. 5.10 The complexity of recognizing different geometric intersection graph classes. Recognition of the green classes is in P, ochre is NP-complete, and blue is $\exists\mathbb{R}$-complete

if we require that disks have unit radius, the recognition problem is known to be NP-hard, but not known to be $\exists\mathbb{R}$-hard [BK96, HK01] (CG-Open3).

5.3.6 *Visibility Graphs*

Visibility graphs are graphs whose vertices can be mapped to geometric objects so that two vertices are adjacent if and only if the two corresponding objects are visible from each other, hence if there exists a "line of sight" that connects the two objects and does not hit any obstacle. Typically, the objects are subsets of the plane, and lines of sight are straight line segments. A generalized notion of visibility graphs was suggested by Develin et al. [DHM03], and we refer to O'Rourke [O'R18] for a recent extensive bibliographical treatment. An older survey on algorithmic aspects of visibility was written by Ghosh [Gho07].

When the objects are points in the plane, and the obstacles are the points themselves, the graphs are called *point visibility graphs* [KPW05, Pfe08]. The recognition problem for these graphs (CG36) has been shown ∃ℝ-complete by Cardinal and Hoffmann [CH17].

Polygon visibility graphs are also the topic of an extended literature, and are closely related to the art gallery problem. A graph on the vertex set V is a polygon visibility graph if there exists a closed polygon whose vertices are in bijection with V, and such that two vertices of V are adjacent if and only if the line segment between the corresponding vertices of the polygon is completely contained in the polygon. Lin and Skiena showed [LS95] that a realizing polygon can require exponential area. Hoffmann and Merckx [HM18] proved that visibility graphs of polygons with holes are hard to recognize (CG37). Boomari et al. [BOZ18] proved that if one is given two graphs on the same vertex set, it is ∃ℝ-complete to decide whether they correspond to the interior and exterior visibility graphs of a polygon (CG38). Still, the computational complexity of the recognition problem for simple polygon visibility graphs remains open, even for the variant in which the ordering of the vertices along the polygon is given. It is also not clear whether segment visibility graphs are ∃ℝ-complete (CG-Open37).

Polygon visibility graphs are closely related to the famous *Art gallery problem* [O'R87]: In an art gallery, given as a simple polygon with n vertices, what is the minimum number of guards, or points, who together can see the whole gallery? When the guards are restricted to lie on the vertices of the polygon, and only vertices of the gallery have to be watched, the problem amounts to finding a dominating set in the polygon visibility graphs, and it is known that $\lfloor n/3 \rfloor$ guards are always sufficient and sometimes necessary [Chv75]. When the guards can be placed anywhere in the polygon, and the whole polygon has to be watched then the problem is ∃ℝ-complete (CG16), as shown by Abrahamsen, Adamaszek, and Miltzow [AAM18]. A version where the guards lie on the boundary of the polygon has also been shown ∃ℝ-complete by Stade [Sta23], see (CG17).

O'Rourke and Streinu [OS98] introduced the bipartite *vertex-edge visibility graphs* in which there is an edge between a vertex and a side of the polygon if the vertex can see some interior points of the side. Similarly, the *edge-edge visibility graphs* is a bipartite graph on pairs of sides of the polygon, and there is an edge between two sides if there are interior points of the two sides that can see each other. O'Rourke and Streinu [OS98] showed that from a vertex-edge visibility graph one can construct the edge-edge visibility graph of a polygon in polynomial time and vice versa, so the two recognition problems have the same complexity, and were both shown to lie in NP [OS97].

O'Rourke and Streinu [OS97] also defined *polygon pseudo-visibility graphs*, in which, roughly, one relaxes straight-lines to pseudolines. Polygon pseudo-visibility graphs were shown to be distinct from visibility graphs by Streinu [Str05], and to be recognizable in polynomial time, as recently shown by Ameer et al. [AGLKW22].

Terrain visibility graphs are defined as the visibility relations of vertices of an x-monotone polygonal line, where the lines of sights must remain above the line [AGLK+20]. The complexity of the recognition problem for these graphs remains unknown as well. A related problem, on so-called triangulated irregular networks (CG39) was studied recently by Boomari et al. [BOZ21].

Table 5.7 $\exists\mathbb{R}$-complete problems about the existence of Nash equilibria, adapted from Bilò et al. [BHM23]

Problem	Condition	Source
NE in a ball (GT1)	$\sigma_i(s_i) \le u\ \forall i$ and $s_i \in \Sigma_i$	Schaefer and Štefankovič [SŠ17]
Second NE (GT4)	σ is not the only NE	Garg et al. [GMVY18]
NE with large payoffs (GT2)	$\mathsf{U}_i(\sigma) \ge u\ \forall i$	Garg et al. [GMVY18]
NE with small payoffs (GT2)	$\mathsf{U}_i(\sigma) \le u\ \forall i$	
NE with large total payoff (GT5)	$\sum_i \mathsf{U}_i(\sigma) \ge u$	Bilò and Mavronicolas [BM21]
NE with small total payoff (GT5)	$\sum_i \mathsf{U}_i(\sigma) \ge u$	
NE with large supports (GT6)	$\lvert\mathsf{Supp}(\sigma_i)\rvert \ge k\ \forall i$	Bilò and Mavronicolas [BM21]
NE with small supports (GT6)	$\lvert\mathsf{Supp}(\sigma_i)\rvert \le k\ \forall i$	
NE with restricting supports (GT3)	$T_i \subseteq \mathsf{Supp}(\sigma_i)\ \forall i$	Garg et al. [GMVY18]
NE with restricted supports (GT3)	$\mathsf{Supp}(\sigma_i) \subseteq T_i\ \forall i$	
Pareto-optimal NE (GT11)	σ is Pareto-optimal	Berthelsen and Hansen [BH22]
Non-Pareto-optimal NE (GT11)	σ is not Pareto-optimal	
Strong NE (GT12)	σ is a strong NE	Berthelsen and Hansen [BH22]
Non-strong NE (GT12)	σ is not a strong NE	

5.3.7 Nash Equilibria and Games

The computational complexity of finding a Nash equilibrium of a game, the existence of which is guaranteed by the classical result of Nash, is the topic of an extensive and relatively recent literature. Daslakakis et al. [DGP09] and Cheng et al. [CDT09] proved that for two-player games, the problem is complete for the class PPAD. Etessami and Yannakakis later proved that for $r \ge 3$ the problem is FIXP-complete [EY10].

It is of interest to consider Nash equilibria in which some additional constraints are satisfied. For instance, does there exist a Nash equilibrium in which every payoff is bounded from above by some constant (GT2), or in which the supports of the mixed strategies are restricted (GT3)? For two players, many of these problems become NP-complete [CS08]. For three players, it has been shown that many of them become $\exists\mathbb{R}$-complete. The first result in this vein is due to Schaefer and Štefankovič [SŠ17], who extended [EY10] to show that 3-player Nash equilibrium with additional constraints on the probabilities is $\exists\mathbb{R}$-complete (GT1). Datta [Dat03] showed that the 3-played Nash equilibrium problem is universal, but the reduction requires an exponential number of strategies.

Table 5.7 lists a collection of problems about deciding the existence of Nash equilibria with some additional properties that are known to be $\exists\mathbb{R}$-complete

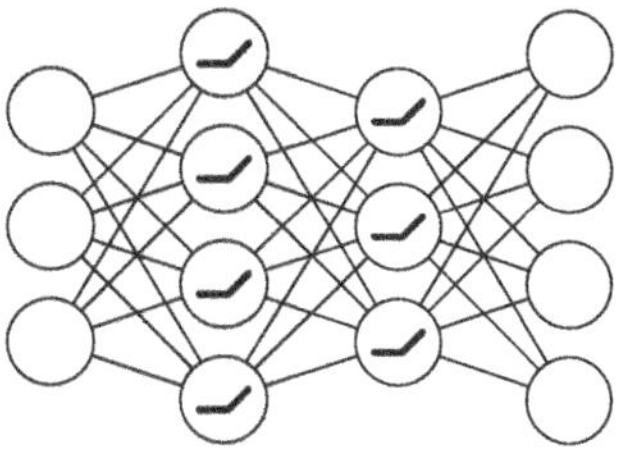

Fig. 5.11 Artificial neural networks are currently among the prominent supervised learning models

[SŠ17, GMVY18, BM21, BH22, BHM23]. In all cases, the number of players can be set to three. In most cases, except for Pareto-optimal and strong Nash equilibria, the game can be assumed to be a win-lose game, hence all payoffs are either 0 or 1. In many cases, variants of the problem in which the game is symmetric, and the sought Nash equilibria is constrained to be symmetric as well, are also ∃ℝ-complete.

A number of ∃ℝ-completeness results have been proved for problems on games in extensive form [Han19, GPS20, HS20], including for instance non-negativity of expected payoff for games with imperfect recall (GT16), and the existence of constrained stationary Nash equilibria in acyclic games (GT18). The existence of equilibria in Arrow-Debreu markets has also been shown ∃ℝ-complete by Garg et al. [GMVY17], (GT22).

5.3.8 Machine Learning and Probabilistic Reasoning

Supervised learning consists of developing systems that can solve certain tasks by leveraging example data instead of handcrafting a specific algorithm. Given a model with the parameters $x \in \mathbb{R}^n$, and some training data, the *training problem* consists of finding a good choice of the parameters using some optimization function indicating the fitting of the training data. Artificial neural networks are popular such models (Fig. 5.11). Although the field of machine learning stems from quite different paradigms and perspectives, we can phrase the training problem as a computational problem and study its computational complexity. It turns out that even the simplest artificial neural networks are ∃ℝ-complete to train to optimality (ML1). As there is a broad range of models with real-valued parameters, we expect that more training problems can be shown ∃ℝ-complete. This section contains also a list of probabilistic reasoning problems that are ∃ℝ-complete (ML3). In probabilistic reasoning, we have a logical formula about probabilities of events, and the question is whether there exist consistent values for those probabilities.

5.3.9 Markov Chains and Decision Processes

A Markov Chain is a stochastic process where the state at time t only depends on the state at time $t-1$. Markov chains offer a rich language to study all types of stochastic phenomenon. Markov decision processes are a bit more complex than Markov chains and are among the foundational models for the reinforcement learning problem and model checking. Here, we have a controller that is allowed to perform an action in each time step. The next state depends probabilistically on the current state and the action taken by the controller, so that once we choose a fixed strategy of the controller the Markov decision process becomes a Markov Chain (Fig. 5.12). Typical questions related to those models are about values of state (MDP2) and (MDP3) or reachability (MDP5) and (MDP7).

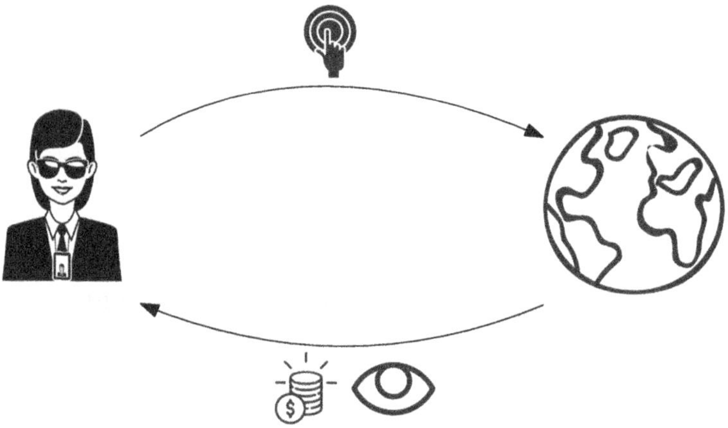

Fig. 5.12 In a Markov Decision Process a controller can perform actions each time step. In return the agent receives some observations about the state of world and a reward. There are many variants with the agent having more or less knowledge about the world, the reward function and the stochastic dynamics

5.3.10 Candidates

As we performed the research for this survey, we came across many problems for which we do not know the complexity, but which are likely to be classifiable using $\exists\mathbb{R}$, $\forall\mathbb{R}$, or some higher level of the hierarchy. We have included these problems throughout the compendium, if they are variants of other listed problems, or otherwise in the final section containing candidate problems.

Dobbins et al. [DKMR23, Sect. 5.2] thoroughly investigate and classify the nature of computational problems which may be complete for higher levels of the hierarchy. Some additional thoughts can be found in the conclusion section of [SŠ23].

5.4 Compendium

Before exploring the compendium, we recommend reading the short introductory section of Part 5.3. Let us add that within each section we have tried to group problems thematically though that has not always been possible; we have therefore included pointers to related problems at the end of many entries. An index at the end of the book may also be helpful in locating problems and getting an overview. Open questions for a problem are separated by the symbol ▼.

5.4.1 Logic

Syntactically, a *(first-order) logic* is specified by its *signature*, consisting of the *predicate*- and *function*-symbols (including constants), together with its logical symbols, such as variables, Boolean connectives, quantifiers and equality. There are then standard rules on how to form well-formed formulas in such a logic. Formulas without free variables are known as *sentences*. If all the quantifiers of a formula (or sentence) occur at the beginning, we say the formula is in *prenex* form. The quantifier-free part of a formula in prenex form is known as the *matrix* of the formula.

For a particular structure matching the signature of the logic, we can define whether a sentence is true or false in that structure, adding *semantics* to the logic. Most questions in this section will be concerned with the complexity of deciding truth in a particular structure, typically the real numbers. But one can also ask whether a set of sentences is *consistent*, that is, we can ask whether there is some structure that satisfies all the sentences; such a structure is often referred to as a *model*.

(L1) Existential Theory of the Reals (ETR)
Given: An existentially quantified sentence over the signature $(0, 1, +, -, \cdot)$ with comparison operators $\{=, <, \leq, \geq, >\}$ and Boolean connectives $\vee$, $\wedge$ and $\neg$.
Question: Is the sentence true over the real numbers?
Complexity: $\exists\mathbb{R}$-complete by definition or by Mnëv [Mnë88] and Shor [Sho91] via a reduction from simple stretchability (CG9). For a fixed number of variables, the problem can be solved in polynomial time, using quantifier elimination.
Comments: Mnëv's [Mnë88] universality theorem can be seen as the earliest $\exists\mathbb{R}$-completeness result; it shows that (simple) stretchability can "express" semialgebraic sets. Shor's [Sho91] paper then draws the connection explicitly, he writes "Mnëv [...] proved an even stronger theorem. When translated into complexity theory terms, his theorem implies that determining the stretchability of pseudoline arrangements is equivalent to the existential theory of the reals". Schaefer and Štefankovič [Sch10, SŠ17] argue for the existential theory of the reals (ETR) as the defining problem

for the complexity class $\exists\mathbb{R}$. In the BSS-model, hardness of ETR is the Cook-Levin theorem of that model, and one of the main results of the Blum-Shub-Smale paper [BSS89], also see [Cuc93]. We want to point out that the encoding is not always defined in the same way [BPR06], yet it is easy to show that all variants that we are aware of are polynomial-time equivalent.
Universality: Mnëv [Mnë88], also see Shor [Sho91].
Also see: Existential theory of totally ordered real variables (L2), feasibility (A1), simple stretchability (CG9), approximate existential theory of the reals (L3), matroid realizability over the reals (CG3).

(L2) Existential Theory of Totally Ordered Real Variables
Given: A set of conditions of the forms $x_i + x_j = x_k$ and $x_i x_j = x_k$, , on the set of real variables $x_1, \dots, x_n$.
Question: Are there $1 = x_1 < x_2 < \cdots < x_n \in \mathbb{R}$ satisfying all conditions?
Complexity: $\exists\mathbb{R}$-complete by Mnëv [Mnë88] and Shor [Sho91] via a reduction from ETR (L1).
Comments: This useful intermediate problem was used by Mnëv [Mnë88] and named by Shor [Sho91]. It can be considered one of the first normal forms of ETR; the ordering is essential in showing stretchability to be $\exists\mathbb{R}$-complete.
Universality: Mnëv [Mnë88], also see Shor [Sho91].
Also see: ETR (L1), stretchability (CG9), continuous CSP (L8).

(L3) Approximate Existential Theory of the Reals (ε-ETR)
Given: An existentially quantified sentence over the signature $(0, 1, +, \cdot)$ with with comparison operators $\{<_\varepsilon, \leq_\varepsilon, \geq_\varepsilon, >_\varepsilon\}$ and logic connectives $\vee, \wedge$ and $\neg$, and $\varepsilon \in \mathbb{Q}_{\geq 0}$. Here $a <_\varepsilon b$ if $a < b + \varepsilon$, and similarly for the other comparison operators.
Question: Is the sentence true over the real numbers?
Complexity: $\exists\mathbb{R}$-complete by Deligkas, Fearnley, Melissourgos, and Spirakis [DFMS22] via a reduction from feasibility (A1) for every fixed $\varepsilon \in \mathbb{Q}_{\geq 0}$.
Also see: ETR (L1), existential theory of totally ordered real variables (L2), feasibility (A1).

(L4) Distinct ETR
Given: A set of conditions of the forms $x_i + x_j = x_k$ and $x_i x_j = x_k$, on the set of real variables $x_1, \dots, x_n$.
Question: Are there *distinct* $x_1, \dots, x_n \in \mathbb{R}$ satisfying all conditions?
Complexity: $\exists\mathbb{R}$-completeness shown in [KdMM23] by Kim, de Mesmay, and Miltzow via a reduction from a variant of strict local positivity (A2). The result also follows from the $\exists\mathbb{R}$-completeness of the existential theory of totally ordered real variables (L2), but that result is more difficult to obtain.
Comments: This useful intermediate problem was used to show that matroid realizability over the reals (CG3) is $\exists\mathbb{R}$-complete.
Also see: ETR (L1), continuous CSP (L8), strict local positivity (A2), matroid realizability over the reals (CG3).

(L5) ETR- INV
Given: A set of conditions/constraints of the forms $x_i + x_j = x_k$, $x_i x_j = 1$ and $x_i = 1$, on the set of real variables $x_1, \ldots, x_n$.
Question: Are there $x_1, \ldots, x_n \in [1/2, 2]$ satisfying all conditions?
Complexity: $\exists\mathbb{R}$-completeness in [AAM18] by Abrahamsen, Adamaszek, and Miltzow via a reduction from feasibility (A1).
Comments: ETR- INV was introduced to show that the art gallery problem (CG16) is $\exists\mathbb{R}$-complete. It has been refined in various ways, for example, see PLANAR-ETR- INV (L6), RANGE- ETR- INV (L7) and the continuous constraint satisfaction problem (L8). The reduction has many small steps that are often only sketched, a detailed explanation can be found in [AM19].
Also see: Art gallery problem (CG16), PLANAR- ETR- INV (L6), RANGE- ETR-INV (L7), continuous constraint satisfaction problem (L8), feasibility (A1).

(L6) PLANAR- ETR- INV
Given: A set of conditions/constraints of the forms $x_i + x_j = x_k$, $x_i x_j = 1$ and $x_i = 1$, on the set of real variables $x_1, \ldots, x_n$. Furthermore, the variable-constraint graph is planar. The *variable-constraint graph* is the bipartite graph that has variable and constraints as vertices, and an edge for each occurrence of a variable in a constraint.
Question: Are there $x_1, \ldots, x_n \in \mathbb{R}$ satisfying all conditions?
Complexity: $\exists\mathbb{R}$-completeness shown by Dobbins, Kleist, Miltzow, and Rzążewski [DKMR18]. Later, Lubiw, Miltzow, and Mondal [LMM22] showed the same result for directed graphs.
Comments: This intermediate problem was used to show $\exists\mathbb{R}$-completeness of Prescribed Area Extension (CG61) and drawability of a graph in a polygonal domain (CG48).
Also see: ETR- INV (L5), graph in polygonal domain (CG48).

(L7) RANGE- ETR- INV
Given: An instance Φ of ETR- INV, a number $\delta = 2^{-\ell}$ for some natural number ℓ and for each variable x_i, an interval $I(x_i)$ of length at most 2δ.
Question: Are there real numbers $x_1, \ldots, x_n \in [1/2, 2]$ satisfying Φ and with each $x_i \in I(x_i)$?
Complexity: $\exists\mathbb{R}$-complete in the first Arxiv version of [AMS20].
Comments: In an updated version, the authors use Continuous Constraint Satisfaction Problems (CCSP) (L8). The problem is also difficult for $\delta = O(n^{-c})$, where n is the number of variables and c is an arbitrary constant.
Also see: ETR- INV (L5).

(L8) Continuous Constraint Satisfaction Problems (CCSP)
Given: Constraints of the form $x_i = \delta$, with $\delta \in \mathbb{Q}$, $x_i \geq 0$, $x_i + x_j = x_k$, and $f(x_i, x_j) = 0$ for a function $f : \mathbb{R}^2 \to \mathbb{R}$, on the set of real variables $x_1, \ldots, x_n$.

(Note that for each function f, we have a different CCSP and thus f is not given as part of the input, but the function f is part of the problem description.)
Question: Are there $x_1, \ldots, x_n \in \mathbb{R}$ satisfying all constraints?
Complexity: If the function f is sufficiently well-behaved and curved then the CCSP is $\exists\mathbb{R}$-complete by Miltzow and Schmiermann [MS24]. The problem remains $\exists\mathbb{R}$-complete if one adds range constraints, i.e., $x_i \in [-\delta, \delta]$ or inequalities, i.e., $f(x_i, x_j) \geq 0$ and $g(x_i, x_j) \geq 0$, assuming f and g meet certain conditions [MS24].
Comments: CCSP generalizes many other $\exists\mathbb{R}$-complete results for technical problems such as ETR- INV (L5). Continuous Constraint Satisfaction problems have also been studied by Jonsson and Thapper [JT14, JT18]. They showed NP-hardness in case the set of relations contains a not essentially convex relation as well as a primitive positively definable BNU (bounded, non-constant, and unary) relation.
Also see: ETR- INV (L5).

(L9) Collinearity Logic

Given: A conjunctive logical formula using the ternary relation COL, three points being collinear and distinct, and its negation.
Question: Are there points in the plane satisfying the constraints of the formula?
Complexity: $\exists\mathbb{R}$-complete by Davis et al. [DGC99] via a reduction from a restricted form of ETR (L1).
Comments: Collinearity logic is a special form of order type realizability (CG1). As an intermediate problem, Davis et al. [DGC99] work with a variant allowing the relation PAR (parallel), which is also $\exists\mathbb{R}$-complete. If we specify *for every triple of points* whether they are collinear or not, the problem becomes equivalent to matroid realizability over the reals (CG3), which is also $\exists\mathbb{R}$-complete.
Open Questions: Is collinearity logic universal? ▼ Does the problem remain $\exists\mathbb{R}$-complete if every subset of collinear points in the realization must have bounded size? A careful analysis of the hardness proof for partial order type realizability (CG1) will probably show that the problem remains $\exists\mathbb{R}$-hard if every maximal subset of collinear points has size at most 6 or 7.
Also see: Order type realizability (CG1), matroid realizability over the reals (CG3), linear hypergraph realizability (CG35).

(L10) $\mathcal{LR}$-Calculus

Given: A conjunctive logical formula using relations left(p, q, r) and right(p, q, r) for points in the Euclidean plane.
Question: Are there points in the plane satisfying the constraints of the formula? We say left(p, q, r) if p and q are distinct, and r lies strictly to the left of the line from p to q, and similarly for right(p, q, r).
Complexity: $\exists\mathbb{R}$-complete by Mnëv [Mnë88] since it is equivalent to simple order type realizability (CG1). Later, Lee [Lee13, Lee14] proved $\exists\mathbb{R}$-completeness for general logical formulas (not necessarily conjunctive) via a reduction from oriented matroid realizability (CG1), which is a slightly weaker result than Mnëv's.
Comments: The full $\mathcal{LR}$-calculus allows nine relations (e.g. see [SL15a]. For further

$\exists\mathbb{R}$-completeness results on spatial calculi, see [Lee13, Lee14].
Universality: Mnëv [Mnë88].
Also see: Order type realizability (CG1).

(L11) Cross-Product Expression Satisfiability
Given: A cross-product term $t(x_1, \ldots, x_n)$; that is, a term that is either one of the variables, or of the form $(r \times s)$, where r and s are cross-product terms.
Question: Are there vectors $v_1, \ldots, v_n \in \mathbb{R}^3$ such that $t(v_1, \ldots, v_n) = v_1$, where $\times$ is the cross-product over $\mathbb{R}^3$?
Complexity: $\exists\mathbb{R}$-complete by Herrmann et al. [HSZ13] via a reduction from Feasibility (A1).
Comments: Herrmann, Sokoli, and Ziegler show that the problem is equivalent to the existential theory of F for any field $F \subseteq \mathbb{R}$, and that the problem remains $\exists\mathbb{R}$-complete over the real projective plane [HSZ13]. They also consider various variants that turn out to be equivalent to polynomial identity testing.
Also see: Dot-product graph (CG25).

(L12) Realizability of Euclidean Embeddings of Rankings (REER)
Given: A conjunctive logical formula of the forms $x_i \preceq_\ell x_j$ and $x_i \prec_\ell x_j$, where $i, j \in [n]$ and $\ell \in [d]$.
Question: Are there points $x_i \in \mathbb{R}^d$, $i \in [n]$ and vectors $v_\ell \in \mathbb{R}^d$ such that all conditions are satisfied? Here $x_i \preceq_\ell x_j$ means that $v_\ell^T(x_i - x_j) \geq 0$, and $x_i \prec_\ell x_j$ means that $v_\ell^T(x_i - x_j) > 0$.
Complexity: $\exists\mathbb{R}$-complete by Schockaert and Lee [SL15a] for any fixed dimension $d \geq 2$ via a reduction from $\mathcal{LR}$-calculus.
Also see: $\mathcal{LR}$-calculus (L10).

(L13) Topological Inference in RCC8 with Convexity
Given: A *constraint network*, that is, a conjunctive logical formula using the predicate "convex", and the eight binary relations of the region-connection calculus (RCC8): DC (disconnected), EC (externally connected), EQ (equal), PO (partially overlapping), TPP (tangential proper part), TPPi (tangential proper part inverse), non-tangential proper part (NTTP), and non-tangential proper part inverse (NTTPi), see [DGC99] for definitions.
Question: Are there regular regions in the plane satisfying the logical condition? A region is *regular* if it is the closure of its interior. It does not have to be connected.
Complexity: $\exists\mathbb{R}$-complete by Davis et al. [DGC99] via a reduction from collinearity logic (L9). The problem remains $\exists\mathbb{R}$-complete if relations are restricted to EC, PP and "convex" [DGC99], even in (fixed) higher dimension [SL15b, Appendix A]. It also remains complete if the set of relations is PO, DC and "convex" and the relation between any two regions has to be specified (a *fully specified contraint network*) [Sch10]; this result is obtained via a reduction from intersection graphs of convex sets (CG32).
Comments: Membership in $\exists\mathbb{R}$ is not trivial, since regions only have to be regular.
Open Questions: Does the topological inference problem in RCC8 with convexity remain in $\exists\mathbb{R}$, if we added the predicate "connected" [Sch10, Remark 4]?
Also see: Convex set intersection graph (CG32).

(L14) Quantum Logic Strong Satisfiability
Given: A propositional formula $\phi(x)$, with $x = (x_1, x_2, \ldots, x_n)$.
Question: Is the formula $\phi(x)$ *strongly satisfiable* over the modular ortholattice $L(\mathbb{R}^d)$, see definition below, for some constant $d \geq 3$, i.e., does there exist an assignment of values $a_i \in L(\mathbb{R}^d)$ to the x_i such that $\phi(a) = \mathbf{1}$ when interpreted in $L(\mathbb{R}^d)$?
Complexity: $\exists\mathbb{R}$-complete by [HZ16].
Comments: The lattice $L(\mathbb{R}^d)$ is the lattice of *linear subspaces of* $\mathbb{R}^d$, with the following three operations, where U and V are two such linear subspaces:

meet: $U \wedge V = U \cap V$,
complement: $\neg U = \{x \in \mathbb{R}^d : \langle x, u \rangle = 0 \ \forall u \in U\}$, the orthogonal subspace,
join: $U \vee V = U + V$, where $+$ denotes the Minkowski sum. Hence the join $U \vee V$ is simply the linear subspace spanned by U and V.

In addition, we define the constants $\mathbf{0} = \{(0, 0, \ldots, 0)\}$ (the origin) and $\mathbf{1} = \mathbb{R}^d$ (the whole space). The lattices $L(\mathbb{R}^d)$ are *modular ortholattices*. They form the basis of the so-called *quantum logic* proposed by Birkhoff and von Neumann [BvN36]. The problem studied by Hermann and Ziegler [HZ16] should not be confused with another quantum satisfiability problem referred to as QSAT. For $d = 2$, the problem is NP-complete.
Also see: Quantum Logic Weak Satisfiability (L15).

(L15) Quantum Logic Weak Satisfiability
Given: A propositional formula $\phi(x)$, with $x = (x_1, x_2, \ldots, x_n)$.
Question: Is the formula $\phi(x)$ *weakly satisfiable* over $L(\mathbb{R}^d)$ for some fixed constant $d \geq 3$, i.e., does there exist an assignment of values $a_i \in L(\mathbb{R}^d)$ to the x_i such that $\phi(a) \neq \mathbf{0}$ when interpreted in $L(\mathbb{R}^d)$?
Complexity: $\exists\mathbb{R}$-complete by [HZ16].
Comments: See quantum logic strong satisfiability (L14) for definitions. For $d = 2$, the problem is NP-complete.
Also see: Quantum Logic Strong Satisfiability (L14).

(L16) Arithmetic Data Exchange Realizability
Given: Two *(database) schemas*, S and T (finite relational signatures over real arithmetic), an *arithmetic schema mapping*, $(S, T, \Sigma_{st}, \Sigma_t)$, where Σ_{st} and Σ_t are finite sets of sentences over $S \cup T$ and T, respectively (with Σ_{st} encoding translation constraints, and Σ_t encoding target constraints), and an *instance* I of S. The target constraints Σ_t are assumed to be *weakly acyclic*. (For detailed definitions, see [tCKO13].)
Question: Is there an instance J of T such that (I, J) satisfies both Σ_{st} and Σ_t?
Complexity: $\exists\mathbb{R}$-complete by ten Cate et al. [tCKO13, Theorem 4.10] via a reduction from rectilinear crossing number (CG50).
Comments: Ten Cate, Kolaitis, and Othman also show that the *certain answer variant* of the data exchange problem is $\forall\mathbb{R}$-complete for conjunctive queries.
Also see: Rectilinear crossing number (CG50).

5.4.2 *Algebra*

Polynomials in this section are *multivariate* unless explicitly stated. Polynomials can be represented in various different ways; our default representation is the *explicit (or sparse)* representation as a sum of monomial terms with integer coefficients. A denser representation of polynomials can be achieved through an *arithmetic (or algebraic) circuit*, which is a directed graph with vertices (*gates*) that represent variables or the mathematical operators $+, -, \cdot$. The difference between the explicit and the circuit representation corresponds to the difference between Boolean formulas (in CNF) and Boolean circuits in logic: circuits can reuse already computed values. Circuits can be extended with additional operators, such as the sign-function or division, which allows them to represent non-polynomial functions [BC09].

In this section we will occasionally encounter $\exists\mathbb{C}$, the existential theory of the complex numbers.

5.4.2.1 Polynomials and Arithmetical circuits

(A1) Feasibility
Given: A family $f = (f_1, \ldots, f_k) : \mathbb{R}^n \to \mathbb{R}^k$ of polynomials $f_i \in \mathbb{Z}[x_1, \ldots, x_n]$.
Question: Is there an $x \in \mathbb{R}^n$ such that $f_i(x) = 0$ for all $i \in [k]$?
Complexity: $\exists\mathbb{R}$-complete, even if the total degree of all polynomials is at most 2 or if $k = 1$ and the single polynomial is nonnegative and has total degree at most 4 [BSS89, BFS99, BCSS98, SŠ17]. Reductions are from ETR (L1). Also remains $\exists\mathbb{R}$-complete if the domain is a compact set like the unit ball or $[-1, 1]^n$ and each polynomial has total degree at most 2 [Sch13, Lemma 3.9]. For a fixed number n of variables, the feasibility problem can be decided in polynomial time (using quantifier elimination). If $k = 1$ and the degree is at most 3, the problem is solvable in polynomial time by Triesch [Tri90].
Comments: In the BSS-model this result traces back to [BSS89, BCSS98]. Buss et al. [BFS99] define the existential theory $\text{ETh}(F)$ of a field F (as a decision problem), and show that the feasibility problem over F is complete for $\text{ETh}(F)$. In the real case, this is true even for $k = 1$. These reductions work with a Tseitin transform, see, for example [Mat14, Proposition 3.2]. Arguably, the $\exists\mathbb{R}$-completeness result is implicit in the normal forms devised by Mnëv [Mnë88] and Shor [Sho91]. Triesch's result was proved in the BSS-model [Tri90], but carries over. See the remark at the beginning of Part 5.3 for some comment on other existential theories.
Open Questions: Does the problem remain $\exists\mathbb{R}$-complete for $n = 1$ and $k = 1$ if the polynomial is represented by a straight-line program? See the candidate problem feasibility of a univariate polynomial.
Also see: ETR (L1), feasibility of univariate polynomials (A-Open1), existential theory of invertible matrices (L-Open2).

(A2) (Strict) Local Positivity
Given: A family $f = (f_1, \ldots, f_k) : \mathbb{R}^n \to \mathbb{R}^k$ of polynomials $f_i \in \mathbb{Z}[x_1, \ldots, x_n]$.

We say that f is *(strictly) locally positive* if $f(x) \geq 0$ ($f(x) > 0$) for some $x \in \mathbb{R}^n$.
Question: Is f (strictly) locally positive?
Complexity: $\exists\mathbb{R}$-complete for family of polynomials of degree 2, by Schaefer and Štefankovič [SŠ17], or a single polynomial of degree 8 by Ouaknine and Worrell [OW17, Theorem 7] via a reduction from Feasibility (A1). The reductions work for both strict and non-strict versions. The result also remains true for bounded domains [OW17, SŠ23]. (Ouaknine and Worrell only prove the bounded case and write that it is "straightforward" to adapt to the case $\mathbb{R}^n$; a proof of that case can be found in [SŠ23], which also lowers the degree to 4.) Kiefer and Tang [KT24] give an independent proof for a single multivariate polynomial of degree 6 on a bounded domain.
Comments: The reductions do not (and cannot) maintain strong forms of universality (such as stable equivalence), since feasibility describes closed sets and positivity describes open sets. Strict local positivity is called STRICT INEQ by Kratochvíl and Matoušek [KM94]. Cucker and Rosselló [CR92] discuss this problem under the name $SPOS_d$.
Also see: Feasibility (A1), global positivity (A3), positivity of linear recurrence (Misc2).

(A3) (Strict) Global Positivity
Given: A polynomial $f : \mathbb{R}^n \to \mathbb{R}$ in $\mathbb{Z}[x_1, \ldots, x_n]$. If $f(x) \geq 0$ ($f(x) > 0$) for all $x \in \mathbb{R}^n$ we say that f is *(strictly) globally positive*.
Question: Is f (strictly) globally positive?
Complexity: $\forall\mathbb{R}$-complete by Ouaknine and Worrell [OW17, Theorem 7] via a reduction from Feasibility (A1). Remains hard for bounded domain and total degree at most 8.
Comments: Strict global positivity is what is typically meant by a function being positive. Since f is strictly globally positive (globally positive) if and only if $-f$ is not locally positive (not strictly locally positive), $\forall\mathbb{R}$-completeness follows from Ouaknine and Worrell [OW17, Theorem 7] showing that (strict) local positivity is $\exists\mathbb{R}$-complete. In the BSS-model, Cucker and Rosselló [CR92] called global positivity *Hilbert's 17th problem*, and classified its complement using the exotic quantifier $\exists^*$ later introduced by Koiran [Koi99, Koi00].
Also see: Feasibility (A1), local positivity (A2).

(A4) Fixed Point
Given: A family $f = (f_1, \ldots, f_n) : \mathbb{R}^n \to \mathbb{R}^n$ of polynomials $f_i \in \mathbb{Z}[x_1, \ldots, x_n]$.
Question: Does f have a *fixed point*, that is an $x \in \mathbb{R}^n$ with $f(x) = x$?
Complexity: $\exists\mathbb{R}$-complete by Schaefer and Štefankovič [SŠ17] via a (trivial) reduction from Feasibility (A1).
Comments: The reduction is trivial.
Also see: Feasibility (A1), Brouwer Fixed Point (A5).

(A5) Brouwer Fixed Point
Given: A family $f = (f_1, \dots, f_n) : [-1, 1]^n \to [-1, 1]^n$ of polynomials in $\mathbb{Z}[x_1, \dots, x_n]$, $q \in \mathbb{Q}^n$, $\delta \in \mathbb{Q}_{\geq 0}$.
Question: Is there a *fixed point* of f within distance δ of q, that is an $x \in [-1, 1]^n$ with $f(x) = x$ and $\|x - q\|_\infty < \delta$?
Complexity: $\exists\mathbb{R}$-complete by Schaefer and Štefankovič [SŠ17] via a reduction from Fixed Point (A4). The problem remains $\exists\mathbb{R}$-complete for other natural compact domains and norms. For fixed dimension n, the problem can be decided in polynomial time (using quantifier elimination).
Comments: Without the restriction $\|x - q\|_\infty < \delta$, there always is a fixed point, by Brouwer's fixed-point theorem. Ettessami and Yannakakis [EY10] introduced the complexity class FIXP to capture the complexity of these types of problems.
Also see: Fixed point (A4).

(A6) Nontrivial Homogeneous Zero (H_2N)
Given: A family $f = (f_1, \dots, f_k) : \mathbb{R}^n \to \mathbb{R}^k$ of homogeneous polynomials $f_i \in \mathbb{Z}[x_1, \dots, x_n]$.
Question: Is there a *non-trivial zero*, that is, an $x \in \mathbb{R}^k$ with $f(x) = 0$ and $x \neq 0$?
Complexity: $\exists\mathbb{R}$-complete by Schaefer [Sch13, Corollary 3.10] from feasibility (A1) with compact domain. Remains $\exists\mathbb{R}$-complete even if the total degree of each polynomial is 2.
Comments: This problem was introduced as H_2N, abbreviating Hilbert's Homogeneous Nullstellensatz, by Koiran and shown NP-hard [Koi00, Theorem 6]. The proof in [Sch13] yields polynomials of degree 4, but that is unnecessary: the term $\sum_{i=1}^n x_i^2$ in the proof does not need to be square, which implies that y_0^4 and x_0^4 can be replaced by y_0^2 and x_0^2, yielding quadratic polynomials. This problem is equivalent to a 3-dimensional tensor having eigenvalue 0. The problem is called Quadratic Feasibility in [HL13], which also contains a sketch of a hardness proof that has issues [SŠ18, Example 2.5].
Also see: Feasibility (A1), ISO (A7), tensor with eigenvalue 0 (A28).

(A7) Isolated Zero (ISO)
Given: A family $f = (f_1, \dots, f_k) : \mathbb{R}^n \to \mathbb{R}^k$ of polynomials $f_i \in \mathbb{Z}[x_1, \dots, x_n]$.
Question: Does f have an *isolated zero*, that is, an $x \in \mathbb{R}^k$ such that $f(x) = 0$ and $f(x') \neq 0$ in an open neighborhood of x?
Complexity: $\forall\mathbb{R}$-complete, by Koiran [Koi00], Schaefer [Sch13] via a reduction from H_2N (A6) due to Koiran. Membership is shown in Schaefer and Štefankovič [SŠ17]. Remains $\forall\mathbb{R}$-complete even if the total degree of each polynomial is 4.
Comments: Koiran [Koi00] introduced the problem ISO as the question whether the algebraic variety $\{x \in \mathbb{R}^n : f(x) = 0\}$ contains an isolated point, and observed that H_2N (A6) reduces to the complement of ISO; since he also showed that H_2N (A6) is NP-hard, this implied coNP-hardness for ISO.
Also see: H_2N (A6), isolated point (A39), semialgebraic set with isolated point (A-Open3).

(A8) Injectivity (polynomial)
Given: A family $f = (f_1, \dots, f_n) : \mathbb{R}^n \to \mathbb{R}^n$ of polynomials $f_i \in \mathbb{Z}[x_1, \dots, x_n]$.
Question: Is f injective on $\mathbb{R}^n$?
Complexity: $\forall\mathbb{R}$-complete (unpublished, see comments). For functions represented by arithmetic circuits, this was shown by Bürgisser and Cucker [BC09, Corollary 9.4].
Comments: Start with the feasibility problem for $g \in \mathbb{Z}[x_1, \dots, x_{n-1}]$. Then g is feasible if and only if $f(x_1, \dots, x_n) := (x_1, \dots, x_{n-1}, x_n g(x_1, \dots, x_{n-1}))$ is not injective. Starting with feasibility on a bounded or open domain implies hardness for testing injectivity on these domains as well.
Also see: Surjectivity (second level) (A45).

(A9) Regularity (polynomial)
Given: A polynomial $f \in \mathbb{Z}[x_1, \dots, x_n]$.
Question: Is f *regular*? That is, there is no common real zero of f and its partial derivatives.
Complexity: $\forall\mathbb{R}$-complete by Cucker and Rosselló [CR92] via a reduction from (non)-feasibility (A1).
Comments: A polynomial $g \in \mathbb{Z}[x_1, \dots, x_n]$ is feasible if and only if $f = g^2$ is not regular.
Also see: Feasibility (A1).

(A10) Totality (arithmetic circuit)
Given: An arithmetic circuit with division defining a (partial) function $f : \mathbb{R}^n \to \mathbb{R}$. The value of the circuit is not *defined* if it requires division by zero.
Question: Is f total on $\mathbb{R}^n$, that is, is f defined for every $x \in \mathbb{R}^n$?
Complexity: $\forall\mathbb{R}$-complete by Bürgisser and Cucker [BC09, Proposition 4.1, Corollary 9.4] via a reduction from Feasibility (A1).
Comments: The problem is hard for a single division gate, since we can let $f = 1/g$, where g is a polynomial. Then f is total if and only if g is not feasible. Without division gate, the problem is trivial.
Also see: Feasibility (A1), injectivity (A8), surjectivity (second level) (A45).

(A11) Continuity (arithmetic circuit)
Given: An arithmetic circuit, division-free but with sign function sgn, defining a function $f : \mathbb{R}^n \to \mathbb{R}$.
Question: Is f continuous on $\mathbb{R}^n$?
Complexity: $\forall\mathbb{R}$-complete by Bürgisser and Cucker [BC09, Corollary 9.4] via a reduction from Feasibility (A1).
Comments: A polynomial g is feasible if and only if $f = \operatorname{sgn}(g^2)$ is not continuous, showing hardness of continuity. The trickier part is showing membership in $\forall\mathbb{R}$, since expressing continuity seems to require a quantifier alternation; Bürgisser and Cucker solve this problem via exotic quantifiers, which can be eliminated in this case.
Also see: Feasibility (A1), continuity at a point (arithmetic circuit) (A12).

(A12) Continuity at a Point (arithmetic circuit)
Given: An arithmetic circuit, division-free but with sign function sgn, defining a function $f : \mathbb{R}^n \to \mathbb{R}$, a point $x \in \mathbb{Q}^n$.
Question: Is f continuous at x?
Complexity: $\forall\mathbb{R}$-complete by Bürgisser and Cucker [BC09, Proposition 6.13, Corollary 9.4] via a reduction from zero belonging to the closure of a semialgebraic set (A38).
Comments: The hardness proof by Bürgisser and Cucker [BC09, Proposition 6.13] starts with a semialgebraic set S for which we know that $0 \notin S$ and testing whether 0 belongs to the closure of S is $\exists\mathbb{R}$-hard. S is given by an arithmetic circuit g. Then 0 belongs to the closure of S if and only if $f = \text{sgn}(g^2)$ is not continuous at 0, and this proof can be adapted for any fixed $x \in \mathbb{Q}^n$. The trickier part is showing membership in $\forall\mathbb{R}$, since expressing continuity seems to require a quantifier alternation; Bürgisser and Cucker solve this problem via exotic quantifiers, which can be eliminated in this case.
Also see: Continuity (arithmetic circuit) (A11), zero in closure of semialgebraic set (A38).

(A13) Domain Density (arithmetic circuit)
Given: An arithmetic circuit (with division gate and sign function), defining a (partial) function $f : \mathbb{R}^n \to \mathbb{R}$.
Question: Is the domain of f dense in $\mathbb{R}^n$?
Complexity: $\forall\mathbb{R}$-complete, by Bürgisser and Cucker [BC09, Proposition 5.7, Corollary 9.4].
Comments: In the Zariski topology, Bürgisser and Cucker [BC09, Proposition 5.7, Corollary 9.4] show that the problem is $\exists\mathbb{R}$-complete.
Also see: Image density (second level) (A46), density of semialgebraic set (A35).

(A14) Polynomial Projection (Linear)
Given: Two polynomials $f, g : \mathbb{R}^n \to \mathbb{R}$ in $\mathbb{Z}[x_1, x_2, \ldots, x_n]$.
Question: Is there a matrix A such that $f(x) = g(Ax)$?
Complexity: $\exists\mathbb{R}$-complete by Bläser et al. [BRS17] via a reduction from the symmetric tensor rank problem (A27) .
Comments: Over an arbitrary integral domain, the problem is complete for the existential theory of that domain [BRS17]. If A is a diagonal, invertible matrix, then the polynomial projection problem can be solved in randomized polynomial time [BRS17]. The general (affine) polynomial projection problem asks whether there is a matrix A and a vector b such that $f(x) = g(Ax + b)$. The special case $A = I$ is known as polynomial equivalence under shifts (A-Open2).

Open Questions: Is the affine polynomial projection problem ∃ℝ-hard? This version is known to be NP-hard by Kayal [Kay12].
Also see: Polynomial equivalence under shifts (A-Open2).

5.4.2.2 Matrices and Tensors

A d-dimension (real) tensor T is an element in $\mathbb{R}^{n_1\times\cdots\times n_d}$. The tensor has rank at most k if it can be written as $\sum_{i=1}^{k} v_1 \otimes \cdots \otimes v_k$, where $v_i \in \mathbb{R}^{n_i}$ for $1 \leq i \leq d$, and $\otimes$ is the outer product of two vectors. Matrices are then the special case of 2-dimensional tensors. There are other notions of tensor- and matrix-rank, some of which we discuss below.

(A15) Matrix Singularity
Given: A square matrix $M(x_1, \ldots, x_k)$ with variable entries $x_1, \ldots, x_k$ and entries in $\mathbb{Q}$.
Question: Is $M(x_1, \ldots, x_k)$ singular, that is, does $M(x_1, \ldots, x_k)$ have determinant 0 for some $(x_1, \ldots, x_k) \in \mathbb{R}^k$?
Complexity: ∃ℝ-complete by Buss et al. [BFS99] via a reduction from Feasibility (A1).
Comments: Buss et al. [BFS99] show that the singularity problem over any commutative ring R is complete for the existential theory of R. The reduction is based on Valiant's universality proof for the determinant [Val79]. Non-singularity, that is, testing whether $M(x_1, \ldots, x_k)$ is non-singular for some $x_1, \ldots, x_k$ lies in randomized polynomial time, so is unlikely to be ∃ℝ-complete [BFS99]. Brijder, Geerts, and Van den Bussche [BGVdBW19] introduce a formal matrix language (MATLANG) in which matrix singularity is expressible, implying that expression evaluation in that language is ∃ℝ-complete.
Also see: Existential theory of invertible matrices (L-Open2), constrained matrix invertibility problem (A-Open7), feasibility (A1).

(A16) Minimum Matrix Rank
Given: A square matrix $M(x_1, \ldots, x_k)$ with variable entries $x_1, \ldots, x_k$ and entries in $\mathbb{Q}$, and an integer r.
Question: Is the rank of $M(x_1, \ldots, x_k)$ at most r for some $(x_1, \ldots, x_k) \in \mathbb{R}^k$?
Complexity: ∃ℝ-complete by Buss et al. [BFS99] via a reduction from Feasibility (A1).
Comments: Buss et al. [BFS99, Lemma 17] show that the minimum matrix rank problems over any field F is complete for the existential theory of F, even if the matrix entries lie in $\{0, 1\}$. A minimum rank problem for matrices with affine entries can be used to prove ∃ℝ-completeness of tensor rank (A27), see [SŠ18]. The corresponding maximum matrix rank problem asking whether there are $x_1, \ldots, x_k$ for which $M(x_1, \ldots, x_k)$ has rank at least r can be decided in randomized polynomial time [BFS99, Theorem 2].
Open Questions: Buss et al. [BFS99] show that approximating the minimum rank of a matrix to within a factor of $1 + \varepsilon$ is NP-hard for sufficiently small ε, is it also ∃ℝ-hard? ▼ How hard is the minimum matrix rank problem if each variable occurs

only once in the matrix? This is the well-known minimum rank matrix completion problem, the maximum rank version of this problem can be solved in randomized polynomial time [BFS99]. Related to this is the problem of computing the minimum rank of a sign pattern matrix (A18). ▼ Is the problem hard for constant k, say $k = 4$?
Also see: Feasibility (A1), tensor rank (A27), minimum rank of sign pattern matrix (A18), minimum graph rank (A20), low rank approximation (matrix) (A-Open8).

(A17) Minimum Rank of Linear Combination of Matrices
Given: A family of matrices $M_1, M_2, \ldots, M_k \in \mathbb{Q}^{m\times n}$ and an integer r.
Question: Is the rank of $\sum_i \lambda_i M_i$ at most r for some nonzero combination $(\lambda_1, \ldots, \lambda_k) \in \mathbb{R}^k$?
Complexity: $\exists\mathbb{R}$-complete by Bläser et al. [BIL+21, Theorem 2.3] via a reduction from H_2N (A6) even for $r = 1$.
Comments: There are interesting connections to various notions of tensor rank, including slice rank and border rank, see [BIL+21]. The result is used to prove $\exists\mathbb{R}$-hardness of the orbit closure containment problem (A29).
Also see: Minimum matrix rank (A16), tensor rank (A27), orbit closure containment (A29).

(A18) Minimum Rank of Sign Pattern Matrix
Given: A *generalized sign pattern matrix*, that is, $M \in \{+, -, 0\}^{n\times n}$, and an integer k.
Question: Is there a matrix $A \in \mathbb{R}^{n\times n}$ of rank at most k so that $\text{sgn}(A) = M$, where sgn is the sign function applied to each entry?
Complexity: $\exists\mathbb{R}$-complete by Bhangale and Kopparty [BK15] via a reduction from simple stretchability (CG9), even for $k = 3$.
Comments: For $k = 2$ the problem is solvable in polynomial time [BK15]. The maximum rank variant is equivalent to a matching problem and can be solved in polynomial time [BK15].
Open Questions: Generalized in the name refers to allowing the entry 0; in a *sign pattern matrix* zero entries are not allowed; in this case Bhangale and Kopparty [BK15] can still show that the problem is NP-hard, but $\exists\mathbb{R}$-completeness is open.
Also see: Minimum matrix rank (A16), feasibility (A1).

(A19) Monotone Matrix Rank
Given: A matrix $M \in \mathbb{Q}^{m\times n}$ and an integer k.
Question: Is the *monotone rank* of M at most k, that is, is there a matrix A of rank at most k, and monotone functions $f_1, \ldots, f_n$ such that $M_{ij} = f_j(A_{ij})$, where $i \in [m]$, $j \in [n]$?
Complexity: $\exists\mathbb{R}$-complete for $k = 2$ by Lienkaemper [Lie23] via a reduction from simple allowable sequence (CG4).
Also see: Minimum matrix rank (A16), minimum rank of sign pattern matrix (A18), (simple) allowable sequence (CG4).

(A20) Minimum Graph Rank
Given: A graph G on n vertices and an integer $k \geq 0$.
Question: Is there a real symmetric matrix $A \in \mathbb{R}^{n\times n}$ of rank at most k so that $v_i v_j \in E(G)$ if and only if $A_{ij} \neq 0$, for $1 \leq i \neq j \leq n$?
Complexity: $\exists\mathbb{R}$-complete even for $k = 3$ by Grace, Hall, Jensen and Lawrence [Law23, Sect. 4] via a reduction from feasibility (A1).
Comments: The smallest k given G is often called the minrank of G, e.g. see [GH22]. Over an arbitrary infinite field, instead of $\mathbb{R}$, the problem (with A being Hermitian instead of symmetric) is equivalent to the existential theory of the field (according to [Law23]).
Also see: Minimum matrix rank (A16), orthogonality dimension (CG-Open18).

(A21) Nonnegative Matrix Factorization (NMF)
Given: A square matrix $M \in \mathbb{Q}^{n\times n}$ with all entries nonnegative and a natural number $k \in \mathbb{N}$.
Question: Are there matrices $A \in \mathbb{R}^{n\times k}$, $B \in \mathbb{R}^{k\times n}$ such that $M = AB$ and all entries of A and B are nonnegative?
Complexity: $\exists\mathbb{R}$-complete by Shitov [Shi16b] via a reduction from feasibility (A1). The reduction also establishes that the nonnegative factorization of a rational matrix into rational (rather than real) factors A, B is $\exists\mathbb{Q}$-complete.
Comments: Vavasis had earlier shown NP-hardness [Vav09]. The smallest k for which M has a nonnegative factorization equals the nonnegative rank of M, that is the smallest number of rank-1 matrices with nonnegative entries, whose sum is M [CR93, Lemma 2.1], so the nonnegative matrix rank problem is also $\exists\mathbb{R}$-complete. Rank problems for matrices with affine entries can be used to prove $\exists\mathbb{R}$-completeness of tensor rank (A27), see [SŠ18].
Universality: Shitov [Shi16b] also shows universality.
Open Questions: Is the problem already hard for some fixed k, say four?
Also see: Symmetric nonnegative matrix trifactorization (A22), nested polytopes (CG72), Hadamard decomposition (A-Open9).

(A22) Symmetric Nonnegative Matrix Trifactorization (SNT)
Given: A symmetric square matrix $M \in \mathbb{Q}^{n\times n}$ with all entries nonnegative and a natural number $k \in \mathbb{N}$.
Question: Are there matrices $A \in \mathbb{R}^{n\times k}$, $B \in \mathbb{R}^{k\times k}$ such that $M = BAB^T$ and all entries of A and B are nonnegative and A is symmetric?
Complexity: $\exists\mathbb{R}$-complete by Shitov [Shi25a] via a reduction from nonnegative matrix factorization (A21) based on [KBŠ23].
Also see: Nonnegative matrix factorization (A21).

(A23) (Real) PSD Rank
Given: A matrix $A \in \mathbb{Q}^{m\times n}$ and a natural number $r \in \mathbb{N}$.
Question: Is psd-rank$(A) \leq r$? That is, are there Hermitian positive semidefinite $r \times r$ matrices $(B_1, \ldots, B_m)$ and $(C_1, \ldots, C_n)$ with $A_{ij} = \text{trace}(B_i C_j)$, for all i, j? If the B_i and C_j are restricted to be real, we obtain the *real PSD rank* of A.

Complexity: Shitov established $\exists\mathbb{R}$-completeness [Shi23, Theorem 3.4] of the (complex) PSD rank. He also established $\exists\mathbb{R}$-completeness [Shi17a] of the real PSD rank. The problem remains hard in case the nominator and denominator of each entry are given in unary.
Comments: The abbreviation PSD stands for positive semidefinite. The literature often does not distinguish between real and complex PSD rank. PSD Rank can be solved in polynomial time for any fixed value of r.
Also see: CPSD Rank (A24).

(A24) (Real) CPSD Rank
Given: A matrix $A \in \mathbb{Q}^{n\times n}$, a natural number $r \in \mathbb{N}$
Question: Is cpsd-rank$(A) \leq r$? That is, are there Hermitian psd matrices $(P_1, \ldots, P_n)$ of size $r \times r$ such that $A_{ij} = \text{trace}(P_i P_j)$ for all i, j? We say A is *completely positive semidefinite*. If the P_i are restricted to be real matrices, we obtain the *real CPSD rank* of A.
Complexity: Shitov established $\exists\mathbb{R}$-completeness [Shi23, Theorem 3.10] of both the CPSD and the real CPSD rank.
Comments: The problem is even hard in case the numerator and denominator of each entry are given in unary. For any fixed integer r, there is a polynomial-time algorithm to determine whether the CPSD rank or real CPSD rank of a given matrix are at most r [Shi23, Theorem 3.14].
Also see: PSD Rank (A23).

(A25) Phaseless Rank
Given: A matrix $A \in \mathbb{Q}^{m\times n}$, a natural number $r \in \mathbb{N}$
Question: Does A have *phaseless rank* at most r, that is, is there a matrix $\varepsilon \in \mathbb{C}^{m\times n}$ with $|\varepsilon_{ij}| = 1$ such that $A \odot \varepsilon$ has rank at most r, where $\odot$ denotes the element-wise matrix product?
Complexity: Shitov established $\exists\mathbb{R}$-completeness [Shi23, Theorem 3.7].
Comments: The problem is hard even if the numerator and denominator of each entry are given in unary.

(A26) Rank-Constrained Optimization
Given: Two matrices $A, B \in \mathbb{Q}^{\ell\times n}$, a natural number k.
Question: Is there an $n \times n$ matrix X in the positive semidefinite cone S^n_+ such that $AX = B$ and rank$(X) \leq k$?
Complexity: $\exists\mathbb{R}$-complete by Bertsimas et al. [BCWP22] via a reduction from linkage realizability (CG66).
Comments: In Bertsimas et al. [BCWP22], the general problem of minimizing an expression of the form $\langle C, X\rangle + \lambda \cdot \text{rank}(X)$ subject to $AX = B$, rank$(X) \leq k$ and $X \in \mathcal{K}$ for some proper cone $\mathcal{K}$ is considered. This problem is already $\exists\mathbb{R}$-hard for $\mathcal{K} = S^n_+$, $k = 2$, $\lambda = 1$ and $C = \mathbf{0}$ by reduction from unit linkage realizability.
Also see: Linkage realizability (CG66).

(A27) Tensor Rank
Given: A 3-dimensional tensor $T = (t_{ijk}) \in \mathbb{Q}^{d_1 \times d_2 \times d_3}$, a natural number k.
Question: Is the *(tensor) rank* of T at most k; that is, are there at most k rank-1 tensors $T_1, \ldots, T_k \in \mathbb{R}^{d_1 \times d_2 \times d_3}$ such that $T = \sum_{i \in [k]} T_i$? A tensor has *rank*-1 if it can be written as $u \otimes v \otimes w$ for real vectors u, v, w, where $\otimes$ is the outer (Kronecker) product of two vectors.
Complexity: $\exists\mathbb{R}$-complete by Schaefer and Štefankovič [SŠ18] via a reduction from a restricted version of the minimum matrix rank problem (A16), and Shitov [Shi16a] via a reduction from feasibility (A1). The problem remains hard for symmetric tensors [Shi16a].
Comments: Tensor rank was famously shown NP-hard by Håstad [Hås90]. Shitov [Shi16a] showed that over an integral domain, the (symmetric) tensor rank problem is as hard as the existential theory of that domain. For a related notion, called *border rank*, whose complexity is still open, see [BIJL18]. For fixed k the problem is fixed-parameter tractable even for higher-dimensional tensors [BSV21, Theorem 1.1], the question is also discussed in [Shi25b].
Universality: There is a universality result [SŠ24] (which fixes the version in [SŠ18]).
Also see: Minimum matrix rank (A16).

(A28) Tensor with Eigenvalue 0
Given: A 3-dimensional tensor $T = (t_{ijk}) \in \mathbb{Q}^{n \times n \times n}$.
Question: Does T have 0 as an *eigenvalue*; that is, is there a non-zero vector $v \in \mathbb{R}^n$ for which $\sum_{i,j \in [n]} t_{i,j,k} v_i v_j = 0$ for all $k \in [n]$?
Complexity: $\exists\mathbb{R}$-complete by Schaefer [Sch13, Corollary 3.10], also see Schaefer and Štefankovič [SŠ18, Example 2.5].
Comments: There also is a proof by Hillar and Lim [HL13], but it appears to have a gap, see the discussion in [SŠ18, Example 2.5]. The problem is equivalent to the non-trivial homogenous zero problem (H_2N) (A6). We should note that there are other, non-equivalent, definitions of tensor eigenvalues.
Open Questions: Is the eigenvalue problem $\exists\mathbb{R}$-complete for symmetric tensors? Hillar and Lim [HL13, Theorem 9.3] show that this problem is NP-hard for eigenvalues in $Q(\sqrt{d})$, where $d \in Q$.
Also see: Tensor rank (A27), H_2N (A6), tensor eigenvector approximation (A-Open11), tensor singular value (A-Open12).

(A29) Orbit Closure Containment
Given: Two 3-dimensional tensors $T, T' \in \mathbb{Q}^{n \times n \times n}$.
Question: Is the orbit closure of T contained in the orbit closure of T' under the $\mathsf{GL}_n \times \mathsf{GL}_n \times \mathsf{GL}_n$ action in $\mathbb{R}$?
Complexity: $\exists\mathbb{R}$-complete by Bläser et al. [BIL+21] via a reduction from minimum rank of linear combination of matrices (A17).
Comments: The orbit of T under G_n is the set $\{gT : g \in \mathsf{G}_n\}$, and we consider the closure of these orbits in the Zariski topology. We refer to Bläser et al. [BIL+21] for detailed definitions and motivations.

Also see: Minimum matrix rank (A16), minimum rank of linear combination of matrices (A17), tensor rank (A27).

5.4.2.3 Semialgebraic and Algebraic Sets

For polynomials and their representations, see Sect. 5.4.2. An *algebraic set* is the zero-set of a family of polynomials: $S = \{x \in \mathbb{R}^n : f_i(x) = 0 \text{ for all } i \in [r]\}$ and is specified by its defining polynomials, f_i, $i \in [r]$. A *basic primary semialgebraic set* is a set of the form $S = \{x \in \mathbb{R}^n : f_i(x) = 0 \text{ for all } i \in [r], g_i(x) > 0 \text{ for all } i \in [s]\}$. A *semialgebraic set* is the finite union of basic primary semialgebraic sets. We will see two ways of specifying a semialgebraic set: via a formula φ as $S = \{x \in \mathbb{R}^n : \varphi(x)\}$, where φ is a quantifier-free formula in the theory of the reals with free variables x; and via an arithmetic circuit C, where $S = \{x \in \mathbb{R}^n : C(x) = 1\}$, where x are the variables of the circuit C. Unless we say otherwise, we assume a semialgebraic set is specified using the formula φ.

(A30) Cardinality of Semialgebraic Set
Given: A semialgebraic set $S \subseteq \mathbb{R}^n$, a natural number $k \geq 1$ or $k = \infty$.
Question: Does S contain fewer than k elements?
Complexity: $\forall\mathbb{R}$-complete by Cucker and Rosselló [CR92] via a reduction from feasibility (A1) for any fixed k, including $k = \infty$.
Comments: Special (named) cases are $k = 1$, the *emptiness* and $k = \infty$, the *finiteness* problem. Cucker and Rosselló [CR92] work in the BSS-model, but their proof carries over to $\forall\mathbb{R}$. If k is not fixed, the problem is complete for the counting version of $\exists\mathbb{R}$, see [BC06, Proposition 8.5(ii)].
Also see: Dimension of semialgebraic set (A31), feasibility (A1).

(A31) Dimension of Semialgebraic Set
Given: A semialgebraic set $S \subseteq \mathbb{R}^n$, an integer $d \geq 0$.
Question: Does S have (real) dimension at least d; that is, is there a projection $\pi : \mathbb{R}^n \to \mathbb{R}^d$ (dropping $n - d$ coordinates), such that $\pi(S)$ contains a non-empty interior?
Complexity: $\exists\mathbb{R}$-complete by Koiran [Koi99] via reduction from feasibility (A1), even for any fixed $d \geq 0$.
Comments: For $d = 0$ the problem degenerates to the non-emptiness problem of a semialgebraic set (A30). Cucker and Rosselló [CR92] looked at the cases $d = n - 1$, which they call the hypersurface problem, and $d = n$, which they call the NEI (nonempty interior) problem. They showed that both problems can be described using the exotic quantifier $\exists^*$ (not formally introduced until Koiran's papers [Koi99, Koi00]).

The $\exists\mathbb{R}$-hardness reduction is straightforward, the non-trivial part is showing that the problem lies in $\exists\mathbb{R}$. Koiran [Koi99] first shows this for the BSS-model, but also explains how to translate the result into the discrete setting, corresponding to $\exists\mathbb{R}$. Koiran had earlier shown that the *complex* dimension of a semialgebraic set is $\exists\mathbb{C}$-complete (again for the BSS-model) [Koi97b]. For his result Koiran introduces the exotic quantifier $\exists^*$ and shows how to eliminate it.
Also see: Local dimension of semialgebraic set (A32), feasibility (A1).

(A32) Local Dimension of Semialgebraic Set
Given: A semialgebraic set $S \subseteq \mathbb{R}^n$, $x \in \mathbb{Q}^n$, an integer $d \geq 0$.
Question: Does S have local dimension at least d at x; that is, is there a sufficiently small neighborhood U of x such that the dimension of $S \cap U$ is at least d?
Complexity: $\exists\mathbb{R}$-complete by Bürgisser and Cucker [BC09, Corollary 6.6, Corollary 9.4] even for $d = 1$ via a reduction from testing whether 0 lies in the closure of a semialgebraic set (A38).
Comments: $\exists\mathbb{R}$-membership uses the exotic quantifier H, which can be eliminated.
Also see: Dimension of semialgebraic set (A31).

(A33) Convexity of Semialgebraic Set
Given: A semialgebraic set $S \subseteq \mathbb{R}^n$, a natural number $n \geq 2$.
Question: Is S *convex*, that is, does the line segment st belong to S for every two points $s, t \in S$?
Complexity: $\forall\mathbb{R}$-complete by Cucker and Rosselló [CR92]. Also, see Schaefer and Štefankovič [SŠ23].
Comments: The proof by Cucker and Rosselló [CR92] is phrased in the BSS-model, but translates easily to the $\forall\mathbb{R}$-completeness setting, as shown in [SŠ23].
Also see: Star-shapedness of semialgebraic set (A-Open4), convexity through translation (CG-Open33).

(A34) Unboundedness of Semialgebraic Set
Given: A semialgebraic set $S \subseteq \mathbb{R}^n$.
Question: Is S *unbounded*, that is, is it true that for every $r > 0$ there is an $x \in S$ with $\|x\| > r$?
Complexity: $\exists\mathbb{R}$-complete by Cucker and Rosselló [CR92], also see Bürgisser and Cucker [BC09, Proposition 6.4, Corollary 9.4], and Schaefer and Štefankovič [SŠ23].
Comments: Membership in $\exists\mathbb{R}$ is the non-trivial part here, since the definition uses a universal quantifier. Cucker and Rosselló [CR92] work in the BSS-model, but consider a restricted variant which corresponds to $\exists\mathbb{R}$, and obtain membership using an upper bound from real algebraic geometry. Bürgisser and Cucker [BC09] work with an exotic quantifier and show how to eliminate it; Schaefer and Štefankovič [SŠ23] give a reduction from feasibility over a bounded domain.

(A35) Radius of Semialgebraic Set
Given: A semialgebraic set $S \subseteq \mathbb{R}^n$, $d \in \mathbb{Q}_{\geq 0}$.
Question: Does S have radius at most d; that is, is there a point $c \in \mathbb{R}^n$ such that $\|c - s\| \leq d$ for every $s \in S$?
Complexity: $\forall\mathbb{R}$-complete by Schaefer and Štefankovič [SŠ23, SŠ25] via reduction from Feasibility (A1) over a bounded domain.
Comments: The $\forall\mathbb{R}$-hardness reduction is straight-forward, the difficulty lies in showing that the problem lies in $\forall\mathbb{R}$ by eliminating the quantification over the center.
Also see: Diameter of semialgebraic set (A36).

(A36) Diameter of Semialgebraic Set
Given: A semialgebraic set $S \subseteq \mathbb{R}^n$, $d \in \mathbb{Q}_{\geq 0}$.
Question: Does S have diameter at most d; that is, do every two points in S have distance at most d from each other?
Complexity: $\forall\mathbb{R}$-complete by Schaefer and Štefankovič [SŠ23, SŠ25] via reduction from Feasibility (A1) over a bounded domain.
Comments: The strict version, in which we require the diameter to be strictly less than d is also $\forall\mathbb{R}$-complete; the only issue is $\forall\mathbb{R}$-membership, but that can be handled using the exotic H-quantifier as we saw in Sect. 5.2.2.4.
Also see: Radius of semialgebraic set (A35).

(A37) Density of Semialgebraic Set
Given: A semialgebraic set $S \subseteq \mathbb{R}^n$.
Question: Is S dense in $\mathbb{R}^n$? A set $S \subseteq T$ is *dense* in T if for every $t \in T$ and every $\varepsilon > 0$ there is an $s \in S$ such that $\|s - t\| < \varepsilon$.
Complexity: In $\forall\mathbb{R}$ by Koiran [Koi99]. $\forall\mathbb{R}$-complete by Bürgisser and Cucker [BC09, Proposition 5.2, Corollary 9.3], Schaefer and Štefankovič [SŠ23, SŠ25] via reduction from strict positivity over a bounded domain.
Comments: The tricky part here is membership in $\forall\mathbb{R}$, that follows using Koiran's method [Koi99]. Bürgisser and Cucker [BC09] showed the problem complete for semialgebraic sets defined by arithmetic circuits, and Schaefer and Štefankovič [SŠ23] strengthened the result to semialgebraic sets defined by formulas. The directed Hausdorff distance between S and T (A44) is 0 if and only if $S \cap T$ is dense in T, so replacing $\mathbb{R}^n$ with another semialgebraic set leads to a $\forall\exists\mathbb{R}$-complete problem.
Also see: distance between semialgebraic sets (A41), (directed) Hausdorff distance between semialgebraic sets (second level) (A44), zero in closure of semialgebraic set (A38), Image density (second level) (A46), Domain density (arithmetic circuit) (A13).

(A38) Zero in Closure of Semialgebraic Set
Given: A semialgebraic set $S \subseteq \mathbb{R}^n$.
Question: Does 0 belong to the closure of S? That is, are there $s_n \in S$ such that $\|s_i\| \to 0$ for $n \to \infty$?
Complexity: $\exists\mathbb{R}$-complete by Cucker and Rosselló [CR92] via a reduction from feasibility.
Comments: The following hardness proof is due to Bürgisser and Cucker [BC09, Proposition 5.1]: a polynomial $f : \mathbb{R}^n \to \mathbb{R}$ (of total degree at most 4) is feasible if and only if the semialgebraic set $S = \{(x, y) \in \mathbb{R}^{n+1} : y^4 f(x/y) = 0 \wedge y \neq 0\}$ is not empty and 0 belongs to the closure of S. Membership is the non-trivial result here, it requires an upper bound from real algebraic geometry. Cucker and Rosselló [CR92] call this the *(Euclidean) adherence* problem. Bürgisser and Cucker [BC09, Proposition 5.1, Corollary 9.4] study both the Euclidean and Zariski adherence problem.
Open Questions: How hard is this problem in the Zariski topology [BC09, Table 2]?
Also see: distance between semialgebraic sets (A41).

(A39) Isolated Point of Semialgebraic Set
Given: A semialgebraic set $S \subseteq \mathbb{R}^n$.
Question: Is 0 an isolated point in S, that is, does S not contain other points arbitrarily close to 0?
Complexity: $\forall\mathbb{R}$-complete by Bürgisser and Cucker [BC09, Corollary 6.7, Corollary 9.4] via a reduction from testing whether 0 lies in the closure of a semialgebraic set (A38).
Comments: The problem remains hard for algebraic sets, see the isolated zero problem (A7). The complexity of testing whether a semialgebraic set contains an isolated point is open (A-Open3).
Also see: Isolated zero (A7), semialgebraic set with isolated point (A-Open3), zero in closure of semialgebraic set (A38).

(A40) Closedness of Basic Semialgebraic Set
Given: A basic Semialgebraic set $B \subseteq \mathbb{R}^n$ and a natural number $n \geq 2$.
Question: Is B *closed*, that is, does it contain all its limit points?
Complexity: $\forall\mathbb{R}$-complete by Bürgisser and Cucker [BC09, Lemma 6.20, Corollary 9.4] via a reduction from Feasibility (A1).
Comments: Hardness is easy: let $B = \{(x, y) \in \mathbb{R}^{n+1} : f(x) = 0, y > 0\}$, then f is not feasible if and only if the basic semialgebraic set B is closed; membership in $\forall\mathbb{R}$ is much trickier, and is open for semialgebraic sets in general. The problem remains $\forall\mathbb{R}$-hard for compactness instead of closedness.
Open Questions: Does the problem lie in $\forall\mathbb{R}$ for semialgebraic sets?
Also see: Zero in Closure of semialgebraic sets (A38).

(A41) Distance between Semialgebraic Sets
Given: Two semialgebraic sets $S, T \subseteq \mathbb{R}^n$.
Question: Do S and T have distance 0, that is, are there $s_n \in S$ and $t_n \in T$ such that $\|s_i - t_t\| \to 0$ for $n \to \infty$?
Complexity: $\exists\mathbb{R}$-complete by Schaefer and Štefankovič [SŠ17] via a reduction from feasibility (A1).
Comments: $\exists\mathbb{R}$-hardness is easy, more interesting is membership in $\exists\mathbb{R}$, since there is a universal quantifier that has to be eliminated. The special case in which T consists of a single point amounts to determining whether a point lies in the closure of a semialgebraic set (A38).
Also see: Zero in closure of semialgebraic set (A38), Hausdorff distance (A44).

(A42) Hyperplane Local Support of Semialgebraic Set
Given: A semialgebraic set $S \subseteq \mathbb{R}^n$ and a hyperplane H.
Question: Does H locally support S; that is, some open subset of H belongs to S and, close to that subset, all of S lies on the same side of H?
Complexity: $\exists\mathbb{R}$-complete by Bürgisser and Cucker [BC09], Schaefer and Štefankovič [SŠ23] via a reduction from feasibility (A1).

Comments: Hardness was shown by Bürgisser and Cucker [BC09] for arithmetic circuits for a (potentially) larger class. Schaefer and Štefankovič [SŠ23] consider semialgebraic sets defined by formulas.
Also see: distance between semialgebraic sets (A41).

5.4.3 Computational Geometry

5.4.3.1 Realizability of Configurations

A rank-d *chirotope* is a function $\chi : \{1, 2, \ldots, n\}^d \to \{-1, 0, 1\}$ with the following properties: (i) exchanging two components i and j changes the sign: $\chi(\tau_{ij}(\lambda)) = -\chi(\lambda)$, hence it is an *alternating sign map*; (ii) the function $|\chi|$ is the characteristic function of independent sets of a *matroid*; (iii) the following *Grassmann-Plücker relation* holds: for every $\lambda \in [n]^{d-2}$ and $a, b, c, d \in [n]\backslash\lambda$, the set

$$\{\chi(\lambda, a, b) \cdot \chi(\lambda, c, d), -\chi(\lambda, a, c) \cdot \chi(\lambda, b, d), \chi(\lambda, a, d) \cdot \chi(\lambda, b, c)\}$$

either contains $\{-1, +1\}$ or equals $\{0\}$. We usually identify a chirotope χ with its opposite $-\chi$. A chirotope defines an oriented matroid of rank d. In rank three, chirotopes are often referred to as *order types*. It can be checked that for a given set of n vectors $\{v_1, v_2, \ldots, v_n\}$ in $\mathbb{R}^d$, the function χ that returns $\text{sgn}(\det(v_{\lambda_1}, v_{\lambda_2}, \ldots, v_{\lambda_d}))$ satisfies the above properties. This function indicates the orientation of the simplex formed by any d-tuple of vectors. In the plane, we can associate to every point $p \in \mathbb{R}^2$ the vector $(p_1, p_2, 1)$. Then the rank-three chirotope given by the determinant indicates whether a given ordered triple of points forms a clockwise, counter-clockwise, or flat triangle. It is not true that every rank-three chirotope corresponds to a set of points in the plane, in that sense. Deciding this is one of the fundamental $\exists\mathbb{R}$-complete problem.

(CG1) (Simple) Order Type Realizability
Given: An order type $\chi : [n]^3 \to \{-1, 0, 1\}$.
Question: Are there points $p_1, \ldots, p_n$ in the plane realizing this order type, hence such that p_k lies to the left of $p_i p_j$ if $\chi(p_i, p_j, p_k) = -1$, to the right if $\chi(p_i, p_j, p_k) = 1$, and the three points are collinear otherwise?
Complexity: $\exists\mathbb{R}$-complete by Mnëv [Mnë88, Sho91] even if the order type is *simple* (also called *uniform*), that is, does not take on value 0.
Comments: This is one of the original $\exists\mathbb{R}$-complete problems, dual to simple stretchability (CG9), and a special case of (rank-3) oriented matroid realizability. For a detailed proof of $\exists\mathbb{R}$-completeness, see Matoušek [Mat14]. In spatial reasoning, the order type is captured by the $\mathcal{LR}$-calculus (left/right calculus, (L10)), which is therefore $\exists\mathbb{R}$-complete [WL10]. There is a partial variant in which χ is only required to be defined on subsets of $[n]$ of fixed size and fixed overlap; that version remains $\exists\mathbb{R}$-complete [Sch21b], which is useful in establishing results for problems with fixed parameters, such as segment intersection graphs of bounded degree (CG26), pseudo-segment stretchability (CG10), and the rectilinear local crossing number (CG52).
Universality: Mnëv [Mnë88], also see Shor [Sho91].

Also see: Collinearity logic (L9), $\mathcal{LR}$-calculus (L10), directional walk (CG6), pseudo-segment stretchability (CG10), rectilinear local crossing number (CG52).

(CG2) Oriented Matroid Realizability Over the Reals
Given: A rank-d oriented matroid, for some fixed d, given for example as a chirotope $\chi : [n]^d \to \{-1, 0, 1\}$.
Question: Can this oriented matroid be realized over the reals, hence are there vectors $v_1, \ldots, v_n \in \mathbb{R}^d$ such that $\chi(\lambda) = \text{sgn}(\det(v_{\lambda_1}, v_{\lambda_2}, \ldots, v_{\lambda_d}))$?
Complexity: $\exists\mathbb{R}$-complete from order type realizability (CG1).
Also see: Order type realizability (CG1).

(CG3) Matroid Realizability Over the Reals
Given: A family $F \subseteq 2^{[n]}$ and a dimension $d \in \mathbb{N}$.
Question: Are there vectors $v_1, \ldots, v_n \in \mathbb{R}^d$ such that for every set $S \in 2^{[n]}$, we have $S \in F$ if and only if the vectors $\{v_i : i \in S\}$ are linearly independent?
Complexity: $\exists\mathbb{R}$-complete by Kim et al. [KdMM23]; the paper contains two reductions: one from order type realizability (CG1), the other from distinct ETR (L4).
Comments: The proof follows from a careful reading of the $\exists\mathbb{R}$-completeness proofs of order type realizability (CG1) [Mat14], but the result was first explicitly stated by Kim et al. [KdMM23] in an arguably simpler and self-contained way. Matroids can be represented in various ways, affecting the complexity status of the realizability problem. Typically matroids are given by a circuit or black-box model. Compared to that, the explicit representation via independent sets may appear wasteful. However, representing the problem explicitly can only make it easier, but it is still $\exists\mathbb{R}$-hard. Moreover, matroid realizability is $\exists\mathbb{R}$-complete for every fixed $d \geq 3$, so at most $O(n^d)$ sets have to be specified.

Sturmfels also studied the problem of matroid realizability over the rational numbers and showed similar equivalences with finding rational solutions to Diophantine equations [Stu87]. However Sturmfels' reduction is not polynomial time, so it does not establish an $\exists\mathbb{R}$-completeness result.
Also see: Order type realizability (CG1), distinct ETR (L4), collinearity logic (L9).

(CG4) (Simple) Allowable Sequence
Given: A sequence $\pi_1, \ldots, \pi_k$ of permutations of $[n]$.
Question: Is the given sequence *allowable*, that is, is there a set of n points in the plane so that the permutations $\pi_1, \ldots, \pi_k$ are obtained, in that order, by projecting the points onto a line rotating by a half turn around some point of the plane?
Complexity: $\exists\mathbb{R}$-complete by Hoffmann and Merckx [HM18] even if the sequence is *simple*, that is, every two consecutive permutations differ by transposition of two adjacent elements, via a reduction from order type realizability (CG1).
Comments: Allowable sequences are one of the oldest tools in the computational geometers tool set, first introduced in the 19th century, see [GP80].
Universality: Hoffmann and Merckx [HM18].
Also see: Visibility graph of polygon with holes (CG37), Order type realizability (CG1).

(CG5) Radial System Realizability
Given: An (abstract) radial system, that is a family $(\pi_i)_{i \in [n]}$ such that π_i is a permutation of $[n] - \{i\}$, for each i.
Question: Are there points $p_1, \ldots, p_n$ in the plane so that for each i, the points seen from p_i occur in radial clockwise order as described by π_i?
Complexity: $\exists\mathbb{R}$-complete by Cardinal and Kusters [CK15] via a reduction from order type realizability (CG1).
Comments: There is a partial variant in which the radial system only needs to be specified for point sets of bounded size and overlap, see [Sch21b]; the partial variant can be used to establish hardness for problems with fixed parameters, such as existence of simultaneous geometric embeddings for a fixed number of graphs, and straight-line realizability of bounded-degree graphs with rotation system (CG55).
Also see: Order type realizability (CG1), straight-line realizability of graph with rotation system (CG55), simultaneous geometric embedding (CG56).

(CG6) Directional Walk
Given: A walk W on n vertices, and a direction left/right for every triple of consecutive points along W.
Question: Are there n points in the plane so that for all triples (p, q, r) of consecutive points along the walk, r lies to the left or right or pq as specified and distinct vertices are mapped to distinct locations?
Complexity: $\exists\mathbb{R}$-complete [Sch21b] via a reduction from a partial version of order type realizability (CG1). Remains $\exists\mathbb{R}$-complete even if each edge occurs at most once on the walk [KR23].
Comments: The directional walk problem can be viewed as a sparsified version of order type realizability, or a special case of the left/right calculus [WL10].
Open Questions: Does the problem remain $\exists\mathbb{R}$-complete if the realizing walk has to be crossing-free? ▼ The proof in [Sch21b] implies that the problem remains $\exists\mathbb{R}$-complete even if each vertex occurs only a finite number of times along the walk. Is the directional walk problem $\exists\mathbb{R}$-complete even if each vertex occurs at most twice (once being trivial)?
Also see: Order type realizability (CG1).

(CG7) Finite Convex Geometry Realizability
Given: A *finite convex geometry* $(V, \mathcal{C})$, that is a set $\mathcal{C} \subseteq 2^V$ that contains $\emptyset$, V, is closed under intersection, and for every $C \in \mathcal{C} \backslash \{V\}$ there is an $v \in V$ such that $C \cup \{v\} \in \mathcal{C}$.
Question: Are there n points in $\mathbb{R}^d$ that realize the given finite convex geometry? Letting $\text{conv}(C)$ denote the convex hull of the points in C, a point set P *realizes* the finite convex geometry $(P, \{C \subseteq P : \text{conv}(C) \cap P = C\})$.
Complexity: $\exists\mathbb{R}$-complete by Adaricheva and Wild [AW10, HM18] for $d = 2$, via a reduction from order type realizability (CG1), and by Hoffmann and Merckx [HM18] for $d \geq 2$ via a reduction from allowable sequence (CG4).
Also see: Allowable sequence (CG4), order type realizability (CG1), convex neural codes (CG-Open10).

(CG8) Free Space Matrix Realizability
Given: A matrix $M_\varepsilon \in \{0, 1\}^{n\times n}$, $\varepsilon \in \mathbb{Q}_{>0}$. M_ε is the *free space matrix* for two polygonal curves $P = (p_1, \ldots, p_n)$ and $Q = (q_1, \ldots, q_m)$ in $\mathbb{R}^d$, if M_ε has a 1 in cell (i, j) if and only if $\|p_i - q_j\| < \varepsilon$.
Question: Are there two polygonal curves P and Q so that M_ε is the free space matrix for P and Q?
Complexity: $\exists\mathbb{R}$-complete for every $d \geq 2$ by Akitaya et al. [ABM+23, Theorem 14] via a reduction from oriented hyperplane realizability (CG13).
Comments: In $\mathbb{R}^1$, the problem can be decided in polynomial time [ABM+23, Theorem 16].
Also see: Free space diagram realizability (CG-Open5), oriented hyperplane realizability (CG13).

5.4.3.2 Arrangements and Stretchability

Arrangements of vectors or points can be dualized into arrangements of hyperplanes. From the topological representation theorem [FL78, BKMS05], oriented matroids can always be realized as arrangements of pseudo-hyperplanes. In the planar case, we obtain *pseudoline arrangements*. In a projective setting, a pseudoline arrangement is defined on a two-sphere, but a pseudoline arrangement can also be defined in the plane as a collection of x-monotone curves that pairwise intersect exactly once. For our purpose, the only relevant information about a pseudoline arrangement is the underlying combinatorics, or equivalence class under homeomorphisms of the plane. A convenient way to specify this information is in the form of a *wiring diagram*, or by specifying, for each pseudoline, the left-to-right order of its intersection points with the other pseudolines.

(CG9) (Pseudoline) Stretchability
Given: A pseudoline arrangement.
Question: Is the arrangement *stretchable*, that is, is it isomorphic to a straight-line arrangement?
Complexity: $\exists\mathbb{R}$-complete, by Mnëv [Mnë88], and Shor [Sho91] via a reduction from the existential theory of totally ordered real variables (L2). The problem remains $\exists\mathbb{R}$-complete if the given arrangement is *simple*, that is, at most two pseudolines intersect in each point [SŠ17]. Since simple stretchability in the plane is equivalent to simple stretchability in the hyperbolic plane, hyperbolic simple stretchability is also $\exists\mathbb{R}$-complete [Bie22, BBDJ23].
Comments: Stretchability is the original $\exists\mathbb{R}$-complete problem in the Mnëv/Shor tradition. Mnëv [Mnë88] showed that pseudoline stretchability is universal, and Shor [Sho91] recast the result from a computer scientist's point of view. A nice introductory presentation of the result is due to Matoušek [Mat14]. Shor [Sho91] also explicitly shows that pseudoline stretchability is NP-hard.
Universality: Both pseudoline stretchability and simple pseudoline stretchability are universal, as shown by Mnëv [Mnë88], also see Shor [Sho91].
Also see: Existential theory of totally ordered real variables (L2), pseudo-segment stretchability (CG10).

(CG10) Pseudo-segment Stretchability
Given: An arrangement of *pseudo-segments*: arcs of x-monotone curves, each pair of which intersects exactly once.
Question: Is the arrangement *stretchable*, that is, is it isomorphic to an arrangement of straight-line segments?
Complexity: $\exists\mathbb{R}$-complete by Mnëv [Mnë88], and Shor [Sho91], since pseudo-line stretchability (CG9) can be viewed as a special case. The problem remains $\exists\mathbb{R}$-complete if each pseudo-segment is involved in at most 72 crossings and the arrangement is simple, by Schaefer [Sch21b] via a reduction from a partial version of order type realizability (CG1). Förster, Kindermann, Miltzow, Parada, Terziadis, and Vogtenhuber [FKM+24] show that the problem remains $\exists\mathbb{R}$-hard even if the intersection graph of the arrangement has chromatic number at most 30.
Comments: This result can be used to show that segment intersection graphs (CG26) of degree at most 216 are $\exists\mathbb{R}$-complete to recognize [Sch21b] and that the geometric thickness of multigraphs (CG57) is $\exists\mathbb{R}$-complete [FKM+24].
Open Questions: What is the smallest number of crossings on a pseudo-segment for which the pseudo-segment stretchability remains $\exists\mathbb{R}$-complete? ▼ Does the problem remain $\exists\mathbb{R}$-complete if the intersection graph of the arrangement has chromatic number at most 2? [FKM+24].
Also see: Order type realizability (CG1), geometric thickness of multigraphs (CG57), segment intersection graph (CG26).

(CG11) Arrangement Graph
Given: A graph G.
Question: Is G an arrangement graph? The vertices of an *arrangement graph* are intersection points of an arrangement L of lines in the plane, and two vertices are adjacent if and only if they lie on the same line of L and the line segment between them does not intersect any of the other lines in L.
Complexity: $\exists\mathbb{R}$-complete, by Bose et al. [BEW03] via a reduction from simple stretchability (CG9).
Also see: Segment number (CG44).

(CG12) (Pseudocircle) Circularizability
Given: An *arrangement* of *pseudocircles*, that is, a collection of closed curves, any two of which are either disjoint or intersect exactly twice.
Question: Is the arrangement *circularizable*, that is, is there a homeomorphism of the plane taking it to an arrangement of circles?
Complexity: $\exists\mathbb{R}$-complete by Kang and Müller [KM14] (see also Felsner and Scheucher [FS20]) via a reduction from simple stretchability (CG9). The problem remains $\exists\mathbb{R}$-hard when restricted to arrangements of great pseudocircles [FS20].
Also see: Simple stretchability (CG9).

(CG13) Oriented Hyperplane Realizability in $\mathbb{R}^d$

Given: A set $S \subseteq \{-,+\}^n$ with $\{(-,\ldots,-),(+,\ldots,+)\} \subseteq S$.

Question: Are there n oriented hyperplanes in $\mathbb{R}^d$ such that for every vector $s \in S$ there is a point in $\mathbb{R}^d$ whose characteristic vector with respect to the hyperplane arrangement is s?

Complexity: $\exists\mathbb{R}$-complete by Tanenbaum et al. [TGS95], Basri et al. [BFG+09], Kang and Müller [KM12] for every $d \geq 2$ via a reduction from simple stretchability (CG9).

Comments: This problem is related to oriented matroid realizability (CG2), where the oriented matroid is given in terms of its cocircuits. Tanenbaum et al. [TGS95] cast the problem slightly differently, as that of recognizing point-halfspace orders (CG14). For $d = 2$ this is very close to the simple stretchability problem (CG9), but with the additional condition that $\{(-,\ldots,-),(+,\ldots,+)\} \subseteq S$. Kang and Müller [KM12] also show that specifying a realizing hyperplane arrangement may require $2^{\Theta(n)}$ bits. Basri et al. [BFG+09] introduce oriented hyperplane realizability under the name matrix visibility problem and show it is $\exists\mathbb{R}$-hard (they only claim NP-hardness, but reduce from oriented matroid realizability (CG1)). For $d = 2$, McDiarmid and Müller study this as the oriented line arrangement problem.

Also see: Point-halfspace orders (CG14), dot-product graph (CG25), unit ball graph (CG24), Euclidean preference (Misc1), oriented matroid realizability (CG2).

(CG14) Point-halfspace orders in $\mathbb{R}^d$

Given: A bipartite poset $(B, T, \leq)$.

Question: Are there injective maps f from B to $\mathbb{R}^2$ and g from T to the set of halfplanes in $\mathbb{R}^2$ such that $f(x) \in g(y)$ if and only if $x \leq y$?

Complexity: $\exists\mathbb{R}$-complete by Tanenbaum et al. [TGS95] via a reduction from simple stretchability (CG9). Also $\exists\mathbb{R}$-complete in any fixed higher dimension d.

Comments: This is a different but essentially equivalent encoding of the oriented hyperplane realizability problem (CG13).

Also see: Oriented hyperplane realizability (CG13).

(CG15) Circle Orders

Given: A poset $P = (S, \leq)$.

Question: Is P the containment order of a set of closed disks in the plane, hence does there exist an injective map f from S to the set of closed disks in $\mathbb{R}^2$ such that $f(x) \subseteq f(y)$ if and only if $x \leq y$?

Complexity: $\exists\mathbb{R}$-complete by Cardinal [Car24] via a reduction from pseudocircle circularizability (CG12). Also $\exists\mathbb{R}$-complete if P is a bipartite poset.

Also see: Point-halfspace orders (CG14).

5.4.3.3 Covering and Packing

(CG16) Art Gallery Problem
Given: A simple polygon P with vertices in $\mathbb{Q}^2$ and a number $k \in \mathbb{N}$.
Question: Does there exist a set $G \subseteq P$ of k guards such that every point inside the polygon is visible from G? Two points a, b are said to be visible to each other in case that the line segment ab is completely contained in the polygon P. The polygon is considered to be closed.
Complexity: $\exists\mathbb{R}$-complete, by Abrahamsen et al. [AAM18, AAM22], and Stade [Sta23]
Comments: Also known as the minimum star cover problem [O'R87]. The original $\exists\mathbb{R}$-completeness proof in [AAM22] is fairly involved, but Stade found a shorter proof in 2022 [Sch23a]. It is known that irrational coordinates can already be necessary for two and three guardable galleries [AAM17, MM22]. There is a series of papers that try to find practical algorithms for the art gallery problem. Interestingly some of them come with provable performance guarantees under some mild assumptions [HM21].
Universality: The art gallery problem admits topological universality [ST22, BEMM+22, BEMM+23].
Open Questions: Does the problem lie in $\exists\mathbb{R}$ for semialgebraic art galleries, even in $\mathbb{R}^2$?
Also see: Polygon convex cover (CG20), star-shapedness of semialgebraic set (A-Open4).

(CG17) Boundary-Guarding
Given: A simple polygon P with vertices in $\mathbb{Q}^2$ and a number $k \in \mathbb{N}$.
Question: Does there exist a set $G \subseteq \partial P$ of k guards such that every point inside the polygon is visible by at least one guard in G?
Complexity: $\exists\mathbb{R}$-complete, by Stade [Sta23] via a reduction from RANGE- ETR-INV (L7).
Comments: The boundary guarding problems differs from the art gallery problem (CG16) in that the guards must be on the boundary.
Open Questions: It is open if boundary-boundary guarding is $\exists\mathbb{R}$-complete; in the boundary-boundary variant, all guards must be placed on the boundary, and only the boundary has to be guarded [Sta23]. This is related to the terrain guarding problem, which turns out to be NP-complete [KK11, FHKS16].
Also see: Art gallery problem (CG16), ETR- INV (L5).

(CG18) Geometric Packing
Given: A simple polygon C, the *container*, and polygonal pieces $p_1, \ldots, p_m$. All polygons have vertices in $\mathbb{Q}^2$.
Question: Is it possible to move (translate and rotate) all the pieces so they lie in the container without overlapping?
Complexity: This problem is $\exists\mathbb{R}$-complete, by Abrahamsen et al. [AMS20], even if the container is a square and all pieces are convex. The reduction is from ETR- INV (L5).

Comments: The problem is in NP in case that only translations are allowed and pieces are polygonal. The problem is $\exists\mathbb{R}$-complete in case that only translations are allowed and pieces or the container are allowed to contain curved parts. When considering translation and rotation, the proof from [AMS20] needed to consider *non-convex* polygonal pieces. This could be simplified in a paper by Miltzow and Schmiermann [MS24] about continuous constraint satisfaction problems (L8), which implies that geometric packing remains $\exists\mathbb{R}$-complete even if all the pieces are convex and polygonal, a simplified proof of that result can also be found in [AMS24].
Open Questions: Does the problem remain $\exists\mathbb{R}$-hard if the pieces are restricted to a fixed number of shapes (one, for example) [Sch21b]? ▼ Is the problem fixed-parameter tractable in the number m of pieces? ▼ How hard is the packing problem if the pieces to be packed are unit disks? Hardness may be easier to show for polygons with holes [Abr24]. ▼ How hard is it to pack triangles in a rectangle [Mut24]?
Also see: Packing triangles with shifts and rotations (CG19), tiling the plane periodically (CG-Open34).

(CG19) Packing Triangles with Shifts and Rotations
Given: A family of triangles each given by three vertices in $\mathbb{Q}^2$, and, for each triangle either a range $[a_x, b_x] \times [a_y, b_y] \in (\mathbb{Q}^2)^2$ of allowable shifts for the triangle, or a center of rotation in $\mathbb{Q}^2$ together with a range $[\alpha, \beta] \in \mathbb{Q}^2$ of allowable rotation angles for the triangle around the center of rotation.
Question: Is it possible to move the triangles (shift or rotate each within its given bounds) so they do not overlap?
Complexity: This problem is $\exists\mathbb{R}$-complete, by Mutlu [Mut24] via a reduction from PLANAR- ETR- INV (L6). The problem remains $\exists\mathbb{R}$-complete even if all the triangles are right isosceles, the shifts are nonnegative, and the center of rotation lies on the boundary of the triangle.
Comments: Mutlu [Mut24] calls this the *real triangular door placement problem*, since the problem can be viewed as placing doors such that the areas swept by the doors do not overlap; in this version, the doors sweep a triangular area.
Open Questions: Does the problem remain $\exists\mathbb{R}$-complete if the pieces are quarter circles, instead of triangular (corresponding to the usual area swept by a door) [Mut24]?
Also see: Geometric packing (CG18), tiling the plane periodically (CG-Open34)

(CG20) Polygon Convex Cover
Given: A simple polygon, a natural number $k \in \mathbb{N}$.
Question: Can the polygon be written as the union of at most k convex polygons?
Complexity: $\exists\mathbb{R}$-complete by Abrahamsen [Abr22] via a reduction from RANGE-ETR- INV (L7). The problem remains $\exists\mathbb{R}$-complete even if all the convex polygons have to be triangles.
Comments: Also known as the *minimum convex cover problem*, it was previously known to be NP-hard [CR94] and to lie in $\exists\mathbb{R}$ [O'R82]. Replacing convex sets with star-shaped sets one obtains the art gallery problem (CG16).
Open Questions: Abrahamsen [Abr24] asks whether the polygon cover problem remains $\exists\mathbb{R}$-hard if a set of convex polygons is given and the original polygon has to be covered by rotated and translated copies of the convex polygons.

Also see: ETR- INV (L5), RANGE- ETR- INV (L7), art gallery problem (CG16), tiling the plane periodically (CG-Open34)

5.4.3.4 Geometric Representations of Graphs and Hypergraphs

(CG21) Unit Disk Graph
Given: A graph G.
Question: Is G the intersection graph of unit disks in the plane?
Complexity: $\exists\mathbb{R}$-complete, by McDiarmid and Müller [MM10, MM13] via a reduction from oriented hyperplane realizability (CG13). The problem remains $\exists\mathbb{R}$-complete for unit balls in d-dimensional space [KM12] again via a reduction from oriented hyperplane realizability (CG13).
Comments: This problem was earlier shown to be NP-hard in [BK98].
Open Questions: Does the problem remain hard for graphs of bounded degree [Sch21b]?
Also see: Disk graph (CG22), hyperbolic unit disk graph (CG23), unit ball graph (CG24), unit disk contact graph (CG-Open3), oriented hyperplane realizability (CG13).

(CG22) Disk Graph
Given: A graph G.
Question: Is G the intersection graph of disks in the plane?
Complexity: $\exists\mathbb{R}$-complete, by McDiarmid and Müller [MM13] via a reduction from oriented hyperplane realizability (CG13).
Open Questions: Does the problem remain $\exists\mathbb{R}$-complete for balls in d-dimensional space [MM13]? ▼ Does the problem remain hard for graphs of bounded degree?
Also see: Unit disk graph (CG21), unit ball graph (CG24).

(CG23) Hyperbolic Unit Disk Graph
Given: A graph G.
Question: Is G the intersection graph of unit disks in the hyperbolic plane?
Complexity: $\exists\mathbb{R}$-complete, by Bieker et al. [BBDJ23] via a reduction from hyperbolic simple stretchability (CG9).
Comments: The proof uses the result that every unit disk graph is also a hyperbolic unit disk graph [BFKS23].
Also see: Unit disk graph (CG21), hyperbolic simple stretchability (under Stretchability (CG9)).

(CG24) Unit Ball Graph in $\mathbb{R}^d$
Given: A graph G.
Question: Is G the intersection graph of unit balls in $\mathbb{R}^d$?
Complexity: $\exists\mathbb{R}$-complete for every fixed $d \geq 2$ by Kang and Müller [KM12] via a reduction from oriented hyperplane realizability (CG13). This generalizes the result for unit disk graphs (CG21), the case $d = 2$, by McDiarmid and Müller [MM10].
Comments: Kang and Müller [KM12] call this the *d-sphericity* problem, and they

establish exponential lower and upper bounds on the centers of a family of balls representing G. Fiduccia et al. [FSTZ98] asked whether the problem lies in NP. For $d = 1$ we get the *unit interval problem*, which is solvable in linear time [LO93].
Open Questions: Does recognizing unit ball graphs remain $\exists\mathbb{R}$-complete in d-dimensional hyperbolic space [BBDJ23, Gro24]?
Also see: Unit disk graph (CG21), oriented hyperplane realizability (CG13), dot-product graph (CG25).

(CG25) Dot-Product Graph in $\mathbb{R}^d$
Given: A graph $G = (V, E)$.
Question: Are there vectors $v_1, \ldots, v_n \in \mathbb{R}^d$ such that $ij \in E$ if and only if $v_i^T v_j \geq 1$ for all $1 \leq i, j \leq n$?
Complexity: $\exists\mathbb{R}$-complete, by Kang and Müller [KM12] for every fixed $d \geq 2$ via a reduction from oriented hyperplane realizability (CG13). This generalizes the result for unit disk graphs (CG21), the case $d = 2$, by McDiarmid and Müller [MM10].
Comments: The problem was introduced by Fiduccia et al. [FSTZ98]. The smallest d for which G has a dot-product representation is the *dot-product dimension* of the graph. Kang and Müller [KM12] also establish exponential upper and lower bounds on the precision of the vectors v_i.
Also see: Oriented hyperplane realizability (CG13), unit ball graph (CG24), unit disk graph (CG21), orthogonality dimension (CG-Open18), cross-product expression satisfiability (L11).

(CG26) Segment Intersection Graph (SEG)
Given: A graph G.
Question: Is G the intersection graph of line segments in the plane?
Complexity: $\exists\mathbb{R}$-complete, by Kratochvíl and Matoušek [KM94], reduction from simple stretchability (CG9). The problem remains $\exists\mathbb{R}$-hard for graphs of bounded degree [Sch21b] via a reduction from pseudo-segment stretchability (CG10).
Comments: A slightly simplified $\exists\mathbb{R}$-hardness proof can be found in [Sch10] and, in full detail, in [Mat14]. Kratochvíl and Matoušek [KM94] show that there are graphs on n vertices that require segment representations of size $2^{2^{\Omega(\sqrt{n})}}$, and that there are always representations of size at most $2^{2^{O(n)}}$.
Open Questions: The problem is known to be NP-hard for graphs of girth at least k [KP07], does it remain $\exists\mathbb{R}$-hard?
Also see: Unit segment intersection graph (CG27), 3D segment intersection graph (CG30), planar slope number (CG41), pseudo-segment stretchability (CG10), k-polyline intersection graph (CG28), incidence graph of geometric 3-configuration (CG-Open17), linear hypergraph realizability (CG35).

(CG27) Unit Segment Intersection Graph
Given: A graph G.
Question: Is G the intersection graph of unit line segments in the plane?
Complexity: $\exists\mathbb{R}$-complete, Hoffmann et al. [HMWW24], reduction from simple stretchability (CG9).
Also see: Segment intersection graph (CG26).

(CG28) k-Polyline Intersection Graph
Given: A graph G.
Question: Is G the intersection graph of k-polylines in the plane? A *k-polyline* is a polygonal chain consisting of $k - 1$ line segments.
Complexity: $\exists\mathbb{R}$-complete, by Hoffmann et al. [HMWW24] for any fixed k via a reduction from simple stretchability (CG9).
Also see: Segment intersection graph (CG26).

(CG29) (Downward) Ray Intersection Graph
Given: A graph G.
Question: Is G the intersection graph of (downward) rays in the plane?
Complexity: The recognition of both ray and downward ray intersection graphs is $\exists\mathbb{R}$-complete by Cardinal, Felsner, Miltzow, Tompkins, and Vogtenhuber [CFM+18] via a reduction from stretchability (CG9).
Comments: Downward ray intersection graphs are equivalent to grounded segment graphs [CFM+18]. The paper [CFM+18] also shows $\exists\mathbb{R}$-completeness of outer segment graph recognition. Outer segments have one endpoint on a common circle and the other endpoint in the interior of the circle.
Also see: Segment intersection graph (CG26).

(CG30) 3D Segment Intersection Graph
Given: A graph G.
Question: Is G the intersection graph of line segments in $\mathbb{R}^3$?
Complexity: $\exists\mathbb{R}$-complete, Evans et al. [ERS+19] via a reduction from simple stretchability (CG9).
Comments: The authors remark that the problem remains $\exists\mathbb{R}$-hard in any fixed dimension.
Also see: Segment intersection graph (CG26).

(CG31) 3D Line Intersection Graph
Given: A graph G.
Question: Is G the intersection graph of lines in $\mathbb{R}^3$?
Complexity: $\exists\mathbb{R}$-complete, by Cardinal [Car24] via a reduction from matroid realizability over the reals (CG3).
Also see: Segment intersection graph (CG26), 3D segment intersection graph (CG30).

(CG32) Convex Set Intersection Graph
Given: A graph G.
Question: Is G the intersection graph of a family of convex sets in the plane?
Complexity: $\exists\mathbb{R}$-complete, by Schaefer [Sch10] by a reduction from simple stretchability (CG9). Remains $\exists\mathbb{R}$-complete if the graph has bounded degree [Sch21b].
Comments: While not explicitly claimed in [Sch10], $\exists\mathbb{R}$-hardness also holds for any family of convex sets that can arbitrarily closely approximate any line segment;

this includes ellipses, triangles and convex polygons. It was shown by Müller et al. [MvLvL13] that the intersection graph recognition problem for translates (or homothets) of any fixed convex polygon can be solved on an exponential-size grid. This implies that the corresponding recognition problems lie in NP.
Open Questions: Bieker et al. [BBDJ23] ask whether the problem remains $\exists\mathbb{R}$-complete in the hyperbolic plane.
Also see: Segment intersection graph (CG26), topological inference in RCC8 with convexity (L13).

(CG33) Nerves of Convex Set
Given: A simplicial complex S of dimension k.
Question: Is S the k-skeleton of the nerve of c-dimensional convex sets in $\mathbb{R}^d$? Given a family A of sets, the *nerve* of A is a simplicial complex that has one vertex for each member of A, and any subset $B \subseteq A$ forms a simplex if their intersection is non-empty.
Complexity: $\exists\mathbb{R}$-complete for many parameter combinations with $k \geq 2$, in particular $k \geq d$ and $c = d - 1$ or $c = d$, by Schnider and Weber [SW23].
Comments: This defines a family of problems. This family is parameterized by the dimension of the convex sets c, the ambient dimension d containing the convex sets and the dimension k of the dimension of the input simplicial complex. Note that for $k = 1$, this is a question about graphs. For example, $(k, c, d) = (1, 1, 1)$ asks if the underlying graph is an interval graph, which is known to be polynomial-time solvable [BL76]. Another illustrative example is $(k, c, d) = (1, 1, 2)$. Here, we ask if the underlying graph is the intersection graph of segments, which is known to be $\exists\mathbb{R}$-complete [KM94]. The hardness results stem from the hardness results of graphs combined with a lifting lemma due to Tancer [Tan10]. Some other parameter combinations are always possible to realize (specifically $d, c \geq 2k + 1$ [Weg67]) and thus the computational problem becomes trivial.
Open Questions: All remaining parameter combinations are still open.
Also see: Segment intersection graph (CG26), convex set intersection graph (CG32), convex neural codes (CG-Open10), finite convex geometry realizability (CG7).

(CG34) Hypergraph Induced by Halfspaces and Points
Given: A hypergraph $H = (V, E)$, an integer $d \geq 0$.
Question: Is there a set of halfspaces and points in $\mathbb{R}^d$ such that the induced hypergraph equals H? The hypergraph has one vertex per point, and a hyperedge consists of all points belonging to the same halfspace.
Complexity: $\exists\mathbb{R}$-complete by Tanenbaum et al. [TGS95] via a reduction from stretchability (CG9) for every fixed $d \geq 2$. The problem remains $\exists\mathbb{R}$-complete if halfspaces are replaced by balls, ellipsoids and, so-called, bi-curved, difference-separable, and computable convex sets as shown by Bertschinger et al. [BEMK+23].
Open Questions: Does the problem remain $\exists\mathbb{R}$-complete for any other natural geometric objects? The authors of [BEMK+23] note that they are not aware of any such examples.
Also see: Stretchability (CG9), point-halfspace orders (CG14), oriented hyperplane realizability (CG13).

(CG35) Linear Hypergraph Realizability
Given: A *linear hypergraph* H, that is, a hypergraph in which every two edges share at most one point.
Question: Is there a *straight-line realization* of H; that is, can every hyperedge be drawn as a straight-line segment containing all vertices of the hyperedge and no other vertices of H? The segments are allowed to cross.
Complexity: $\exists\mathbb{R}$-complete, by Dobler et al. [DKL+24] via a reduction from matroid realizability over the reals (CG3)
Comments: Linear hypergraphs of degree at most 2 are always realizable in the sense defined above. The *rank* of a hypergraph is the maximum size of a hyperedge. The straight-line realizability problem becomes non-trivial already for hypergraphs of rank and degree at most 3 [DKL+24]; this corresponds to incidence graphs of geometric 3-configurations (CG-Open17)
Open Questions: Dobler, Kobourov, Lenhart, Mchedlidze, Nöllenburg, and Symvonis ask whether the problem remains $\exists\mathbb{R}$-hard if the hypergraph has bounded rank, of even rank 3. The answer for bounded rank is likely yes, see the comments on a bounded version of collinearity logic (L9).
Also see: Matroid realizability over the reals (CG3), collinearity logic (L9), segment intersection graph (CG26), incidence graph of geometric 3-configuration (CG-Open17).

(CG36) (Point) Visibility Graph
Given: A graph G.
Question: Is G the visibility graph of a set of points in the plane? The visibility graph of a set of points in the plane has one vertex per point, and two vertices are adjacent if and only if the line segment between the two points does not contain any other point of the set.
Complexity: $\exists\mathbb{R}$-complete, by Cardinal and Hoffmann [CH17].
Comments: If the points in the realization have to lie on a grid, the problem becomes $\exists\mathbb{Q}$-complete [CH17].
Open Questions: The complexity of recognizing visibility graphs of polygons is open. In the visibility graph of a polygon, two vertices of the polygon are adjacent if and only if the line segment between them is contained in the polygon.
Also see: Visibility graph of polygon with holes (CG37), unit spanning ratio (CG62), segment visibility graph (CG-Open37), obstacle number (CG-Open38), hidden set in polygon (CG-Open25).

(CG37) Visibility Graph of Polygon with Holes
Given: A graph G.
Question: Is there a polygon with holes such that G is the visibility graph of the vertices of G; that is, the vertices of G are the vertices of the polygon and its holes, and there is an edge between two vertices if the segment connecting the two vertices lies inside the polygon?
Complexity: $\exists\mathbb{R}$-complete by Hoffmann and Merckx [HM18] via a reduction from allowable sequence (CG4) and Boomari et al. [BOZ18] from simple stretchability (CG9).

Comments: For convex polygons with a single hole the visibility graphs are circular arc graphs and therefore recognizable in polynomial time [CE95].
Open Questions: What is the complexity of recognizing visibility graphs of polygons without holes? Boomari et al. [BOZ18] conjecture that the problem remains $\exists\mathbb{R}$-complete.
Also see: Allowable sequence (CG4), point visibility graph (CG36), internal/external polygon visibility graph (CG38).

(CG38) Internal/External Polygon Visibility Graph
Given: A graph $G = G_{int} \cup G_{ext}$.
Question: Is there a simple polygon such that G_{int} is the visibility graph of the vertices of G inside the polygon, and G_{ext} is the visibility graph of the vertices of G outside the polygon? Two vertices can see each other if the line segment connecting them does not cut through the side of the polygon.
Complexity: $\exists\mathbb{R}$-complete by Boomari et al. [BOZ18] from simple stretchability (CG9). The problem remains $\exists\mathbb{R}$-complete even if the blocking vertex for each pair of vertices that cannot see each other is specified (see [BOZ18] for definition of blocking vertex).
Open Questions: What is the complexity of recognizing polygon visibility graphs, that is, if just G_{int} is given?
Also see: Visibility graph of polygon with holes (CG37), point visibility graph (CG36).

(CG39) Triangulated Irregular Network Visibility Graph
Given: A graph G, a plane triangulation T.
Question: Is there a triangulated irregular network (TIN) with visibility graph G and xy-projection isomorphic to T? The *triangulated irregular network (TIN)* assigns coordinates in $\mathbb{R}^3$ to the vertices of T. Two vertices in the TIN can *see* each other if the straight-line segment connecting them lies above the TIN.
Complexity: $\exists\mathbb{R}$-complete by Boomari et al. [BOZ21] from triangulation stretchability (CG47).
Also see: Triangulation stretchability (CG47), boundary guarding (CG17).

(CG40) Generalized Transmission Graph
Given: A directed graph $G = (V, E)$, a family C of geometric objects (e.g. lines in $\mathbb{R}^2$).
Question: Are there objects $C(v) \in C$ and points $p(v) \in C(v)$ for every vertex $v \in V$, such that $p(u) \in C(v)$ if and only if $uv \in E$ for all $u, v \in V$?
Complexity: $\exists\mathbb{R}$-complete for C being lines in $\mathbb{R}^2$ by Klost and Mulzer [KM18] via a reduction from simple stretchability (CG9). The problem is also $\exists\mathbb{R}$-complete for C being a restricted type of circular sectors instead of lines [KM18].
Open Questions: For $\exists\mathbb{R}$-completeness of circular sectors, Klost and Mulzer [KM18] require a somewhat technical condition (they call wide-spread); they conjecture that that condition can be removed. ▼ What about other natural families of geometric objects? Are there general conditions on C that would guarantee $\exists\mathbb{R}$-hardness?
Also see: Segment intersection graph (CG26).

5.4.3.5 Graph Drawing

Graphs in this section are simple (no multiple edges, no loops) and undirected, unless we say otherwise. A *drawing* of a graph assigns a different location to each vertex of the graph, typically in $\mathbb{R}^2$, and connects the endpoints of each edge by a (Jordan) curve. We assume that there are no vertices in the interior of edges, edges intersect finitely often, and no more than two edges cross in a point. In a *geometric/straight-line* drawing, all edges are represented by straight-line segments. We refer to crossing-free drawings as *embeddings* or *planar*; a graph given with a crossing-free drawing is *plane*.

A *rotation* at a vertex is a cyclic permutation of the edges incident on the vertex (meant to represent the order in which the edges leave the vertex in a drawing or embedding). A *rotation system* assigns a rotation to each vertex of a graph.

We will occasionally encounter $\exists\mathbb{Q}$, the existential theory of the rationals.

(CG41) Planar Slope Number
Given: A graph G, an integer $k \geq 0$.
Question: Is there a crossing-free straight-line drawing of G in which the edges are drawn using at most k different slopes?
Complexity: $\exists\mathbb{R}$-complete, by Hoffmann [Hof17], via a reduction from stretchability (CG9). For fixed k, the problem lies in NP[Hof17] using a result from [KM94] that shows that intersection graphs of segments with a fixed number of slopes can be recognized in NP.
Comments: A drawing realizing the planar slope number of the graph may not be possible on a grid, and it is $\exists\mathbb{Q}$-complete to decide whether a grid realization exists, and there is a $2^{2^{\Omega(n^{1/3})}}$ lower bound on the grid size [Hof17].
Open Questions: How hard is it to determine the (non-planar) slope number of a graph [Hof17]?
Also see: Upward planar slope number (CG42), upward outerplanar slope number (CG-Open15), segment intersection graph (CG26), line cover number (CG43), segment number (CG44), faithful drawability of edge-colored graph (CG-Open20).

(CG42) Upward Planar Slope Number
Given: A directed graph G, an integer $k \geq 0$. A (straight-line) drawing of G is *upward* if u lies below v for every directed edge $uv \in E(G)$.
Question: Is there an upward planar straight-line drawing of G in which the edges are drawn using at most k different slopes?
Complexity: $\exists\mathbb{R}$-complete, by Quapil [Qua24], reduction from stretchability (CG9).
Comments: A drawing realizing the planar slope number of the graph may not be possible on a grid, and it is $\exists\mathbb{Q}$-complete to decide whether a grid realization exists, and there is a $2^{2^{\Omega(n^{1/3})}}$ lower bound on the grid size [Hof17].
Open Questions: Does the problem turn NP-complete for fixed k, as is the case for planar slope number (CG41)? It is known to be NP-hard in that case [KZ23]. ▼ Is the problem $\exists\mathbb{Q}$-complete if the vertices have to lie on a grid?
Also see: Planar slope number (CG41), upward outerplanar slope number (CG-Open15).

(CG43) Line Cover Number in $\mathbb{R}^d$
Given: A graph G, an integer $k \geq 0$
Question: Are there k lines in $\mathbb{R}^d$ whose union contains a crossing-free straight-line drawing of G? We say the lines *cover* G.
Complexity: $\exists\mathbb{R}$-complete by Chaplick et al. [CFL+23] for $d \in \{2, 3\}$.
Comments: Chaplick et al. [CFL+23] define the *affine cover number* as the smallest number of ℓ-dimensional planes in $\mathbb{R}^d$ that cover a straight-line drawing of G. Since the affine cover number of a graph is the same for $d > 3$ and $d = 3$, it is sufficient to study line cover numbers in $\mathbb{R}^2$ and line and plane cover numbers in $\mathbb{R}^3$.
Open Questions: Chaplick et al. [CFL+23] show that the plane cover number is NP-hard in $\mathbb{R}^3$. Is it $\exists\mathbb{R}$-complete? Is it $\exists\mathbb{R}$-complete for other values of (ℓ, d)?
Also see: Planar slope number (CG41), circle cover number under spherical cover number (CG-Open13), minimum line number (CG-Open14), planar stick number of drawing (CG46).

(CG44) Segment Number
Given: A graph G, an integer $k \geq 0$.
Question: Is there a crossing-free straight-line drawing of G that is the union of k straight-line segments? The smallest such k is known as the *(planar) segment number*.
Complexity: $\exists\mathbb{R}$-complete by Okamoto et al. [ORW19] via a reduction from arrangement graph recognition (CG11) even if the maximum degree of the graph is 4 and by Erickson [Eri19] via a reduction from simple stretchability (CG9) if the graph is 4-regular and given with a fixed embedding, which has to be realized. Several variants are $\exists\mathbb{R}$-complete as well, including the case that polyline drawings are allowed (*bend segment number*), that crossings are allowed (*crossing segment number*), or that the straight-line drawing lies in $\mathbb{R}^3$ (*3D segment number*) [ORW19].
Comments: Erickson's result is phrased as straightening a closed curve, see the isotopic polygon problem (CG45). The arc number (CG-Open12) was introduced by Schulz [Sch15] as a variation of the segment number.
Also see: Arrangement graph (CG11), arc number (CG-Open12), minimum line number (CG-Open14), isotopic polygon (CG45), planar stick number of drawing (CG46).

(CG45) Isotopic Polygon
Given: A closed curve γ, that is, a 4-regular plane graph (each self-intersection of γ becomes a degree-4 vertex), an integer $k \geq 0$.
Question: Is there a polygon with k vertices isotopic to γ?
Complexity: $\exists\mathbb{R}$-complete by Erickson [Eri19] via a reduction from simple stretchability (CG9).
Comments: The isotopic polygon problem is a special case of the planar stick number of drawing (CG46), and can be viewed as the fixed embedding variant of the segment number (CG44), and the
Universality: Universal [Eri19].
Also see: Segment number (CG44), planar stick number of drawing (CG46).

(CG46) Planar Stick Number of Drawing
Given: A graph G with topological drawing D (not necessarily crossing-free), an integer $k \geq 0$.
Question: Is there a poly-line drawing of G isomorphic to D with at most k segments in total?
Complexity: $\exists\mathbb{R}$-complete via a reduction from isotopic polygon (CG45).
Comments: This problem was suggested by the work of Khandhawit et al. [KPW23] who introduce the stick number of a spatial diagram of a graph; since this notion requires knot equivalence, there is currently no chance to relate it to $\exists\mathbb{R}$, but the simplified version presented here should still be of interest, as the underlying computational problem. Erickson [Eri24] observed that hardness follows from the isotopic polygon problem (CG45), even if $G = K_3$: start with the closed curve γ from that problem and view it as a topological drawing of K_3.
Also see: Segment number (CG44), isotopic polygon (CG45), line cover number (CG43), minimum line number (CG-Open14).

(CG47) Triangulation Stretchability
Given: A plane triangulation T, and a collection of paths in T.
Question: Is there a straight-line drawing of T in which each path in the collection lies on a line?
Complexity: $\exists\mathbb{R}$-complete by Boomari et al. [BOZ21] from simple stretchability (CG9).
Also see: Triangulated irregular network visibility graph (CG39), simple stretchability (CG9).

(CG48) Graph in a Polygonal Domain
Given: A graph G and a polygon P, the *domain*. Some of the vertices of G are specified to be vertices of P.
Question: Does there exists a straight line drawing D of G such that D lies inside the domain P?
Complexity: $\exists\mathbb{R}$-complete [LMM22] via a reduction from PLANAR- ETR- INV (L6).
Comments: The problem is closely related to partial drawing extensibility (PDE) (CG-Open7). In the PDE problem, we are given a partial straight-line drawing of a graph, and we are asked whether this drawing can be extended to a straight-line drawing of the full graph [Pat06]. This differs from the graph in a polygonal domain problem in that in PDE we are not allowed to place vertices on existing edges, whereas in the current problem we are allowed to place vertices on the already existing polygon.
Open Questions: Patrignani [Pat06] showed that the partial drawing extensibility problem (CG-Open7) is NP-hard, is it $\exists\mathbb{R}$-complete?
Also see: Partial drawing extensibility (CG-Open7), anchored graph drawing (CG-Open9).

(CG49) Drawing Graph on Segments with Obstacles
Given: A plane graph G, a line-segment ℓ_v (open or closed) for every $v \in V(G)$, and a set S of line-segments, all in the plane.
Question: Is there a straight-line drawing of G, isomorphic to the given embedding, in which every vertex v lies on ℓ_v, and edges do not cross the line-segments in S?
Complexity: $\exists\mathbb{R}$-complete by Bieker [Bie20] via a reduction from a variant of PLANAR- ETR- INV (L6).
Comments: Since open line segments can be points, this problem can be turned into a restricted version of the partial drawing extensibility problem (CG-Open7), whose complexity is open.
Open Questions: Segments are allowed to be both open and closed. Does the problem remain $\exists\mathbb{R}$-hard if all segments are open, or all segments are closed? ▼ The proof requires segments with four different slopes, does the problem become easier with fewer slopes? ▼ Does the problem remain $\exists\mathbb{R}$-hard if S is empty? ▼ Does the problem remain $\exists\mathbb{R}$-hard if the embedding of the graph is not fixed?
Also see: Graph in a polygonal domain (CG48), partial drawing extensibility (CG-Open7), pseudo-segment stretchability (CG10).

(CG50) Rectilinear Crossing Number
Given: A graph G, an integer $k \geq 0$.
Question: Does G have a straight-line drawing with at most k crossings? The smallest such k is the *rectilinear crossing number* of G.
Complexity: $\exists\mathbb{R}$-complete. $\exists\mathbb{R}$-hardness was shown by Bienstock [Bie91] via a reduction from simple stretchability (CG9) and Dean [Dea02] showed that the problem lies in $\exists\mathbb{R}$.
Comments: For $k \leq 2$ it is known that the rectilinear crossing number equals the standard crossing number [BD93], so the problem can be decided in polynomial time for those values. (The case $k = 3$ is also claimed, but only sketched, in the paper.)
Universality: Bienstock's reduction yields algebraic universality [Bie91].
Open Questions: Does the problem remain $\exists\mathbb{R}$-complete for graphs of bounded degree? ▼ Does the problem remain $\exists\mathbb{R}$-complete for fixed k? What is the complexity of the problem for $k = 4$? ▼ Does the problem remain $\exists\mathbb{R}$-complete if each edge consists of t segments, for some fixed t? What about $t = 2$? This question is based on Bienstock [Bie91] who named the corresponding crossing number the *t-polygonal crossing number*, and pointed out that realizing an optimal t-polygonal crossing number drawing may require exponential precision for every fixed t.
Also see: Rectilinear local crossing number (CG52), pseudolinear vs. rectilinear crossing number (CG51), segment intersection graph (CG26), maximum rectilinear crossing number (CG-Open6).

(CG51) Pseudolinear versus Rectilinear Crossing Number
Given: A graph G.
Question: If G has a pseudolinear drawing with at most k crossings, does G have a straight-line drawing with k crossings, for every k? In other words, does the

pseudolinear crossing number of G equal its rectilinear crossing number (CG50)? A *pseudolinear drawing* of a graph is a pseudoline arrangement in which every edge is drawn as part of a pseudoline.
Complexity: $\exists\mathbb{R}$-complete by Hernández-Vélez, Lea nos, and Salazar [HVLnS17] via a reduction from stretchability (CG9).
Also see: Rectilinear crossing number (CG50), stretchability (CG9).

(CG52) Rectilinear Local Crossing Number
Given: A graph G, an integer $k \geq 0$.
Question: Does G have a straight-line drawing with at most k crossings *per edge*? The smallest such k is the *rectilinear local crossing number* of G.
Complexity: $\exists\mathbb{R}$-complete [Sch21a]. Remains $\exists\mathbb{R}$-complete for fixed $k \geq 867$ [Sch21a].
Comments: $\exists\mathbb{R}$-complete [Sch21a] via a reduction from simple pseudo-segment stretchability (CG10). NP-complete for $k = 1$ [Sch14], and NP-hard for every fixed $k \geq 2$ [Sch21a].
Universality: Reduction yields algebraic universality [Sch21a].
Open Questions: Does the problem remain $\exists\mathbb{R}$-complete for graphs of bounded degree? ▼ What is the largest k for which the problem lies in NP? What is the smallest k for which the problem is $\exists\mathbb{R}$-complete?
Also see: Rectilinear crossing number (CG50), segment intersection graph (CG26), pseudo-segment stretchability (CG10).

(CG53) Geometric Partial 1-Planarity
Given: A graph $G = (V, E)$, a subset of edges $F \subseteq E$.
Question: Is there a straight-line drawing of G in which every edge of F has at most one crossing?
Complexity: $\exists\mathbb{R}$-complete by Schaefer [Sch14] via a reduction from stretchability (CG9).
Comments: The special case of geometric 1-planarity in which $F = E$ is only NP-complete, a folklore result [Sch14].
Open Questions: What is the complexity of geometric partial planarity (CG-Open8), the variant of the problem in which edges in F have to be crossing-free?
Also see: Geometric partial planarity (CG-Open8), rectilinear local crossing number (CG52).

(CG54) Weak Rectilinear Realizability
Given: A graph $G = (V, E)$, a set $R \subseteq \binom{E}{2}$.
Question: Is there a straight-line drawing of G such that $\{e, f\} \in R$ for every two edges e and f that cross in the drawing?
Complexity: $\exists\mathbb{R}$-complete, even for complete graphs by Kynčl [Kyn11] via a reduction from simple stretchability (CG9). Remains $\exists\mathbb{R}$-complete in the *strict* case, in which R contains exactly the pairs of edges that cross in the drawing, even if G is the complete graph.
Comments: The strict case for a matching is the segment intersection graph problem (CG26), which was known to be $\exists\mathbb{R}$-complete [KM94]; the fact that the problem

remains hard for complete graphs is novel here [Kyn11]. For topological drawings, the problem in NP-complete [Kra91b, SSŠ03].
Also see: Segment intersection graph (CG26), straight-line realizability of graph with rotation system (CG55).

(CG55) Straight-line Realizability of Graph with Rotation System
Given: A graph $G = (V, E)$ with *rotation system* π, that is, a function that assigns to every vertex $v \in V$ a cyclic permutation of the edges incident on v.
Question: Is there a straight-line drawing of G such that the cyclic permutation of the edges incident to each vertex v agrees with $\pi(v)$?
Complexity: $\exists\mathbb{R}$-complete for complete graphs [Kyn11] via a reduction from strict rectilinear realizability (CG54). Remains $\exists\mathbb{R}$-complete for graphs of bounded degree [Sch21b] via a reduction from the partial version of radial system realizability (CG5) [Sch21b].
Comments: For complete graphs, the pairs of intersecting edges completely determine the rotation system of a straight-line drawing and vice versa; this implies that for complete graphs, the problem is equivalent to the strict rectilinear realizability problem.
Open Questions: How hard is the problem for cubic graphs? ▼ Is it NP-complete to test whether a graph with rotation system has a good (topological) drawing? For complete graphs, this can be decided in polynomial time [Kyn20].
Also see: Weak rectilinear realizability (CG54), simultaneous geometric embedding (CG26), radial system realizability (partial version) (CG5).

(CG56) Simultaneous Geometric Embedding (SGE)
Given: A family $(G_i)_{i\in[k]}$ of graphs on a shared vertex set.
Question: Is there a straight-line drawing of $G = \bigcup_{i\in[k]} G_i$ in which each G_i by itself is planar?
Complexity: $\exists\mathbb{R}$-complete by Cardinal and Kusters [CK15]. Remains $\exists\mathbb{R}$-complete even if k is fixed and all graphs are paths [Sch21b] and in the case that there are no *public* edges (edges belonging to more than one graph), the *empty sunflower case* [KR23] even for fixed $k = 31$ [FKM+24]. A weak hardness result for SGE was first presented in [EBGJ+08].
Comments: Explicitly shown $\exists\mathbb{R}$-complete in [CK15] via a reduction from simple stretchability (CG9); the result can also be proved by reduction from strict rectilinear realizability of complete graphs (CG54) [Kyn11], but the reduction from simple stretchability leads to stronger precision bounds: a simultaneous geometric embedding may require a grid of size $2^{2^{\Omega(n)}}$, where $n = V(G)$. The reductions for fixed k start with pseudo-segment stretchability (CG10).
Open Questions: Does the empty sunflower case remain $\exists\mathbb{R}$-hard if all graphs are trees? ▼ Does the problem remain $\exists\mathbb{R}$-complete for $k = 2$ [Car15]?
Also see: Strict rectilinear realizability (CG54), geometric thickness of multigraphs (CG57), geometric thickness (CG-Open1), pseudo-segment stretchability (CG10).

(CG57) Geometric Thickness of Multigraph
Given: A multigraph G, a natural number $k \in \mathbb{N}$.
Question: Does G have a straight-line drawing in which the edges can be colored with k colors in such a way that no crossing in the drawing involves two edges of the same color?
Complexity: $\exists\mathbb{R}$-complete by Förster, Kindermann, Miltzow, Parada, Terziadis, and Vogtenhuber [FKM+24] for every fixed $k \geq 57$ via a reduction from pseudo-segment stretchability (CG10). The problem remains $\exists\mathbb{R}$-complete even if G has fixed local crossing number [FKM+24].
Comments: Geometric thickness was introduced by Kainen [Kai73]. It has similarities to the simultaneous geometric embedding problem (CG56), which is $\exists\mathbb{R}$-complete. For a brief discussion, see [Sch21b, Remark 7]. The complexity of the geometric thickness of (simple) graphs (CG-Open1) remains open.
Also see: Simultaneous geometric embedding (CG56) geometric thickness (CG-Open1).

(CG58) Right-Angle Crossing Drawability
Given: A graph G.
Question: Does G have a straight-line drawing in which all crossings are right-angle crossings? Such drawings are known as right-angle crossing drawings, or *RAC-drawings*, for short.
Complexity: $\exists\mathbb{R}$-complete by Schaefer [Sch23b] via a reduction from feasibility (A1) (using straight-line drawability with junctions as an intermediate problem). Remains $\exists\mathbb{R}$-complete even if the topological embedding of the graph is given and each edge is involved in at most eleven crossings, and crossing angles are doubly exponentially close to being right-angle [Sch23b].
Comments: Determining whether G has a RAC-drawing on the integer grid is $\exists\mathbb{Q}$-complete [Sch23b]. The *resolution* at a crossing is the smaller of the two angles incident to a crossing; the *crossing resolution* of a straight-line drawing is the largest resolution at any crossing, so RAC-drawings have maximal crossing resolution. The maximal resolution at vertices is known as the angular resolution (CG60).
Open Questions: In a RAC_k drawing each edge is allowed to have up to k bends. Are RAC_1- and RAC_2-drawability $\exists\mathbb{R}$-hard? ▼ Does RAC-drawbility remain $\exists\mathbb{R}$-hard for graphs of bounded degree? ▼ Does RAC-drawability remain $\exists\mathbb{R}$-hard for 1-planar drawing, that is, drawings with at most one crossing per edge? See geometric partial 1-planarity (CG53). ▼ Does RAC-drawability remain $\exists\mathbb{R}$-complete if all right-angle crossings have to occur between axis-parallel edges? This variant is introduced as apRAC-drawability in [ABK+23] who showed that the problem is NP-hard. ▼ What is the complexity of RAC-drawability for *outer drawings*, that is, drawings in which all vertices have to lie on the outer face (this notion seems to have been introduced only recently [FKO24])?
Also see: Angular resolution (CG60), geometric partial 1-planarity (CG53).

(CG59) Lombardi-Drawability With Rotation System
Given: A graph $G = (V, E)$ with *rotation system* π, that is, a function that assigns to every vertex $v \in V$ a cyclic permutation of the edges incident on v.

Question: Does G have a Lombardi drawing in the plane respecting π? In a *Lombardi drawing* all edges must be drawn as circular arcs (crossings are allowed), and every vertex must have *perfect angular resolution*, that is, all angles between consecutive ends at the vertex must be the same. The drawing *respects* the rotation system if the cyclic permutation of the edges incident to each vertex v agrees with $\pi(v)$?
Complexity: $\exists\mathbb{R}$-complete by Jungeblut [Jun23a] via a reduction from simple stretchability (CG9).
Comments: Lovász introduced a somewhat similar notion of realizing a graph by a system of circles; he conjectured that every simple connected 4-regular planar graph admits such a realization. The conjecture was disproved by Bekos and Raftopoulou [BR15].
Open Questions: How hard is Lombardi drawability if no rotation system is given [Jun23a, Open Problem 1]? ▼ How hard is Lombardi drawability if we require the drawing to be crossing-free (with or without rotation system)n [Jun23a, Open Problem 2]? ▼ Does Lombardi drawability remain hard for graphs of bounded degree?
Also see: Straight-line realizability of graph with rotation system (CG55), angular resolution (CG60), arc number (CG-Open12).

(CG60) Angular Resolution
Given: A graph G, a natural number $k \geq 2$.
Question: Does G have a straight-line drawing with *angular resolution* at least $\pi/(2k)$, that is, a drawing in which all angles between incident edges are at least $\pi/(2k)$? The *angular resolution* of G is the supremum of the angular resolution of all straight-line drawings of G.
Complexity: $\exists\mathbb{R}$-complete by Schaefer [Sch23a] via a reduction from feasibility (A1) (using straight-line drawability with junctions as an intermediate problem). Remains $\exists\mathbb{R}$-complete even if a crossing-free topological embedding of the graph is given [Sch23a].
Comments: There is a subtle difference between a graph having angular resolution at least α and there being a drawing of the graph with angular resolution at least α, since angular resolution of a graph is defined as a supremum, as observed in [FHH+93]. Both versions are $\exists\mathbb{R}$-complete.
Open Questions: Testing for angular resolution at least $\pi/4$ is NP-complete. Is the angular resolution problem $\exists\mathbb{R}$-complete for $\pi/3$ [Sch23a]? ▼ Does the problem remain $\exists\mathbb{R}$-complete if G is planar, but the angular-resolution drawing does not have to be [Sch23a]?
Also see: RAC-drawability (CG58), isosceles graph drawing (CG-Open16).

(CG61) Prescribed Area Extension
Given: A plane graph G, an area assignment for G, where an *area assignment* for G assigns a positive real number to every face of G; locations for some of the vertices of G are fixed in the plane.
Question: Does G have a straight-line drawing realizing the area assignment, that is, each face having the area prescribed by the area assignment, with the fixed vertices at their prescribed location?

Complexity: $\exists\mathbb{R}$-complete, Dobbins et al. [DKMR23] via a reduction from PLANAR- ETR- INV (L6).
Open Questions: Does the problem remain $\exists\mathbb{R}$-hard without fixing vertex locations?
Also see: Prescribed area (CG-Open35), area universality (CG-Open36), area universality for triples with partial area assignment (CG76).

(CG62) Straight-line Realizability with Unit Spanning Ratio
Given: A graph G.
Question: Is there a *proper* straight-line drawing of G in the plane with spanning ratio one? Here proper means that no two vertices are drawn on the same point, and no point corresponding to a vertex lies on a segment representing an edge, apart from the two endpoints. The *spanning ratio* of a drawing is the maximum ratio, over all pairs of vertices, of the length of a shortest path in the drawing and the Euclidean distance.
Complexity: $\exists\mathbb{R}$-complete, by Aichholzer et al. [ABB+22] via a reduction from the point visibility graph problem (CG36).
Comments: Aichholzer et al. [ABB+22] show that a graph has a proper drawing with spanning ratio 1 if and only if it is the visibility graph of a set of points in the plane (CG36). They also show that the problem becomes linear-time solvable if the drawing has to be planar (crossing-free).
Also see: Point visibility graph (CG36).

(CG63) Graph Realizability
Given: A graph G with weight function $\ell : E(G) \to \mathbb{Q}_{\geq 0}$.
Question: Is there a straight-line drawing of G in which every edge e of G has length $\ell(e)$?
Complexity: $\exists\mathbb{R}$-complete by Schaefer [Sch13] via a reduction from stretchability (CG9). The problem remains $\exists\mathbb{R}$-complete even if $\ell(e) = 1$ for all $e \in E(G)$.
Comments: Graph drawings are assumed to be non-degenerate in that vertices do not overlap with other vertices or edges; relaxing that condition leads to the linkage realizability problem (CG66). The problem was first shown NP-hard by Saxe [Sax79] and Yemini [Yem79]; Eades and Wormald [EW90] showed NP-hardness if the drawing has to be crossing-free (and all edges have the same length), and that result was strengthened by Cabello et al. [CDR07]. The problem is known under different names to different communities, e.g. as the Euclidean distance matrix completion problem [Lau09], and it has many variants and applications [LLMM14]. A related problem is the reconstruction problem: given a multiset of distances (of the form $\sqrt{q}$ for $q \in \mathbb{Q}$) are there n points in $\mathbb{R}^d$ such that the multiset contains exactly their pairwise distances. This problem turns out to be NP-complete [LSS03].
Open Questions: It is open whether the problem remains $\exists\mathbb{R}$-complete for embedded graphs. (A claimed $\exists\mathbb{R}$-hardness proof [ADD+16] may not be complete. We were informed by one of the authors that the PhD thesis of Abel [Abe16] contains more details, "but is still lacking some detail and edits".)
Also see: Unit distance graph (CG65), linkage realizability (CG66), graph flattenability (CG-Open19).

(CG64) Graph with Long and Short Edges
Given: A graph $G = (V, E)$ with a partition of the edges $E = E_s \cup E_\ell$ into *short* and *long* edges.
Question: Is there a straight-line drawing of G in which exactly the short edges have length at most 1?
Complexity: $\exists\mathbb{R}$-complete by Angelini, Cornelsen, Haase, Katsanou, Montecchiani, and Symvonis [ACH+24] via a unit disk graph (CG21).
Comments: Graphs with long and short edges were introduced in [ACH+24] under the name *dichotomous ordinal graphs*. The same paper suggests a reduction from Euclidean preference (Misc1), but there also is a direct reduction from unit disk graphs (CG21): $G = (V, E)$ is a unit disk graph if and only if the graph H on vertex set V can be realized with short edges E and long edges $\binom{V}{2} - E$.
Open Questions: Does the problem remain $\exists\mathbb{R}$-complete if the realization has to be crossing-free? If the short edges have to induce a plane graph, then, as [ACH+24] point out, the problem is NP-hard [AKPT16]. ▼ Does the problem remain $\exists\mathbb{R}$-complete if the underlying graph is a complete bipartite graph? That is the case of most interest in [ACH+24].
Also see: Unit disk graph (CG21), Euclidean preference (Misc1), graph realizability (CG63).

(CG65) (Strict) Unit Distance Graph
Given: A graph G.
Question: Is G a (strict) subgraph of $E_2 = (\mathbb{R}^2, \{uv \in (\mathbb{R}^2)^2 : |u - v| = 1\})$?
Complexity: Recognition of both unit distance graphs and strict unit distance graphs is $\exists\mathbb{R}$-complete by Schaefer [Sch13] via a reduction from graph realizability (CG63).
Comments: By definition, different vertices of G correspond to distinct points in E_2; if we relax this condition, we get the degenerate unit distance graph problem [HKP11], which we discuss as part of the linkage realizability problem (CG66). Erdős, Harary, and Tutte [EHT65] introduced the *Euclidean dimension* of a graph, which is the smallest d for which G is a subgraph of $E_d = (\mathbb{R}^d, \{uv \in (\mathbb{R}^d)^2 : |u - v| = 1\})$; so testing whether a graph has dimension at most 2 is already $\exists\mathbb{R}$-complete.
Universality: Maehara [Mae91] showed that the distances that occur in rigid unit distance graphs are exactly the algebraic numbers.
Open Questions: Is the Euclidean dimension $\exists\mathbb{R}$-complete for every fixed d?
Also see: Linkage realizability (CG66), unit disk contact graph (CG-Open3), graph realizability (CG63).

(CG66) Linkage Realizability
Given: A graph G with a function $\ell : E(G) \to \mathbb{Q}_{\geq 0}$.
Question: Can we assign a point $p_u \in \mathbb{R}^d$ to every $u \in V(G)$ such that $|f(u) - f(v)| = \ell(uv)$ for every edge uv?
Complexity: $\exists\mathbb{R}$-complete by Abbott [Abb08] via a reduction from Feasibility (A1) for every $d \geq 2$, The problem remains $\exists\mathbb{R}$-complete even for unit linkages, that is if $\ell(e) = 1$ for all $e \in E(G)$, in $\mathbb{R}^2$ by Schaefer [Sch13] via a reduction from stretchability (CG9).
Comments: The main difference to graph realizability (CG63) is that vertices can

overlap with other vertices and edges. In some definitions, linkages also come with points with pre-assigned locations. Unit linkages are called *degenerate unit distance graphs* by Horvat, Kratochvíl, and Pisanski [HKP11] who showed that the problem is NP-hard.
Universality: The universality of linkage realizability is Kempe's universality theorem [KM02, Abb08].
Open Questions: Is realizability of unit linkages $\exists\mathbb{R}$-complete in arbitrary dimension? There is a gadget, due to Saxe [Sax79], that increases the dimension by one, but it is not a unit linkage.
Also see: Unit distance graph (CG65), graph realizability (CG63), linkage rigidity (CG67).

(CG67) Linkage Rigidity
Given: A graph G with a function $\ell : E(G) \to \mathbb{Q}_{\geq 0}$, and a *configuration*, that is, $p_u \in Q^d$ for every $u \in V(G)$ such that $|p_u - p_v| = \ell(uv)$ for every edge uv.
Question: Is the configuration rigid, that is, do small changes to all p_u so that the length conditions are still satisfied lead to the same configuration modulo rigid motions of the plane?
Complexity: $\forall\mathbb{R}$-complete by Abbott [Abb08], Schaefer [Sch13] for $d \geq 2$ via a reduction from ISO (A7). Abbott [Abb08] shows that ISO reduces to linkage rigidity for $d = 2$, and $d - 1$-dimensional linkage rigidity reduces to d-dimensional linkage rigidity. Together with the $\exists\mathbb{R}$-completeness of ISO, this implies $\exists\mathbb{R}$-completeness of d-dimensional linkage rigidity for every fixed $d \geq 2$. Membership of linkage rigidity in $\forall\mathbb{R}$ is shown in [Sch13].
Comments: The main difference to graph realizability (CG63) is that vertices can overlap with other vertices and edges.
Also see: Unit distance graph (CG65), linkage realizability (CG66), graph realizability (CG63).

5.4.3.6 Polytopes and Simplicial Complexes

A polytope is the convex hull of a finite point set in $\mathbb{R}^d$. The *face lattice* of a polytope is the lattice of its faces: the *join* of two faces is the smallest face containing both faces, and the *meet* is the intersection of the faces. Given a lattice, that is, a partial order with join and meet operations, one can ask whether it is *polytopal*, namely whether it is the face lattice of a polytope. We refer to the face lattice of P as its *combinatorial type*, and to P itself as a *realization* of this combinatorial type. The realizability question is known as the *algorithmic Steinitz problem*, due to a famous theorem of Steinitz. Let the *graph of a polytope* be the graph formed by the vertices and edges of the polytope. The graph is also referred to as the 1-*skeleton*, and the k-skeleton for $k \geq 1$ is the complex of all faces of dimension up to k. Steinitz theorem characterizes the graphs of 3-dimensional polytopes: they are precisely the simple, planar, 3-connected graphs [Zie95, Sect. 4], hence easy to recognize. For 3-polytopes the facial lattice is determined by the graph of the polytope, so the algorithmic Steinitz problem for 3-polytopes is tractable.

Note that the complete combinatorial type of a polytope in $\mathbb{R}^d$ can be deduced from its set of $(d-1)$-dimensional faces, or *facets*: the lower-dimensional faces are simply obtained by intersecting subsets of facets. The input to the problems can therefore be restricted to be a list of facets.

We obtain an interesting variant if we restrict the problem to *simple* polytopes, the skeletons of which are regular graphs. In that case the facial lattice is determined by the graph of the polytope (Blind and Mani with an easy proof by Kalai, see [Zie95, Sect. 3.4] and [KP03, Problem 15]).

A polytope is said to be *simplicial* if all its facets are simplices.

Subdivisions are closely related to polytopes. Given a set of points, a subdivision is a covering of their convex hull by polytopes, or *cells*, the vertices of which are points of the set, such that any two intersect in a common face. If all cells are simplices, then the subdivision is called a *triangulation*. A *Delaunay subdivision* is such that the cells are defined by an empty sphere condition: A cell C is part of the Delaunay subdivision of a set of points in $\mathbb{R}^d$ if and only if there exists a $(d-1)$-sphere passing through the vertices of C but containing no other point of the set in its interior. For generic point sets, such cells are always simplices, and we talk of *Delaunay triangulations*.

An *abstract simplicial k-complex* is a collection of sets of size at most k that is closed under taking subsets. More generally, an *(abstract) polyhedral complex* is a collection of (combinatorial types of) polytopes, such that every face of every polytope is also part of the collection, and any two polytopes in the collection intersect in a common face.

(CG68) Polytope Realizability
Given: A combinatorial type, given as a list of facets.
Question: Does there exist a polytope of this combinatorial type?
Complexity: $\exists\mathbb{R}$-complete, by Mnëv [Mnë88] via a reduction from existential theory of totally ordered real variables (L2).
Comments: This problem is also known as the *algorithmic Steinitz problem*. The problem is known to be hard in particular for n-dimensional polytopes with $n+4$ vertices. The input is then a collection of subsets of size at least n of the $n+4$ vertices.
Universality: Mnëv [Mnë88], also see Richter-Gebert [RG99].
Also see: 4-dimensional Polytope Realizability (CG69), Simplicial Polytope Realizability (CG70), Inscribed Simplicial Polytope Realizability (CG71), treetope graph (CG-Open42), existential theory of totally ordered real variables (L2).

(CG69) 4-dimensional Polytope Realizability
Given: A combinatorial type, given as a list of facets.
Question: Does there exist a 4-dimensional polytope of this combinatorial type?
Complexity: $\exists\mathbb{R}$-complete, by Richer-Gebert and Ziegler [RGZ95] and Richter-Gebert [RG96] via a reduction from existential theory of totally ordered real variables (L2).
Universality: Richter-Gebert and Ziegler [RGZ95], and Richter-Gebert [RG96, RG99].
Open Questions: The complexity of the problem for simplicial polytopes of constant dimension $d \geq 4$ is open, see Richter-Gebert [RG96]. ▼ The complexity of the related problem in dimension 4 in which we are only given the graph (1-skeleton) of the polytope is also open.

Also see: Polytope Realizability (CG68), 4-dimensional simplicial polytope realizability (CG-Open40), 4-dimensional polytope graph realizability (CG-Open41), treetope graph (CG-Open42).

(CG70) Simplicial Polytope Realizability
Given: A combinatorial type, given as a list of facets.
Question: Does there exist a simplicial polytope of this combinatorial type?
Complexity: $\exists\mathbb{R}$-complete, by Mnëv [Mnë88], Adiprasito and Padrol [AP17] even in even dimension via a reduction from oriented matroid realizability (CG2).
Universality: Adiprasito and Padrol [AP17].
Comments: Universality of simplicial polytopes was announced by Mnëv in [Mnë88], but with an incomplete proof. Adiprasito and Padrol prove the universality result for *neighborly polytopes*, the d-dimensional polytopes whose $\lfloor d/2 \rfloor$-skeleton is complete. Neighborly polytopes in even dimension are simplicial, so this implies the current result, but we can further assume neighborliness. The problem is known to be hard in particular for n-dimensional polytopes with $n + 4$ vertices.
Open Questions: Encouraged by the results of Richter-Gebert [RG96], Bokowski and Guedes de Oliveira [BGdO90], Adiprasito and Padrol make the "daring conjecture" that the result holds in dimension 4.
Also see: Inscribed Simplicial Polytope Realizability (CG71), 4-dimensional simplicial polytope realizability (CG-Open40).

(CG71) Inscribed Simplicial Polytope Realizability
Given: A combinatorial type, given as a list of facets.
Question: Does there exist a simplicial polytope of this combinatorial type, all vertices of which lie on a sphere?
Complexity: $\exists\mathbb{R}$-complete, by Adiprasito et al. [APT15].
Universality: Adiprasito et al. [APT15].
Comments: Delaunay subdivisions are obtained by stereographic projections of inscribed polytopes. The problem is known to be hard in particular for n-dimensional polytopes with $n + 4$ vertices.
Open Questions: The complexity of the problem for inscribed polytopes (not necessarily simplicial) of constant dimension $d \geq 4$ is open as well. ▼ In dimension 3, a result by Rivin implies that inscribed polytopes can be recognized in polynomial time [Riv96], although it is not clear how to efficiently produce a realization.
Also see: Delaunay Triangulation (CG73).

(CG72) Nested Polytopes
Given: Two polytopes $A \subseteq B \subseteq \mathbb{R}^d$ and a number $k \in \mathbb{N}$. The polytope A is described by vertices and the polytope B is described by its facets.
Question: Does there exist a polytope X on at most k vertices with $A \subseteq X \subseteq B$?
Complexity: $\exists\mathbb{R}$-complete, by Dobbins et al. [DHM19] via a direct reduction from ETR- INV (L5).
Universality: Universality follows from [DHM19].
Comments: Vavasis [Vav09] showed that there is a reduction from nonnegative matrix factorization (A21) to nested polytope, for the special case $k = d + 1$. (In

the case $k = d - 1$ the intermediate polytope is a simplex, and thus they call this problem the intermediate simplex problem.)

Shitov showed in turn $\exists\mathbb{R}$-completeness and universality for nonnegative matrix factorization and thus also implied the nested polytope result [Shi16b].
Also see: ETR- INV (L5).

(CG73) Delaunay Triangulation
Given: A collection F of subsets of size $n - 4$ of $[n]$.
Question: Does there exist a $(n - 5)$-dimensional Delaunay triangulation on n vertices, such that the vertices of the cells of the triangulations are exactly the subsets in F?
Complexity: $\exists\mathbb{R}$-complete, by Adiprasito et al. [APT15] via a reduction from Inscribed Simplicial Polytope Realizability (CG71).
Universality: Adiprasito et al. [APT15].
Comments: The result is directly reduced from the complexity of Inscribed Simplicial Polytope Realizability (CG71): An inscribed polytope in dimension d is mapped to a Delaunay subdivision in dimension $d - 1$ via a stereographic projection.
Open Questions: The complexity of the problem in constant dimension $d \geq 3$ is open. In dimension two, it is known from a result by Rivin that Delaunay triangulations can be recognized in polynomial time [Riv96], although it is not clear how to efficiently produce a realization. Adiprasito, Padrol, and Theran conjecture that universality already holds for 3-dimensional Delaunay subdivisions [APT15]. Padrol and Theran [PT14] give examples of Delaunay triangulations with disconnected realization spaces.
Also see: Inscribed Simplicial Polytope Realizability (CG71), Gabriel graph (CG-Open24).

(CG74) Simplicial Complex
Given: An abstract simplicial k-complex $C = (V, E)$ and a natural number $d \in \mathbb{N}$.
Question: Does C have a geometric embedding in $\mathbb{R}^d$? A *geometric embedding* of C is a function $\phi : V \to \mathbb{R}^d$ such that for every $e \in E$, $\phi(e) = \text{conv}(\{\phi(v) : v \in e\})$ is a simplex of dimension $|e| - 1$, and for every pair $e, e' \in E$, we have that $\phi(e) \cap \phi(e') = \phi(e \cap e')$.
Complexity: $\exists\mathbb{R}$-complete for all $d \geq 3$ and $k \in \{d - 1, d\}$, by Abrahamsen et al. [AKM23] via a reduction from stretchability (CG9).
Comments: For piecewise linear embedding, the problem has been shown to be NP-hard even in dimension 3, see [MTW11, dMRST20].
Open Questions: For which combinations of k and d is the problem $\exists\mathbb{R}$-complete? An interesting case is when $k = 2$ and $d = 4$.
Also see: Polyhedral complex (CG75).

(CG75) Polyhedral Complex
Given: An abstract polyhedral complex C composed of quadrangles and triangles.
Question: Does C have a geometric embedding in $\mathbb{R}^3$ in which the image of every polygon is a convex polygon, and their relative interiors are disjoint?

Complexity: $\exists\mathbb{R}$-complete, by Brehm (unpublished), via a reduction from collinearity (L9).
Comments: The result has been announced and presented, but never published formally. It is hinted at by Ziegler [Zie08]. Wilson [Wil12] also showed that the same result holds for polyhedral complexes composed of convex three-dimensional polyhedra.
Also see: Collinearity logic (L9), simplicial complex (CG74).

5.4.4 Game Theory

5.4.4.1 Games in Strategic Form

We briefly recall the definitions related to games given in *strategic form*. An r-player game Γ in strategic form is a pair $\Gamma = (\Sigma_i, \mathsf{U}_i)$, for $i \in 1, 2, \ldots, r$, where Σ_i is the set of strategies available to the ith player, and U_i is the payoff function of player i, of the form $\mathsf{U}_i : \Sigma_1 \times \Sigma_2 \times \ldots \times \Sigma_r \to \mathbb{R}$.

A *mixed strategy* for the ith player, as opposed to a *pure strategy* $s \in \Sigma_i$, is a probability distribution σ_i on the pure strategies in Σ_i. Let $\sigma = (\sigma_i)_{i=1,2,\ldots,r}$ be a collection of mixed strategies for each of the r players. The expected payoff of the ith player under these probability distributions is:

$$\mathsf{U}_i(\sigma) = \sum_{s \in \prod_{i=1}^{r} \Sigma_i} \mathsf{U}_i(s) \cdot \prod_{i=1}^{r} \sigma_i(s_i) \cdot$$

A Nash equilibrium is a collection of mixed strategies such that no player can improve her expected payoff by deviating from her current mixed strategy. Hence it consists of a collection of mixed strategies σ such that for every $i \in \{1, 2, \ldots, r\}$ and every σ' obtained by replacing σ_i by another mixed strategy σ_i', we have $\mathsf{U}_i(\sigma) \geq \mathsf{U}_i(\sigma')$. The classical result of Nash is that such an equilibrium exists for any game. A Nash equilibrium is said to be *pure* if it consists only of pure strategies.

Many of the problems in this section are discussed in Bilò and Mavronicolas [BM21], and Bilò et al. [BHM23].

(GT1) Nash Equilibrium in a Ball
Given: A 3-player game Γ in strategic form, $u \in \mathbb{Q}$.
Question: Does Γ have a Nash equilibrium σ such that $\sigma_i(s) \leq u$ for all $i = 1, 2, 3$ and all $s \in \Sigma_i$?
Complexity: $\exists\mathbb{R}$-complete by Schaefer and Štefankovič [SŠ17] via a reduction from Brouwer Fixed Point (A5) due to Ettessami and Yannakakis [EY10].
Comments: The result holds for a fixed $u = 1/2$ [SŠ17].
Also see: Brouwer Fixed Point (A5).

(GT2) Nash Equilibrium with Bounded Payoffs
Given: A 3-player game Γ in strategic form, $u \in \mathbb{Q}$.
Question: Does Γ have a Nash equilibrium σ such that $\mathsf{U}_i(\sigma) \geq u$ for all $i \in \{1, 2, 3\}$?

Complexity: $\exists\mathbb{R}$-complete by Garg et al. [GMVY18] via a reduction from Nash Equilibrium in a ball (GT1).
Comments: The problem in which we require that $\mathsf{U}_i(\sigma) \leq u$ instead is also $\exists\mathbb{R}$-complete. Both results hold for symmetric win-lose games as well, as shown by Bilò et al. [BHM23]. Berthelsen and Hansen [BH22] also show that for zero-sum games, deciding the existence of a Nash equilibrium in which $\mathsf{U}_i(\sigma) = 0$ for all $i \in \{1, 2, 3\}$ is $\exists\mathbb{R}$-complete. The proofs of the latter are related to those in Hansen [Han19].

(GT3) Nash Equilibrium with Restricted Support
Given: A 3-player game Γ in strategic form, a subset $T_i \subseteq \Sigma_i$ for each $i = 1, 2, 3$.
Question: Does Γ have a Nash equilibrium σ such that the set of strategies $s \in \Sigma_i$ such that $\sigma_i(s) > 0$ is contained in T_i for every $i \in \{1, 2, 3\}$?
Complexity: $\exists\mathbb{R}$-complete by Garg et al. [GMVY18] via a reduction from Nash Equilibrium in a ball (GT1).
Comments: The problem in which we require that the set contains T_i instead is also $\exists\mathbb{R}$-complete [GMVY18, BM21]. The two problems are respectively called *Superset* and *Subset* in [GMVY18]. Both results hold for win-lose games as well [BHM23].

(GT4) Second Nash Equilibrium
Given: A 3-player game Γ in strategic form.
Question: Does Γ have at least two distinct Nash equilibria?
Complexity: $\exists\mathbb{R}$-complete by Garg et al. [GMVY18] via a reduction from Nash Equilibrium with Bounded Payoffs (GT2). The problem is referred to as *NonUnique* in [GMVY18].

(GT5) Nash Equilibrium with Bounded Total Payoff
Given: A 3-player game Γ in strategic form, $u \in \mathbb{Q}$.
Question: Does Γ have a Nash equilibrium σ such that $\sum_{i\in\{1,2,3\}} \mathsf{U}_i(\sigma) \geq u$?
Complexity: $\exists\mathbb{R}$-complete by Bilò and Mavronicolas [BM21] via a reduction from Nash Equilibrium with Restricted Support (GT3).
Comments: The problem in which we require that $\sum_i \mathsf{U}_i(\sigma) \leq u$ instead is also $\exists\mathbb{R}$-complete. An alternative hardness proof is also given by Berthelsen and Hansen [BH22]. Both results hold for symmetric win-lose games as well [BHM23].

(GT6) Nash Equilibrium with Bounded Support
Given: A 3-player game Γ in strategic form, a natural number $k \in \mathbb{N}$.
Question: Does Γ have a Nash equilibrium σ such that the number of nonzero components of σ_i is at least k, for every $i \in \{1, 2, 3\}$?
Complexity: $\exists\mathbb{R}$-complete by Bilò and Mavronicolas [BM21] via a reduction from Nash Equilibrium with Restricted Support (GT3).
Comments: The problem in which we require that the support has size at most k is also $\exists\mathbb{R}$-complete. Both results hold for symmetric win-lose games as well [BHM23].

(GT7) Minmax Value
Given: A 3-player game Γ in strategic form.
Question: Is the minmax value of player 1 at most 0? Here the *minmax value* of player i is the minimum utility $\min_\sigma \max_{s_i} \mathsf{U}_i(\sigma \backslash s_i)$ of a best reply strategy, where $\sigma \backslash s_i$ is the collection of strategies obtained by replacing σ_i by s_i in σ.
Complexity: $\exists\mathbb{R}$-complete by Hansen [Han19] via a reduction from a variant of (A1) similar to that of [Sch13, Lemma 3.9], where the solution of a system of quadratic equations is promised to lie in a polyhedron contained in the standard simplex.
Comments: The problem remains $\exists\mathbb{R}$-complete under various promises, see Hansen [Han19] for details.

(GT8) Trembling Hand Perfect Equilibrium
Given: A 3-player game Γ in strategic form, and a pure Nash equilibrium σ of Γ.
Question: Is σ a trembling hand perfect equilibrium? A collection of strategies σ is a *trembling hand perfect equilibrium* if and only if it is the limit of a sequence of ε-perfect equilibria with $\varepsilon \to 0^+$, where an *ε-perfect equilibrium* is a Nash equilibrium that is fully mixed, and in which only pure strategies that are best replies get probability more than ε, see [Sel75, Mye78].
Complexity: $\exists\mathbb{R}$-complete by Hansen [Han19] via a reduction from Minmax value (GT7).

(GT9) Proper Equilibrium
Given: A 3-player game Γ in strategic form, and a pure Nash equilibrium σ of Γ.
Question: Is σ a proper equilibrium? Here σ is a *proper equilibrium* if it is the limit point of a sequence of ε-proper equilibria with $\varepsilon \to 0^+$, where an *ε-proper equilibrium* is a fully mixed equilibrium such that given two pure strategies k and ℓ of the same player, if k is a worse reply against σ than ℓ, then σ must assign a probability to k that is at most ε times the probability it assigns to ℓ, see [Sel75].
Complexity: $\exists\mathbb{R}$-complete by Hansen [Han19] via a reduction from Minmax value (GT7).

(GT10) Strategy Sets Closed under Rational Behavior
Given: A 3-player game Γ in strategic form, and a collection of subsets $S_i \subseteq \Sigma_i$, $i = 1, 2, 3$.
Question: Is the collection $\{S_i\}$ closed under rational behavior (CURB)? Here sets of pure strategies $\{S_i\}_{i=1}^m$ of an m-player game is CURB if and only if for all pure strategies s of Player i such that s is a best reply to a distribution $S_1 \times S_2 \times \ldots \times S_{i-1} \times S_{i+1} \times \ldots \times S_m$, we have that $s \in S_i$.
Complexity: $\exists\mathbb{R}$-complete by Hansen [Han19] via a reduction from Minmax value (GT7).

(GT11) Pareto Optimal Nash Equilibrium
Given: A 3-player game Γ in strategic form.
Question: Does Γ have a Pareto optimal Nash equilibrium? A collection σ of mixed strategies is *Pareto optimal* if there does not exist any σ' such that $\mathsf{U}_i(\sigma') \geq \mathsf{U}_i(\sigma)$ for all $i \in \{1, 2, 3\}$, and $\mathsf{U}_i(\sigma') > \mathsf{U}_i(\sigma)$ for at least one $i \in \{1, 2, 3\}$.

Complexity: $\exists\mathbb{R}$-complete by Berthelsen and Hansen [BH22] via a reduction from the variant of Nash Equilibrium with Bounded Payoffs (GT2), in which we require $\mathsf{U}_i(\sigma) = 0$.
Comments: The result also holds for zero-sum games. The problem of deciding whether there exists a non-Pareto optimal Nash equilibrium is $\exists\mathbb{R}$-hard for symmetric win-lose games as well [BM21, BHM23].
Open Questions: It is open if the result also holds for win-lose games [BHM23].

(GT12) Strong Nash Equilibrium
Given: A 3-player game Γ in strategic form.
Question: Does Γ have a strong Nash equilibrium? A collection σ of mixed strategies is a *strong Nash equilibrium* if no coalition of players can together increase their payoff, hence if there is no nonempty subset $B \subseteq \{1, 2, 3\}$ such that all players in B can increase their payoff by simultaneously deviating from their strategies in σ.
Complexity: $\exists\mathbb{R}$-complete by Berthelsen and Hansen [BH22] via a reduction from the variant of Nash Equilibrium with Bounded Payoffs (GT2) in which we require $\mathsf{U}_i(\sigma) = 0$.
Comments: The result also holds for zero-sum games. The problem of deciding whether there exists a non-strong Nash equilibrium is $\exists\mathbb{R}$-hard for symmetric win-lose games as well [BHM23].
Open Questions: It is open if the result also holds for win-lose games [BHM23].

(GT13) Nonsymmetric Nash Equilibrium
Given: A symmetric 3-player game Γ in strategic form.
Question: Does Γ have a nonsymmetric Nash equilibrium?
Complexity: $\exists\mathbb{R}$-complete by Berthelsen and Hansen [BH22] via a reduction from the variant of Nash Equilibrium with Bounded Payoffs (GT2) in which we require $\mathsf{U}_i(\sigma) = 0$.

(GT14) Irrational Nash Equilibrium
Given: A 3-player game Γ in strategic form.
Question: Does Γ have a Nash equilibrium σ in which $\sigma_i(s) \notin \mathbb{Q}$ for some $i \in \{1, 2, 3\}$ and some $s \in \Sigma_i$?
Complexity: $\exists\mathbb{R}$-hard by Berthelsen and Hansen [BH22] via a reduction from the variant of Nash Equilibrium with Bounded Payoffs (GT2) in which we require $\mathsf{U}_i(\sigma) = 0$.
Comments: The result also holds for zero-sum games [BH22], and for win-lose games [BHM23]. The problem of deciding whether there exists a rational Nash equilibrium, in which $\sigma_i(s) \in \mathbb{Q}$ for all $i \in \{1, 2, 3\}$ and all $s \in \Sigma_i$ is shown to be $\exists\mathbb{Q}$-hard [BH22]. It is open whether the problem is even decidable.

5.4.4.2 Games in Extensive Form

Games in *extensive form* are defined by a rooted tree with utility vectors at the leaves. Nodes of the tree are controlled either by one of the player, or by a so-called chance player, in which case we associate a probability distribution to the outgoing edges. In games in extensive form with *imperfect information*, strategies can only be based on partial information, or *signals* regarding the moves of the opponent. A behavioral strategy is a mixed strategy based only on the received signals. In games with *imperfect recall*, not only strategies can only be based on partial information, but players can forget about the sequence of signals or actions. The definitions extend to games that are defined on directed acyclic graphs. We refer to Hansen [Han19] and Gimbert et al. [GPS20] for detailed definitions.

(GT15) Trembling Hand Perfect Equilibrium for Games in Extensive Form
Given: A 3-player game Γ in extensive form, and a pure Nash equilibrium σ of Γ.
Question: Is σ a trembling hand perfect equilibrium?
Complexity: $\exists\mathbb{R}$-complete by Hansen [Han19] via a reduction from Minmax value (GT7).
Comments: We refer to [Sel75, vD84] for the definition of trembling hand perfect equilibria for games in extensive form. Hansen [Han19] also gives $\exists\mathbb{R}$-completeness results for other properties of pure Nash equilibria of games in extensive form, in particular that of belonging to a *sequential equilibrium* [KW82] and being a *quasi-perfect equilibrium* [vD84].
Also see: Trembling Hand Perfect Equilibrium (GT8).

(GT16) Nonnegative Expected Payoff for Games in Extensive Form
Given: A one-player game in extensive form, with imperfect recall.
Question: Is the maximum expected payoff, over all behavioral strategies, nonnegative?
Complexity: $\exists\mathbb{R}$-complete by Gimbert et al. [GPS20], via a reduction from feasibility (A1). The result also holds for deciding whether the maximum payoff is strictly positive.
Also see: Feasibility (A1).

(GT17) Nonnegative Maxmin Payoff for Games in Extensive Form
Given: A zero-sum two-player game in extensive form, with imperfect recall.
Question: Is the maxmin payoff, over all behavioral strategies, nonnegative?
Complexity: Both $\exists\mathbb{R}$-hard and $\forall\mathbb{R}$-hard, and belongs to $\exists\forall\mathbb{R}$ by Gimbert et al. [GPS20]. Reductions are from nonnegative Expected Payoff for Games in Extensive form (GT16).
Also see: nonnegative Expected Payoff for Games in Extensive form (GT16).

(GT18) Stationary Nash Equilibrium in Acyclic Games
Given: An acyclic m-player recursive game with payoff demands $L \in \mathbb{Q}^m$.
Question: Does there exist a stationary Nash equilibrium satisfying all the demands? A *stationary equilibrium* is one involving stationary strategies, that only depend on the current node of the graph.

Complexity: $\exists\mathbb{R}$-complete by Hansen and Sølvsten [HS20]. The reduction is from feasibility (A1) with homogeneous quadratic polynomials over a simplex.
Comments: $\exists\mathbb{R}$-completeness holds even for 7-player games with nonnegative rewards.

5.4.4.3 Fair Division Problems

(GT19) Square-Cut Pizza Sharing
Given: A collection of n pizzas, where a *pizza* is a mass distribution (measure) on $[0, 1]^2$ that can be computed for polygonal subsets using arithmetic circuits.
Question: Can all pizzas be halved simultaneously by a *square cut*, that is, a polygonal arc consisting of horizontal and vertical segments and such that the square cut makes at most $n - 2$ turns?
Complexity: $\exists\mathbb{R}$-complete by Deligkas et al. [DFM20, DFM22] via a reduction from consensus halving (GT21). Remains $\exists\mathbb{R}$-complete even if each pizza consists of at most 6 square or triangular pieces of mass (not necessarily uniform), and each point belongs to at most 3 pizzas [DFM20, DFM22].
Comments: The $\exists\mathbb{R}$-hardness result is sharp in the sense that n pizzas can always be halved simultaneously with a square-cut making at most $n - 1$ turns [KRPS16].
Also see: Line-cut pizza sharing (GT20), consensus halving (GT21).

(GT20) Line-Cut Pizza Sharing
Given: A collection of $2n$ pizzas, where a *pizza* is a mass distribution (measure) on $[0, 1]^2$ that can be computed for polygonal subsets using arithmetic circuits.
Question: Can all pizzas be halved simultaneously by making k line cuts (and assembling the resulting pieces appropriately)?
Complexity: $\exists\mathbb{R}$-complete by Schnider [Sch21c] via a reduction from consensus halving (GT21) with $k = n - 1$
Comments: The $\exists\mathbb{R}$-hardness result is sharp in the sense that $2n$ pizzas can always be halved simultaneously with at most n line cuts [HK20].
Also see: Square-cut pizza sharing (GT19), consensus halving (GT21).

(GT21) Consensus Halving
Given: A family of n arithmetic circuits computing functions $F_i : [0, 1] \to \mathbb{R}$, $i \in [n]$, a natural number $k \in \mathbb{N}$. The functions F_i are used to valuate intervals and unions of intervals by defining $F_i([a, b]) = F_i(b) - F_i(a)$, and $F_i(A \cup B) = F_i(A) + F_i(B)$ for disjoint unions of intervals A and B of $[0, 1]$.
Question: Is there a *k-cut*, that is, k points $0 \leq x_1 \leq \cdots \leq x_k \leq 1$, with the following property: if we split $[0, 1]$ into two sets $A_+ = \bigcup_{0 \leq 2i \leq k} [x_{2i}, x_{2i+1}]$ and $A_- = \bigcup_{0 \leq 2i+1 \leq k} [x_{2i+1}, x_{2i+2}]$, where we let $x_0 = 0$ and $x_{k+1} = 1$, then $F_i(A_+) = F_i(A_-)$ for all $i \in [n]$? That is all agents, the holders of the valuation functions, valuate the two "halves" A_+ and A_- equally.
Complexity: $\exists\mathbb{R}$-complete by Deligkas et al. [DFMS21] via a reduction from feasibility (A1) over a compact domain. The problem is hard for $k = n - 1$ and always solvable (using the Borsuk-Ulam Theorem) for $k = n$.

Open Questions: Does the problem remain hard if the F_i are given as polynomials? Deligkas et al. [DFMS22] show that there is a quasi-polynomial-time approximation scheme in this case.
Also see: Square-cut pizza sharing (GT19), line-cut pizza sharing (GT20).

5.4.4.4 Markets

(GT22) Equilibrium in Exchange Markets
Given: An Arrow-Debreu exchange market, consisting of a set $\mathcal{G}$ of g divisible goods, a set $\mathcal{A}$ of agents, a matrix W such that W_{ij} is the amount of good j with agent i, and a nonnegative piecewise-linear concave utility function $U_i : \mathbb{R}^g_+ \to \mathbb{R}_+$.
Question: Is there an equilibrium in the market, that is, $p \in \mathbb{R}^g_+$ and $x_i \in \mathbb{R}^g$ for all $i \in \mathcal{A}$, such that

$$\forall i \in \mathcal{A} : x_i = \arg\max_y \{U_i(y) : \sum_j y_j p_j \leq \sum_j W_{ij} p_j\}$$

$$\forall j \in \mathcal{G} : \sum_i x_{ij} = \sum_i W_{ij}?$$

Complexity: $\exists\mathbb{R}$-complete by Garg et al. [GMVY17] via a reduction from feasibility (A1). The result holds even under the assumption that the utility functions are Leontief utility functions, a special subclass of piecewise-linear concave functions, see [GMVY17] for definitions.
Also see: Clearing recovery rate (GT23)

(GT23) Clearing Recovery Rate
Given: A *financial network* on n banks, consisting of *external assets* $e_i \in \mathbb{Q}_{\geq 0}$, *debt contracts*, $c_{i,j} \in \mathbb{Q}_{\geq 0}$, and *credit default swaps (CDS)*, $c^k_{i,j} \in \mathbb{Q}_{\geq 0}$; *default parameters* $\alpha, \beta \in [0, 1] \cap \mathbb{Q}$.
Question: For a recovery rate vector $r = (r_i)_{i \in [n]} \in [0, 1]^n$ one computes the *liability* of bank i to bank j as $\ell_{i,j}(r) = c_{i,j} + \sum_{k \in [n]}(1 - r_k)c^k_{i,j}$ and the *total liability* of bank i as $\ell_i(r) = \sum_{j \in [n]} \ell_{i,j}(r)$. The *payment* of bank i to bank j is computed as $p_{i,j}(r) = r_i \cdot \ell_{i,j}(r)$. With that bank i has assets $a_i(r) = e_i + \sum_{j \in [n]} p_{j,i}(r)$ *before default costs* and assets $a'_i(r) = \alpha e_i + \beta \sum_{j \in [n]} p_{i,j}(r)$ *after default costs*. The recover rate vector r is *clearing* if $r_i = 1$ when $a_i(r) \geq \ell_i(r)$ (bank i does not default) and $r_i = a'_i(r)/\ell_i(r)$ when $a_i(r) < \ell_i(r)$ (bank i defaults). Is there is a clearing recovery vector?
Complexity: $\exists\mathbb{R}$-complete by Dohn et al. [DHK24] even if $\beta = 1$ and $\alpha < 1$ via a reduction from feasibility (A1).
Comments: Dohn, Hansen, and Klinkby also show that in case that $\alpha = \beta = 1$ it is $\exists\mathbb{R}$-complete to determine whether there is a clearing recovery rate in which a particular bank survives (does not default).
Also see: Equilibrium in Exchange Markets (GT22)

5.4.5 *Machine Learning and Probabilistic Reasoning*

Artificial neural networks come in many variants. For the purpose of this section, we only define fully-connected, feed-forward, ReLU networks. Such a network realizes a parameterized function $N_p : \mathbb{R}^n \to \mathbb{R}^m$ that maps an input $x \in \mathbb{R}^n$ to some output $N_p(x) \in \mathbb{R}^m$ given some parameters p (weights and biases). Given a set of data points (x, y), the goal is to find the right p such that N_p predicts the y-values of the data points to within a given threshold. The function N_p is the composition of two types of functions. The first type is a simple affine transformation $x \mapsto Ax + b$. The second type is an activation function; examples include the identity function and the rectified linear unit activation function RELU. (The name is more complicated than the function itself.) RELU maps $x \in \mathbb{R}^n$ to $y \in \mathbb{R}^n$, where $y_i = \max(x_i, 0)$. An artificial neural network can be described as a repeated composition of affine functions and RELU's. We refer the interested reader to the comprehensive book by Goodfellow, Bengio and Aaron on deep learning [GBC16], for an in-depth introduction of the great variety neural network architectures and training methods.
When we want to reason about probabilities in a formal sense, we can define formal languages $\mathcal{L}$ that can express probabilities of events, where events can again be combined using elementary variables, conditional probabilities and conjunction of events, comparison of probabilities, independence statements, confirmation statements and potentially many more. There is an illustrative example that shows us that irrational numbers might be needed to find a solution. We try to find probabilities such that $\Pr(A \cap B) = 1 - \Pr(A \cap B)$ and $\Pr(A \cap B) = \Pr(B) \cdot \Pr(B)$ implies that $\Pr(B) = \sqrt{1/2}$.

(ML1) Training Neural Networks
Given: A set of data points, a neural network architecture, a cost function, a threshold
Question: Do there exist biases and weights such that the total error is below the threshold?
Complexity: First Abrahamsen et al. [AKM21] showed $\exists\mathbb{R}$-completeness for arbitrary network architectures, two layers, and the identity activation function. Bertschinger et al. [BHJ+22] strengthened the result to fully-connected neural networks with only two input, two output neurons and the ReLU activation function. In case that we allow that some weights are given in the input than the problem was shown to be $\exists\mathbb{R}$-complete already in 1992 by Zhang [Zha92].
Comments: The $\exists\mathbb{R}$-hardness results were extended in part to other activation functions involving non-algebraic functions [HHKV23]. In that case the problems might not be even decidable. To make the problems with non-algebraic activation functions easier, the neural network problem was studied in a regime where incorrect answers are allowed [BDC23].
Also see: Reachability in neural network (ML2).

(ML2) Reachability in Neural Networks
Given: A feed-forward neural network N with weights and biases, seen as a function from $N : \mathbb{R}^n \to \mathbb{R}^m$, and two convex polytopes $A \subseteq \mathbb{R}^n$ and $B \subseteq \mathbb{R}^m$, described by linear inequalities.
Question: Do there exist points $x \in A$ and $y \in B$ such that $N(x) = y$?
Complexity: The complexity of the problem depends on the type of activation functions that are allowed, see Wurm [Wur23, Wur24]. The problem becomes $\exists\mathbb{R}$-hard for many common non-linear activation functions.

Comments: Wurm [Wur23, Wur24] shows a close relation to constraint satisfaction problems (CSPs). Different computational complexities of CSPs translate directly to the corresponding complexity of the neural network reachability problem.
Also see: Training neural network (ML1).

(ML3) Satisfiability of Probabilistic Languages $\mathcal{L}$
Given: A formula φ in the probabilistic language $\mathcal{L}$.
Question: Does there exists an assignment of probabilities that makes φ true?
Complexity: Mossé et al. [MII21] showed that the problem is $\exists\mathbb{R}$-complete for the languages $\mathcal{L}^{\text{cond}}_{\text{prob}}$, $\mathcal{L}^{\text{poly}}_{\text{prob}}$, $\mathcal{L}^{\text{cond}}_{\text{causal}}$, and $\mathcal{L}^{\text{poly}}_{\text{causal}}$. Ibeling et al. [IIMM23] showed $\exists\mathbb{R}$-completeness for the languages $\mathcal{L}^{\text{ind}}$, $\mathcal{L}^{\text{confirm}}$, $\mathcal{L}^{\text{cond}}$, and $\mathcal{L}^{\text{poly}}$ The problem is known to be NP-complete for the languages $\mathcal{L}^{\text{comb}}_{\text{prob}}$, $\mathcal{L}^{\text{lin}}_{\text{prob}}$, $\mathcal{L}^{\text{comp}}_{\text{causal}}$, and or $\mathcal{L}^{\text{lin}}_{\text{causal}}$, also see the paper by Fagin et al. [FHM90]. The reductions are from ETR- INV (L5).
Comments: The results indicate that the difference between NP-completeness and $\exists\mathbb{R}$-completeness is caused by allowing conditional probability statements or the multiplication of probabilities. The satisfiability problem for probabilistic languages that are allowed to contain summation symbols, e.g. abbreviating a sum like $f(1) + f(2) + f(3) + \cdots + f(n)$ as $\sum_{i=1}^{n} f(i)$, is complete for the class succ$\exists\mathbb{R}$, a real version of NEXP [vdZBL23].
Also see: ETR- INV (L5).

(ML4) Gaussian Conditional Independence Inference Problem
Given: A Conditional Independence Inference Rule φ. See [Stu05] for definitions.
Question: Is the rule φ valid for all regular Gaussians?
Complexity: Boege (see [Boe22a, Theorem 5.38] and [Boe21, Theorem 11]) showed that the problem is $\forall\mathbb{R}$-complete.
Comments: The proof is via a reduction from feasibility (A1), using Shor's normal form and an encoding via geometric von Staudt gadgets.
Open Questions: For a conditional independence problem on finite domains that may be succ$\exists\mathbb{R}$-complete, see [Mak24].
Also see: Feasibility (A1).

5.4.6 Markov Chains and Decision Processes

We recommend the book by Sutton and Barto on reinforcement learning [SB18] for the definition of Markov Decision Processes (MDPs) and their variants; the book devotes a whole chapter to illustrating the notion in depth.

(MDP1) Realizability of polynomial parametric Markov Decision Processes
Given: A polynomial, parametric Markov Decision Process with at least two states.
Question: Is the polynomial parametric Markov Decision Process realizable?
Complexity: The problem is first claimed to be $\exists\mathbb{R}$-complete by Junges et al. [JKPW21] via a reduction from local positivity over a bounded domain (A2). They in turn attribute this to earlier work by Lanotte et al. [LMT07].

Comments: Polynomial and parametric in the context of Markov Decision Processes mean that the transition probabilities are polynomial functions $f(x)$ instead of real numbers. The question is whether there exists some x such that f describes a well-defined Markov Decision Process.

(MDP2) Never Worse States in Graph Parametric MDP
Given: A (non-weighted) parametric Markov Decision Process $\mathcal{M}$ with graph $G(\mathcal{M}) = (V, E)$, together with $v \in V$ and $W \subseteq V$.
Question: Is W never worse than v? A set of states W is *never worse* than a state v if, for all valuations of the parameters, there exists $w \in W$ such that the probability of reaching the target state from w is at least as large as that from v [EPR23].
Complexity: Engelen, Pérez, and Rao showed that the problem is $\forall\mathbb{R}$-complete [EPR23] via a reduction from (the complement of) strict local positivity (A2).
Also see: Equally bad states in graph parameteric MDP (MDP3), strict local positivity (A2).

(MDP3) Equally Bad States in Graph Parametric MDP
Given: A (non-weighted) parametric Markov Decision Process $\mathcal{M}$ with graph $G(\mathcal{M}) = (V, E)$, together with $v, w \in V$.
Question: Is v never worse than w and vice versa? A state is *never worse* than another if, for all valuations of the parameters, the probability of reaching the target state from the former state is at least as large as that from the latter [EPR23, EPR25].
Complexity: Engelen, Pérez, and Rao showed that the problem is $\forall\mathbb{R}$-complete [EPR23] via a reduction from (MDP2).
Also see: Never worse states in graph parameteric MDP (MDP2).

(MDP4) Total Variation Distance in MDP
Given: A (labeled) Markov Decision Process and two initial distributions μ and ν.
Question: Is there a (memoryless) strategy such that total variation distance in the induced Markov Chain equals 0?
Complexity: Kiefer and Tang showed that the problem is $\exists\mathbb{R}$-complete [KT20] via a reduction from nonnegative matrix factorization (A21). Remains $\exists\mathbb{R}$-complete for total variation distance less than 1 [KT20].
Comments: The problem is polynomial-time decidable if we ask if the total variation distance can be made larger than 0. For the related notion of probabilistic bisimilarity all studied cases are either NP-complete or in P [KT20].
Open Questions: If we ask for total variation distance equal to one then it is known that the problem is NP-hard and in $\exists\mathbb{R}$ [KT20]. So it remains to settle this case.
Also see: nonnegative matrix factorization (A21).

(MDP5) Reachability for Augmented Interval Markov Chain
Given: An augmented interval Markov Chain $\mathcal{M}$, an initial vertex s, a target vertex t, and a threshold $\tau \in [0, 1]$.
Question: Does there exists a Markov Chain $M \in [\mathcal{M}]$ refining $\mathcal{M}$ such that the probability to go from s to t is at least/most τ?
Complexity: Chonev showed that the problem is $\exists\mathbb{R}$-complete [Cho19].

Comments: The problem is also $\exists\mathbb{R}$-complete, if we consider the approximate version.
Also see: Quantitative reachability for parametric acyclic Markov chain (MDP7).

(MDP6) Model Checking for Interval MC w.r.t. Unambiguous Automata
Given: An interval Markov chain and a regular language E.
Question: Does there exists values for the undetermined transitions of the Markov chain such that E is satisfied?
Complexity: The problem was shown to be $\mathsf{NP}_{\mathbb{R}}$-complete by Benedikt, Lenhardt, and Worrell [BLW13a]. The same proof easily also shows $\exists\mathbb{R}$-completeness.
Comments: They reduce from Local Positivity (A2) on a bounded domain with a single multivariate polynomial of a specific form.

(MDP7) Quantitative Reachability for Parametric Acyclic Markov Chains
Given: A parametric *acyclic* Markov Chain and a target state T.
Question: Does there exists a well-defined and graph preserving valuation of the probability transitions such that the probability to reach T is at least/most $1/2$?
Complexity: The problem was shown to be $\exists\mathbb{R}$-complete by Junges et al. [JKPW21].
Comments: The problem is reminiscent of the Reachability problem for Augmented Interval Markov Chains. Clearly, the $\exists\mathbb{R}$-completeness also carries over to Markov Decision Processes. The authors also use the term *parameter synthesis* in this context to describe the question of finding parameters for some property to hold.
Also see: Reachability for augmented interval Markov chain (MDP5).

(MDP8) Memoryless Distance Minimisation Problem for Markov Decision Processes
Given: A Markov Decision Process with two initial states, a threshold $\theta \in \mathbb{Q}_{\geq 0}$.
Question: Does there exists a memoryless strategy such that the distance between the two initial states is less than θ?
Complexity: The problem was shown to be $\exists\mathbb{R}$-complete by Kiefer and Tang [KT24] via a reduction from strict local Positivity (A2), with bounded domain and a single polynomial of degree 6.
Comments: For $\theta = 1$, the problem is NP-complete [KT20]. For general, non-memoryless strategies, the problem is undecidable [KT24].

5.4.7 Miscellaneous

In this section, we collect problems that we could not fit into any of the other categories.

(Misc1) Euclidean Preference
Given: A finite set A, and a *profile* V, a finite collection of strict orders $\succ_v$ over A, that is, each $\succ_v$ is total, irreflexive, antisymmetric and transitive. The orders are also called *preference relations* or *votes*.
Question: Is there a function $x : V \cup A \to \mathbb{R}^d$ such that $a \succ_v b$ implies that

$\|x(v) - x(a)\|_2 < \|x(v) - x(b)\|_2$ for all $v \in V$ and $a, b \in A$? In that case the profile V is known as *d-Euclidean*.
Complexity: $\exists\mathbb{R}$-complete for any fixed $d \geq 2$ by Peters [Pet17] via a reduction from oriented hyperplane realizability (CG13). Remains $\exists\mathbb{R}$-complete if all orders are *dichotomous*, that is, there is no triple a, b, c with $a \succ_v b \succ_v c$ for any $v \in V$ [Pet16, Pet17].
Comments: For the Manhattan and maximum metrics the problem lies in NP, so it is unlikely to be $\exists\mathbb{R}$-hard in those cases [Pet17]; NP-hardness seems to be open for those metrics.
Also see: Oriented hyperplane realizability (CG13).

(Misc2) (Ultimate) Positivity of Linear Recurrence
Given: Coefficients $a_1, \ldots, a_k \in \mathbb{Q}^k$ and initial values $u_0, \ldots, u_{k-1} \in \mathbb{Q}^k$. With these numbers we define the sequence $u_n = \sum_{i=1}^{k} a_i u_{n-i}$ for $n \geq k$.
Question: Is the sequence $(u_n)_{n\geq 0}$ *(ultimately) positive*; that is, is $u_n > 0$ for (almost) all n?
Complexity: Both positivity and ultimate positivity are $\forall\mathbb{R}$-hard by Ouaknine and Worrell [OW17] via a reduction from strict local positivity (of a polynomial) (A2). For simple recurrences (the characteristic polynomial has no multiple roots), the problems can be decided in $\mathsf{coNP}^{\exists\mathbb{R}}$ [OW14, OW17].
Comments: These problems are related to Skolem's problem, which asks whether $u_n = 0$ for some n. Decidability of Skolem's problem is a famously open question.
Open Questions: Does positivity (or ultimate positivity) lie in $\forall\mathbb{R}$? ▼ What is the complexity of the robust version of the problem (in which each a_i is given with a small neighborhood), as studied in [Vah23]?
Also see: Strict local positivity (A2), zeroness of convolution recursive sequence (A-Open13).

(Misc3) Linear Update Loop Termination
Given: Boolean formula $\varphi(x)$ over polynomial inequalities in $x \in \mathbb{R}^n$, matrix $A \in \mathbb{R}^{n\times n}$, vector $b \in \mathbb{R}^n$.
Question: Does the program **while** $\varphi(x)$ **do** $x = Ax + b$ terminate for every initial value $x \in \mathbb{R}^n$?
Complexity: $\forall\mathbb{R}$-complete by Frohn et al. [FHG23, Theorem 7.5] via a reduction from feasibility (A1). Remains $\forall\mathbb{R}$-complete if $b = 0$, but becomes coNP-complete if φ consists of linear polynomial conditions only [FHG23, Theorem 7.3].
Comments: The non-trivial part in this result is $\forall\mathbb{R}$-membership, the $\forall\mathbb{R}$-hardness is immediate with program **while** $\varphi(x)$ **do** $x = x$. For earlier work on related problems, see [COW15]. For a related problem in discrete probabilistic programming, see [ZMO24, Sect. 4.5.1].
Open Questions: What is the complexity of the problem if instead we ask whether the program terminates for a specific initial value $x \in \mathbb{Q}^n$? For specific types of conditions $\varphi(x)$ this was called the Extended Orbit Problem in [COW15] and shown to lie in $\exists\mathbb{R}$. Another problem that has similarities is the compact escape problem (A43).

Also see: (Ultimate) positivity of linear recurrence (Misc2), compact escape problem (A43), feasibility (A1).

5.4.8 Higher Levels of the Real Hierarchy

5.4.8.1 Logic

(L17) $Q_1 \cdots Q_k$-Fragment of the Theory of the Reals
Given: A sentence over the signature $(0, 1, +, \cdot)$ with alternating quantifier blocks $Q_1, \ldots, Q_k$, with $Q_i \in \{\exists, \forall\}$ comparison operators $\{=, <\leq, \geq, >\}$ and logic connectives $\vee$, $\wedge$ and $\neg$.
Question: Is the sentence true if interpreted over the reals?
Complexity: Complete for $Q_1 \cdots Q_k\mathbb{R}$ by Bürgisser and Cucker [BC09]. The problem remains complete for $Q_1 \cdots Q_k\mathbb{R}$ even if the matrix is of the form $f(x) > 0$ or $f(x) \geq 0$, for a polynomial f, and, under certain conditions, for $f(x) = 0$ as shown by Schaefer and Štefankovič [SŠ23, Tables 1 and 2]. Also, the quantifiers can be restricted to $(-1, 1)$ and, as long as $k \leq 2$, to $[-1, 1]$, see [SŠ25]. There are also some robustness results, mostly on the first and second level, that allow the use of exotic quantifiers, $\exists^*$, $\forall^*$ and H, see [BC09, Jun23b].
Comments: The proofs dealing with bounded domains in [SŠ23, Sect. 2.3] contained gaps; an updated version with corrected proofs was published as [SŠ25], not all results could be recovered though, see Remark 1.1 in the paper.
Open Questions: Are formulas of the type $(\exists x_1 \in [-1, 1]^{n_1})(\forall x_2 \in [-1, 1]^{n_2})(\exists x_3 \in [-1, 1]^{n_3})\, f(x_1, x_2, x_3) > 0$ complete for $\exists\forall\exists\mathbb{R}$? This is the first open case in [SŠ25, Sect. 2.3.3].
Also see: ETR (L1).

5.4.8.2 Algebra

(A43) Compact Escape Problem
Given: A compact semialgebraic set $K \subseteq \mathbb{R}^n$, a matrix $A \in \mathbb{Q}^{n\times n}$.
Question: For all $x \in K$ is there a k such that $A^k x \notin K$?
Complexity: $\forall\exists\mathbb{R}$-complete by D'Costa, Lefaucheux, Neumann, Ouaknine, and Worrell [DLN+21] via a reduction from the $\forall\exists$-fragment of the theory of the reals (L17).
Comments: D'Costa, Lefaucheux, Neumann, Ouaknine, and Worrell [DLN+21] show that the problem is hard for $\forall\exists_<\mathbb{R}$, hardness then follows, since $\forall\exists\mathbb{R} = \forall\exists_<\mathbb{R}$ by Schaefer and Štefankovič [SŠ23].
Also see: $Q_1 \cdots Q_k$-fragment of the theory of the reals (L17), linear update loop termination (Misc3).

(A44) Hausdorff Distance
Given: Two semialgebraic sets $A, B \subseteq \mathbb{R}^n, d \in \mathbb{Q}$. The *Hausdorff distance* is defined as $d_H(A, B) := \max(\sup_{a\in A} \inf_{b\in B} \|a - b\|, \sup_{b\in B} \inf_{a\in A} \|a - b\|)$.
Question: Do A and B have Hausdorff distance at most d, that is, is $d_H(A, B) \leq d$?
Complexity: $\forall\exists\mathbb{R}$-complete by Jungeblut et al. [JKM23] and Schaefer and Štefankovič [SŠ23] via reductions from the $\forall\exists$-fragment of the theory of the reals (L17). The problem remains hard for $d = 0$ [JKM23, SŠ23] or d double-exponentially small [JKM23].
Comments: Jungeblut et al. [JKM23] show that the problem is complete for $\forall\exists_<\mathbb{R}$, Schaefer and Štefankovič [SŠ23] show the full result, based on Bürgisser and Cucker [BC09] who called the $d = 0$ case *Euclidean Relative Denseness*.. The directed Hausdorff distance, $\sup_{a\in A} \inf_{b\in B} \|a - b\|$ is also $\forall\exists\mathbb{R}$-complete [JKM23, SŠ23].
Open Questions: Jungeblut asks whether the problem becomes $\exists\forall\exists\mathbb{R}$-complete if we ask whether there is a translation t in $\mathbb{R}^n$ such that $d_H(t(A), B) \leq d$; this is a restricted case of the Gromov-Hausdorff distance (A-Open6).
Also see: distance between semialgebraic sets (A41), Gromov-Hausdorff distance (A-Open6), Fréchet distance between surfaces (CG-Open31).

(A45) Surjectivity
Given: A polynomial f in $\mathbb{Z}[x_1, \ldots, x_n]$, semialgebraic sets A and B.
Question: Is f surjective as a function from A to B?
Complexity: $\forall\exists\mathbb{R}$-complete by Schaefer and Štefankovič [SŠ23] extending work by Bürgisser and Cucker [BC09, Proposition 4.1]. Hardness can be obtained via a reduction from a constrained version of the $Q_1 \cdots Q_k$-fragment of the theory of the reals (L17). The problem remains $\forall\exists\mathbb{R}$-complete if A and B are Cartesian products of $\mathbb{R}$ and $[-1, 1]$ and the total degree of f is bounded.
Comments: Bürgisser and Cucker [BC09] showed that surjectivity of rational functions given by arithmetic circuits is $\forall\exists\mathbb{R}$-complete, Schaefer and Štefankovič [SŠ23] proved the result for polynomials.
Open Questions: Does the problem remain $\forall\exists\mathbb{R}$-complete if both A and B are of the form $\mathbb{R}^n$?
Also see: Image density (arithmetic circuit) (A46).

(A46) Image Density (arithmetic circuit)
Given: An arithmetic circuit (with sign function), defining a function $f : \mathbb{R}^n \to \mathbb{R}$.
Question: Is the range of f dense in $\mathbb{R}^n$?
Complexity: $\forall\exists\mathbb{R}$-complete, combining results from Bürgisser-Cucker [BC09, Proposition 5.7, Corollary 9.4] and Junginger [Jun23b, Theorem 4.27].
Comments: Bürgisser and Cucker show that this problem is hard for a complexity class involving exotic quantifiers; Junginger shows that that class equals $\forall\exists\mathbb{R}$. In the Zariski topology, Bürgisser and Cucker [BC09, Proposition 5.7, Corollary 9.4] show that the problem is $\exists\mathbb{R}$-complete.
Also see: Surjectivity (second level) (A45), domain density (arithmetic circuit) (A13), density of semialgebraic set (A35).

5.4.8.3 Computational Geometry

(CG76) Area Universality For Triples with Partial Area Assignment
Given: Vertices V, a set of vertex triples $F \subseteq \binom{V}{3}$, and a *partial area assignment*, that is, a function $\alpha' : F' \to \mathbb{R}_{>0}$, where $F' \subseteq F$.
Question: An *area assignment* assigns a positive real number to every triple in F. Can every area assignment α that extends α', that is, agrees with it on F', be realized by placing the vertices of V in the plane so that every vertex triple in F forms a triangle with the area prescribed by α?
Complexity: $\forall\exists\mathbb{R}$-complete, by result of Dobbins et al. [DKMR23] via a reduction from a constrained form of the $\forall\exists$-fragment of the theory of the reals (L17). The problem remains hard if the size of F is linear in the size of V
Comments: This problem is a step towards settling the complexity of the area universality problem (CG-Open36). Dobbins et al. [DKMR23] suggest studying triangulations in a next step.
Also see: Area Universality (CG-Open36), prescribed area extension (CG61).

5.4.9 *Candidates*

5.4.9.1 Logic

(L-Open1) Universality of Unambiguous Context-Free Grammars (UUCFG)
Given: A context-free grammar G over alphabet Σ that is *unambiguous*, that is, every word generated by the grammar has a unique derivation tree.
Question: Is the grammar *universal*, that is, is $L(G) = \Sigma^*$?
Complexity: This problem was shown to lie in $\forall\mathbb{R}$ by Clemente [Cle20] using another interesting intermediate problem, zeroness of convolution recursive sequences (A-Open13), which also lies in $\forall\mathbb{R}$. Clemente also mentions a more direct proof of membership in Stanisław Purgał's master's thesis. Finally, he shows that a slightly generalized problem, the coin-flip measure comparison problem, is as hard as SSQR problem.
Comments: UUCFG is a *promise problem*, since we are assuming that G is unambiguous, an undecidable property of context-free grammars we cannot verify in $\forall\mathbb{R}$. Strictly speaking then UUCFG lies in the promise version of $\forall\mathbb{R}$, which means the answer given by the $\forall\mathbb{R}$-characterization need only be correct in case G is unambiguous and is allowed to be arbitrary otherwise.
Also see: Positivity of linear recurrence (Misc2), zeroness of convolution recursive sequence (A-Open13).

(L-Open2) Existential Theory of Invertible Matrices
Given: A conjunction of conditions of the form $AX_i = X_jB$, where the X_k are (matrix) variables and $A, B \in \mathbb{Q}^{n\times n}$.
Question: Are there invertible matrices $X_k \in \mathbb{R}^{n\times n}$ satisfying all conditions?
Complexity: In $\exists\mathbb{R}$.

Comments: The problem was introduced by Dubut [Dub20] to upper-bound the complexity of the bisimilarity of diagrams (itself a possible candidate for $\exists\mathbb{R}$-completeness?). Without the invertibility contraints, the problem turns into a linear program becoming solvable in polynomial-time. It may be easier to show the problem $\exists\mathbb{R}$-complete if one is given access to the inverses of the X_i, but that's not the case in the original problem.
Also see: Matrix singularity (A15), constrained matrix invertibility problem (A-Open7), feasibility (A1).

5.4.9.2 Algebra

(A-Open1) Feasibility of Univariate Polynomials
Given: A univariate polynomial $f \in \mathbb{Z}[x]$ defined by a straight-line program or arithmetic circuit.
Question: Does f have a root, that is, is there an $x \in \mathbb{R}^n$ such that $f(x) = 0$?
Complexity: Known to be NP-hard (for a history of proofs, see Perrucci and Sabia [PS07]), and in $\exists\mathbb{R}$. Is it $\exists\mathbb{R}$-complete?
Comments: This candidate is a bit surprising; the known NP-hardness proofs are quite intricate. Does the problem lie in NP? If one asks whether f has a *rational* root in a given interval, then the problem is decidable using Eisenstein's criterion, see for example [DGVVG99, Lemma 5.22]; that makes it somewhat less likely that the rational version of the problem is $\exists\mathbb{Q}$-hard. See [BJ24] for complexity results on counting the number of real roots.
Open Questions: What is the computational complexity of determining whether f has a rational root in some given interval?
Also see: Feasibility (A1).

(A-Open2) Polynomial Equivalence under Shifts
Given: Two polynomials $f, g : \mathbb{R}^n \to \mathbb{R}$ in $\mathbb{Z}[x_1, x_2, \dots, x_n]$.
Question: Is there a vector b such that $f(x) = g(x + b)$?
Complexity: In $\exists\mathbb{R}$.
Comments: Problem is discussed in Chillara et al. [CGS23], but the authors switch to a minimality version: given a polynomial f, is there a vector b such that $f(x + b)$ has fewer monomials? For this version the authors show that over an integral domain which is not a field, the minimality version is as hard as the existential theory of the domain. Over $\mathbb{Z}$ this implies that the problem is undecidable, but the reduction does not apply to $\mathbb{R}$ (or $\mathbb{Q}$ or $\mathbb{C}$).
Also see: Polynomial projection (linear) (A14).

(A-Open3) Semialgebraic Set with Isolated Point
Given: A semialgebraic set $S \subseteq \mathbb{R}^n$.
Question: Does S contain an isolated point; that is, is there an $x \in S$ such that S contains no other points close to x?
Complexity: In $\exists\forall\mathbb{R}$ and known to be $\forall\mathbb{R}$-hard by Bürgisser and Cucker [BC09, Corollaries 6.8 and 9.4], but not known to be $\exists\forall\mathbb{R}$-complete.

Comments: Testing whether a specific point belonging to S is isolated, is $\forall\mathbb{R}$-complete, see the Isolated zero problem (A7).
Also see: Isolated point of semialgebraic set (A39), isolated zero (A7).

(A-Open4) Star-Shapedness of Semialgebraic Set
Given: A semialgebraic set $S \subseteq \mathbb{R}^n$.
Question: Is S *star-shaped*? That is, is there a point in S which can *see* all points in S in the sense that the line segment connecting the points belongs to S?
Complexity: In $\exists\forall\mathbb{R}$ by definition.
Comments: Suggested in Schaefer and Štefankovič [SŠ23].
Also see: Convexity of semialgebraic set (A33).

(A-Open5) Vapnik-Červonenkis Dimension of Family of Semialgebraic Sets
Given: A family of semialgebraic sets $\mathcal{F} = \{\{x \in \mathbb{R}^n : \varphi(x, y)\} : y \in \mathbb{R}^m\}$, integer k.
Question: Does $\mathcal{F}$ have Vapnik-Červonenkis Dimension at least k, that is, are there k points P in $\mathbb{R}^n$ such that for every subset $P' \subseteq P$ of those points, there is an $S \in \mathcal{F}$ such that $P' = S \cap P$?
Complexity: In $\exists\forall\exists\mathbb{R}$; probably not complete for that class, since the universal quantifier is discrete, not real.
Comments: Suggested in Schaefer and Štefankovič [SŠ23]. Over discrete sets this problem is complete for the third level of the polynomial-time hierarchy [Sch99].

(A-Open6) Gromov-Hausdorff Distance in Euclidean Space
Given: Two semialgebraic sets $A, B \subseteq \mathbb{R}^n$, $d \in \mathbb{Q}$. The *Hausdorff distance* of two sets A and B is

$$d_H(A, B) := \max(\sup_{a \in A} \inf_{b \in B} \|a - b\|, \sup_{b \in B} \inf_{a \in A} \|a - b\|).$$

The *Gromov-Hausdorff distance*, $d_{GH}(A, B)$, *of* A *and* B *in Euclidean space* is the infimum of $d_H(f(A), g(B))$ over all isometries (distance-preserving functions) f, g from $\mathbb{R}^n$ to $\mathbb{R}^n$.
Question: Do A and B have Gromov-Hausdorff distance at most d, that is, is $d_{GH}(A, B) \leq d$?
Complexity: In $\exists\forall\exists\mathbb{R}$ by definition, and $\forall\exists\mathbb{R}$-hardness follows from the $\forall\exists\mathbb{R}$-hardness of the Hausdorff distance (A44).
Comments: The Gromov-Hausdorff distance has a more general definition than the one given here; the fact that for Euclidean spaces it is equivalent to the definition of d_{GH} above, was shown by Mémoli [Mém08]. To reduce the Hausdorff distance to the Gromov-Hausdorff distance, we can add a set of isolated points to the two sets that forces the isometries to be the identity.
Also see: Hausdorff distance (A44).

(A-Open7) Constrained Matrix Invertibility Problem
Given: Two matrices $A, B \in \{0, *\}^{n\times n}$.
Question: Can the $*$-entries in A and B be replaced by real entries so that $AB = I$, where I is the n-dimensional identity matrix?
Complexity: Over finite fields, the problem is known to be NP-complete, by Bjerkevik et al. [BBK20].
Comments: This problem arises in the study of interleaving and stability of persistence modules [BBK20], there may be further relevant open problems in that area.
Also see: Matrix singularity (A15), existential theory of invertible matrices (L-Open2).

(A-Open8) Matrix ℓ^p-Low Rank Approximation
Given: A matrix $A \in \mathbb{Q}^{m\times n}$, an integer $k \geq 0$, $\varepsilon \in \mathbb{Q}_{>0}$.
Question: Is there a matrix $B \in \mathbb{R}^{m\times n}$ such that $\|A - B\|_p < \varepsilon$ and $\operatorname{rank}(B) \leq k$? Here, $\|\|_p$ denotes the ℓ^p*-norm*, which is $(\sum_{i\in[m],j\in[n]} |A_{i,j} - B_{i,j}|^p)^{1/p}$ for $p \geq 1$, and $|\{(i, j) \in [m] \times [n] : A_{ij} \neq B_{ij}\}|$ for $p = 0$.
Complexity: In $\exists\mathbb{R}$, and known to be NP-hard for $p \in \{0, 1\}$ even for $k = 1$ [GV18].
Also see: Minimum matrix rank (A16).

(A-Open9) Hadamard Decomposition
Given: A matrix $M \in \mathbb{Q}^{m\times n}$, natural numbers $k_1, k_2 \in \mathbb{N}$.
Question: Are there matrices M_1, M_2 of rank k_1 and k_2 such that $M = M_1 \odot M_2$, where $\odot$ is the component-wise product, known as the *Hadamard product*, of two matrices?
Complexity: In $\exists\mathbb{R}$ as observed by Ciaperoni et al. [CGM24] who asked the question.
Also see: nonnegative matrix factorization (A21).

(A-Open10) Partial Sign Pattern of Principally Regular Matrix
Given: An integer n, and a *partial sign pattern*, that is a family of sets $K_i \subseteq [n]$ and values $s_i \in \{-1, 1\}$. For $K \subseteq [n]$, we denote by $A[K]$ the submatrix of A in which rows and columns are restricted to K. A real matrix is *principally regular* if it is symmetric, and $\operatorname{sgn}\det A[K] \neq 0$ for all non-empty $K \subseteq [n]$.
Question: Is there a principally regular matrix $A \in \mathbb{R}^{n\times n}$ such that $\operatorname{sgn}\det A[K_i] = s_i$?
Complexity: In $\exists\mathbb{R}$ by Boege et al. [BSZ24].
Comments: Boege et al. [BSZ24] also consider variants in which only leading minors are restricted; that is, each K is of the form $\{1, \ldots, \ell\}$, and not all minors have to be non-zero.

(A-Open11) Tensor Eigenvector Approximation
Given: A 3-dimensonal tensor $T = (t_{ijk}) \in \mathbb{Q}^{d\times d\times d}$, $\lambda \in \mathbb{Q}$, $\varepsilon \in \mathbb{Q}_{>0}$.
Question: Is there an $x \in \mathbb{R}^d$ within distance ε of an eigenvector x' of T for eigenvalue λ; that is, $\sum_{i,j} t_{i,j,k} x'_i x'_j = \lambda x'_k$, and $\|x - x'\| < \varepsilon$?
Complexity: Known to be NP-hard by Hillar and Lim [HL13] and in $\exists\mathbb{R}$ by definition. Is it $\exists\mathbb{R}$-complete?
Also see: Tensor with eigenvalue 0 (A28), tensor rank (A27).

(A-Open12) Tensor Singular Value
Given: A 3-dimensional tensor $T = (t_{ijk}) \in \mathbb{Q}^{d \times d \times d}$, $\sigma \in \mathbb{Q}_{>0}$.
Question: Is σ a singular value of T? (See Hillar and Lim [HL13, Sect. 6] for relevant definitions.)
Complexity: Known to be NP-hard even for $\sigma = 0$, by Hillar and Lim [HL13, Theorem 6.2] and in $\exists\mathbb{R}$ by definition. Is it $\exists\mathbb{R}$-complete?
Comments: It is also open whether testing if σ is an upper bound on the spectral norm of T is $\exists\mathbb{R}$-complete; it is known to be NP-hard by Hillar and Lim [HL13, Theorem 1.10].
Also see: Tensor with eigenvalue 0 (A28).

(A-Open13) Zeroness of Convolution Recursive Sequence
Given: A convolution recursive sequence f. A sequence $f : \mathbb{N} \to \mathbb{R}$ is *convolution recursive* if there is a family $g = (g_1, \ldots, g_k) : \mathbb{R}^k \to \mathbb{R}^k$ of polynomials $g_i \in Q[x_1, \ldots, x_k]$ and k sequences of real numbers $f_1 = f, f_2, \ldots, f_k$ such that $\sigma f_i = g_i(f_1, \ldots, f_k)$, where the polynomials g_i are interpreted as follows: addition is addition of sequences and multiplication is convolution of sequences; the operator σ shifts the sequence by one position.
Question: Is $f(n) = 0$ for all n?
Complexity: This problem lies in $\forall\mathbb{R}$ by Clemente [Cle20] who uses it to show that universality of an unambiguous context-free grammar (L-Open1) lies in $\forall\mathbb{R}$.
Also see: Universality of unambiguous Context-Free Grammars (L-Open1), positivity of linear recurrence (Misc2).

5.4.9.3 Computational Geometry

(CG-Open1) Geometric Thickness
Given: A graph G, a natural number $k \in \mathbb{N}$.
Question: Does G have a straight-line drawing in which the edges can be colored with k colors in such a way that no crossing in the drawing involves two edges of the same color?
Complexity: Known to be NP-hard even for $k = 2$ [DGM16], but no better upper bound than $\exists\mathbb{R}$ is known, including the case $k = 2$. For multigraphs the problem is $\exists\mathbb{R}$-complete when $k = 30$ [FKM+24].
Comments: Geometric thickness was introduced by Kainen [Kai73]. It has similarities to the simultaneous geometric embedding problem (CG56), which is $\exists\mathbb{R}$-complete. For a brief discussion, see [Sch21b, Remark 7].
Also see: Geometric thickness of multigraphs (CG57), simultaneous geometric embedding (CG56).

(CG-Open2) Colin de Verdière Invariant
Given: A graph $G = (V, E)$ and a number $k \in \mathbb{N}$.
Question: Is the Colin de Verdière invariant $\mu(G)$ at least k? For a definition see the comment section below.
Complexity: Lies in $\exists\mathbb{R}$ as shown by Kaluza and Tancer [KT19].

Comments: The graph parameter was introduced by Colin de Verdière [CdV90, in French] and is connected to well-studied graph classes. For example $\mu(G) \leq 2$ if and only if G is an outerplanar graph and $\mu(G) \leq 3$ if and only if G is a planar graph [vdHLS99]. The parameter $\mu(G)$ is the largest corank of a symmetric matrix $M \in \mathbb{R}^{n \times n}$ such that

- $ij \in E$ implies $M_{ij} < 0$ and $M_{ij} = 0$ otherwise;
- M has exactly one negative eigenvalue, of multiplicity 1;
- There is a matrix X such that $MX = 0$ and X has only zero entries on the diagonal and at the non-zero entries of M.

For fixed k the invariant μ can be computed in polynomial-time, since it is minor-closed.
Also see: minimum graph rank (A20), minimium matrix rank (A16).

(CG-Open3) Unit Disk Contact Graph
Given: A graph G.
Question: Is G the *contact graph* of a collection of disjoint (open) unit disks in the plane, where two disks form a contact if they share exactly one point on their boundaries?
Complexity: Known to be NP-hard [BK96, KNP15] even if G is outerplanar.
Comments: Unit disk contact graphs are planar, and by Koebe's theorem all planar graphs are contact graphs of disks. There also is a weak version, which allows contacts between disks whose corresponding vertices are not adjacent; weak unit disk contact graphs thus are the subgraphs of unit disk contact graphs.
Also see: Unit disk graph (CG21), ply number (CG-Open4), kissing number (CG-Open21).

(CG-Open4) Ply Number
Given: A graph G, an integer $k \geq 0$.
Question: Is there a straight-line drawing of G with the following property: If at each vertex we draw an open disk with radius half the length of the longest edge incident to the vertex, then no more than k disks intersect at any point in the plane? The smallest such k is the *ply number* of G.
Complexity: The problem is NP-hard, since having ply number at most 1 is equivalent to being a unit disk contact graph (CG-Open3), see [DGDH+15], and that problem is NP-hard [KNP15].
Comments: The ply number of a graph was introduced as a new metric for measuring the quality of graph drawings in [DGDH+15].
Also see: Unit distance graphs (CG65), unit disk contact graphs (CG-Open3).

(CG-Open5) Free Space Diagram Realizability
Given: A *free space diagram*, that is, a rectangle $[0, \ell_1] \times [0, \ell_2]$ in $\mathbb{R}^2$ that is subdivided into a grid of rectangular cells; each cell may be associated with an ellipse (which is cropped to the area of the cell), and a parameter $\varepsilon \in \mathbb{Q}_{\geq 0}$.
Question: Are there two polygonal curves P_1, P_2 of length ℓ_1 and ℓ_2 such that

the union of the cropped ellipses in the free space diagram corresponds to $\{(x, y) : \|P_1(x) - P_2(y)\|_2 \leq \varepsilon\}$? Here, $P(z)$ denotes the parameterization of polygon P by length.
Complexity: In $\exists\mathbb{R}$.
Comments: The paper [ABM+23] shows that planar linkage realizability (CG66) reduces to free space diagram realizability, but an earlier claim that planar linkage realizability is $\exists\mathbb{R}$-complete may have been erroneous, leaving the complexity of the free space diagram realizability problem in limbo as well.
Also see: Free space matrix realizability (CG8), linkage realizability (CG66), Fréchet distance between surfaces (CG-Open31).

(CG-Open6) Maximum Rectilinear Crossing Number
Given: A graph G, an integer $k \geq 0$.
Question: Does G have a straight-line drawing with at least k crossings? The largest such k is the *maximum rectilinear crossing number* of G.
Complexity: In $\exists\mathbb{R}$. Bald et al. [BJL16] showed that the problem is NP-hard.
Open Questions: What is the complexity of the maximum rectilinear local crossing number?
Also see: Rectilinear crossing number (CG50), rectilinear local crossing number (CG52).

(CG-Open7) Partial Drawing Extensibility
Given: A graph G, prescribed locations in $\mathbb{Q}^2$ for a subset of the vertices of G.
Question: Is there a planar straight-line drawing of G respecting the prescribed vertex locations?
Complexity: Known to be NP-hard by Patrignani [Pat06], which implies that the problem is NP-complete on a grid; in $\exists\mathbb{R}$ by definition.
Comments: Closely related to the graph in a polygonal domain problem (CG48). The graph on segments with obstacles problem (CG49) can be considered a restricted variant.
Also see: Graph in a polygonal domain (CG48), graph on segments with obstacles (CG49), anchored graph drawing (CG-Open9).

(CG-Open8) Geometric Partial Planarity
Given: A graph $G = (V, E)$, a subset of edges $F \subseteq E$.
Question: Is there a straight-line drawing of G in which all edges in F are crossing-free?
Comments: If we replace "are crossing-free" with "have at most one crossing", then the problem is $\exists\mathbb{R}$-complete, see geometric partial 1-planarity (CG53).
Also see: Geometric partial 1-planarity (CG-Open8), rectilinear local crossing number (CG52).

(CG-Open9) Anchored Graph Drawing
Given: A graph G, and a point $p(v) \in \mathbb{Q}^2$ and distance $d(v) \in \mathbb{Q}_{\geq 0}$ for every vertex $v \in V(G)$.
Question: Is there a planar straight-line drawing of G such that the location of each vertex v is within distance at most $d(v)$ of $p(v)$?
Complexity: Known to be NP-hard by Angelini et al. [ADLDB+14], in $\exists\mathbb{R}$ by definition.
Comments: The problem remains NP-hard for maximum and infinity norm as well.
Also see: Graph in a polygonal domain (CG48), graph on segments with obstacles (CG49).

(CG-Open10) Convex Neural Codes
Given: A *combinatorial neural code* C: a set of subsets of $[n]$.
Question: Does there exist a collection of convex sets $X_1, X_2, \ldots, X_n$ in R^d, for some fixed constant $d \geq 2$, such that $C = \{I(x) : x \in \mathbb{R}^d\}$, where $I(x) = \{i \in [n] : x \in X_i\}$ is the *intersection pattern* of x?
Complexity: $\exists\mathbb{R}$-hard even for $d = 2$, from [KLR23], and decidable if $d = 2$, from [BJ23].
Open Questions: Does the problem lie in $\exists\mathbb{R}$ for $d = 2$? Is the problem decidable for $d = 3$?
Also see: Convex set intersection graph (CG32), nerves of convex sets (CG33), finite convex geometry realizability (CG7).

(CG-Open11) Convex Clustered Planarity
Given: A graph $G = (V, E)$, a hierarchy $C \subseteq 2^V$. In a *hierarchy* if any two sets, *clusters*, intersect, one most be a subset of the other.
Question: Is there a straight-line drawing of G such that for every $C \in C$ the convex hull of the vertices in C contains no vertices not in C, and any edge that crosses the boundary of the convex hull of C may do so at most once? This is a *clustered convex* straight-line embedding of G.
Comments: This is a variant of the clustered planarity problem, which is now known to be solvable in P [FT22]. The traditional variant does not allow vertices on the boundary of a cluster. This may not be the most likely candidate for $\exists\mathbb{R}$-hardness, is there a stretchability result like Fáry's theorem instead?
Also see: Simultaneous geometric embedding (CG56), convex set intersection graph (CG32).

(CG-Open12) Arc Number
Given: A graph G, an integer $k \geq 0$.
Question: Is there a crossing-free circular-arc drawing of G that is the union of k circular arcs? The smallest such k is known as the *arc number*.
Comments: The arc number was introduced by Schulz [Sch15] as a variation of the segment number (CG44).
Also see: Segment number (CG44), Lombardi drawability (CG59), spherical cover number (CG-Open13).

(CG-Open13) Spherical Cover Number
Given: A graph G, an integer $k \geq 0$.
Question: Are there k spheres in $\mathbb{R}^d$ whose union contains a crossing-free circular-arc drawing of G? We say the spheres *cover G*.
Complexity: Lies in $\exists\mathbb{R}$.
Comments: Introduced by Kryven et al. [KRW19], the case $d = 2$ is also known as the *circle cover number*.

(CG-Open14) Minimum Line Number
Given: A graph G, an integer $k \geq 0$.
Question: Is there a straight-line drawing of G with minimum segment number, and line-cover number at most k? (See entries for segment number (CG44) and line-cover number (CG43) for definitions.)
Complexity: In $\exists\mathbb{R}$.
Comments: This problem is very similar to both segment number and line cover number, but differs from both [KRW19]. It was introduced in [DMNW13].
Also see: Segment number (CG44), line cover number (CG43), planar stick number of drawing (CG46).

(CG-Open15) Upward Outerplanar Slope Number
Given: A directed graph G, an integer $k \geq 0$. A (straight-line) drawing of G is *upward* if u lies below v for every directed edge $uv \in E(G)$.
Question: Is there an upward, outerplanar straight-line drawing of G in which the edges are drawn using at most k different slopes?
Complexity: NP-hard by Klawitter and Zink [KZ23] for $k = 3$.
Comments: A drawing realizing the planar slope number of the graph may not be possible on a grid, and it is $\exists\mathbb{Q}$-complete to decide whether a grid realization exists, and there is a $2^{2^{\Omega(n^{1/3})}}$ lower bound on the grid size [Hof17].
Open Questions: How hard is it to determine the (non-planar) slope number of a graph [Hof17]?
Also see: Planar slope number (CG41), upward planar slope number (CG42).

(CG-Open16) Isosceles Graph Drawing
Given: A maximal planar graph G. Call a straight-line drawing of G *isosceles* if all interior faces are isosceles triangles, that is, have at least two sides of the same length.
Question: Does G have a crossing-free, isosceles drawing?
Complexity: In $\exists\mathbb{R}$.
Comments: Malkevich asked whether all maximal planar graphs have a crossing-free, isosceles drawing [DMO03]. Frati [Fra10] showed that this is not the case, and asked about the computational complexity of this problem, also see [DMO03].
Open Questions: Do all maximal planar graphs have an isosceles drawing (not necessarily crossing-free)? See [DMO03, Fra10].
Also see: Angular resolution (CG60).

(CG-Open17) Incidence Graph of Geometric Configuration
Given: A bipartite cubic graph $G = (L \cup P, E)$.
Question: Is G the incidence (Levi) graph of a geometric 3-configuration, that is, are there points for P, and lines for L such that E has an edge $p\ell$ if and only if p is incident to ℓ?
Complexity: In $\exists\mathbb{R}$.
Comments: Asked by Pisanski on mathoverflow [Pis]. This is a special case of the linear hypergraph realizability problem (CG35).
Also see: Pseudo-segment stretchability (CG10), linear hypergraph realizability (CG35), segment intersection graph (CG26).

(CG-Open18) Orthogonality Dimension
Given: A graph G on n vertices, an integer $k \geq 0$.
Question: Can we assign unit vectors in $\mathbb{R}^k$ to each vertex in G such that vectors assigned to adjacent vertices are orthogonal?
Complexity: For sufficiently large k, a gap-version of this problem is NP-hard by Golovnev and Haviv [GH22]; the problem lies in $\exists\mathbb{R}$ by Arends et al. [AOW11].
Comments: The orthogonality dimension of a graph was introduced by Lovasz [Lov79] as its *orthonormal representation* in the study of the θ-function. Arends, Ouaknine, and Wampler study a closely related problem in $\mathbb{R}^3$, which they call the "embeddability problem". They arrive at it in their search for small Kochen-Specker vector systems, which play a role in quantum computing. They consider additional restrictions on G that are all of an NP nature, so the orthogonality dimension seems to capture the computational core of the embeddability problem. The *local orthogonality dimension* [AH23] may also be of interest.
Also see: Dot-product graph (CG25), unit ball graph (CG24).

(CG-Open19) Graph Flattenability
Given: A graph G, integer $d \geq 0$.
Question: For every straight-line embedding of G in $\mathbb{R}^{d'}$, with arbitrary $d' \geq d$, is there a straight-line embedding in $\mathbb{R}^d$ with the same distances between all pairs of vertices?
Complexity: The problem lies in $\forall\exists\mathbb{R}$ [BC07] (the authors observe that it is equivalent to testing the equality of two semialgebraic sets). For fixed d, graph flattenability is minor-closed and can therefore be solved in polynomial time; for hardness results it will therefore be essential that d is part of the input. Lists of excluded minors are known for $d \leq 3$ [BC07].
Comments: The problem was introduced under the name d-realizability by Belk and Connelly [BC07], but it had been studied earlier.
Open Questions: Is the problem NP- or coNP-hard?
Also see: Graph realizability (CG63), area universality (CG-Open36), unit distance graph (CG65).

(CG-Open20) Faithful Drawability of Edge-Colored Graph
Given: A graph G and a proper edge-coloring of G, that is, any two adjacent edges differ in color.
Question: Is there a *faithful* drawing of G, that is, a drawing of G in which all edges of the same color lie on concurrent lines, and no two edges lie on the same line?
Complexity: In $\exists\mathbb{R}$, but otherwise open even for 4-regular graphs.
Comments: Richter shows that faithful drawability can be recognized in polynomial time for cubic graphs and conjectures that the problem is $\exists\mathbb{R}$-complete for 4-regular graphs [Ric11, Conjecture 9]. By moving the common points of the bundles of concurrent lines to infinity, we obtain a drawing in which all edges of the same color are parallel; this seems to suggest a connection to the planar slope number (CG41).
Also see: Planar slope number (CG41).

(CG-Open21) Kissing Number
Given: A compact semialgebraic set $S \subseteq \mathbb{R}^n$, an integer $k \geq 0$.
Question: Are there k non-overlapping congruent copies of S each touching S? The largest such k is the *kissing number* of S.
Complexity: Open.
Comments: For many natural geometric sets, such as polygons or spheres, there are known upper bounds on the kissing number, which places the problem in $\exists\mathbb{R}$, so solvers can, in principle, be used to find the kissing number in those instances. A similar problem, the largest number of pairwise touching non-overlapping congruent copies of cylinders, was tackled using a similar, but more sophisticated approach, see [BLR15].
Also see: Unit disk contact graph (CG-Open3).

(CG-Open22) Euclidean Minimum Spanning Tree
Given: A tree T.
Question: Is T the Euclidean minimum spanning tree of a set of points in the plane? In the Euclidean minimum spanning tree the weights of the edges are the distances between the points in the plane.
Complexity: In $\exists\mathbb{R}$ and known to be NP-hard by Eades and Whitesides [EW96b].
Comments: The maximum degree of T has to be 6, since every tree of maximum degree 5 is a Euclidean Minimum Spanning Tree; the construction showing this requires a double-exponential grid; as far as we know it is open whether this can be improved [MS92].
Also see: Nearest neighbor graph (CG-Open23), Gabriel graph (CG-Open24).

(CG-Open23) Nearest Neighbor Graph
Given: A graph G.
Question: Is G the nearest neighbor graph of a set of points in the plane? The *nearest-neighbor graph* of a point set has a vertex for each point, and there is an edge between two vertices if they are nearest neighbors to each other (mutually). Here p is a *nearest neighbor* of q if there is no point closer to p than q.
Complexity: In $\exists\mathbb{R}$ and known to be NP-hard by Eades and Whitesides [EW96a].

Comments: There are several variants of this problem, the variant we presented here is called the *mutual* nearest neighbor graph in [EW96a].
Open Questions: Kitching and Whitesides [KW05] proved that recognizing nearest neighbor graphs in three dimensional space remains NP-hard, is it ∃ℝ-complete there? Is there a reduction from two to three dimensions?
Also see: Euclidean minimum spanning tree (CG-Open22), Gabriel graph (CG-Open24).

(CG-Open24) Gabriel Graph
Given: A graph G.
Question: Is G a Gabriel graph in the plane? The *Gabriel graph* of a point set has a vertex for each point, and there is an edge between two vertices pq if the disk with diameter pq contains no other vertices of the graph.
Complexity: In ∃ℝ and known to be NP-hard by Eades and Whitesides [EW96a].
Comments: The realization of a Gabriel graph contains the Euclidean minimum spanning tree (CG-Open22) and the nearest neighborhood graph (CG-Open23) of the point set, and is itself a subset of the Delaunay triangulation (CG73).
Also see: Euclidean minimum spanning tree (CG-Open22), nearest neighbor graph (CG-Open23), Delaunay triangulation (CG73).

(CG-Open25) Hidden Set in Polygon
Given: A simple polygon P, natural number $k \geq 0$.
Question: Are there k points inside the region enclosed by P that cannot see each other pairwise?
Complexity: Known to be NP-hard by Eidenbenz [Eid02] and to lie in ∃ℝ [Fan20].
Comments: As with visibility graphs one can consider variants in which the hidden points must lie on the boundary or the polygon has holes. There are also variants based on terrains rather than polygons. (This problem is also sometimes called the *bird nesting box problem*. The idea is that you want to place as many bird nesting boxes as possible. However, if two bird nesting boxes see each other they may be avoided by birds as many bird species avoid other birds.)
Also see: Visibility graphs (CG36).

(CG-Open26) Watchman Route Problem
Given: A polygon with holes, an integer $k \geq 0$.
Question: Is there a polygonal curve of length at most k such that every point of the polygon can be seen from some point on the polygonal curve?
Complexity: In ∃ℝ, as shown by Volk [Vol23]. NP-hardness is shown by Dumitrescu and Tóth [DT12], who explain why the earlier NP-hardness proof by Chin and Ntafos [CN88], in the paper that introduced the problem, does not work. The problem is polynomial-time solvable for simple polygons, see [DT12] for more background.
Comments: The definition of the problem only gives ∃∀ℝ-membership, showing the problem lies in ∃ℝ requires additional analysis. Volk [Vol23] also considers the variant in which the polygonal curve has at most k bends.
Open Questions: What is the complexity of the problem in unbounded dimension? In that case, ∃∀ℝ is the best upper bound.

Also see: Lawn moving problem (CG-Open27) art gallery problem (CG16), visibility graph of polygon with holes (CG37).

(CG-Open27) Lawn Mowing Problem
Given: A polygon with holes, an integer $k \geq 0$.
Question: Is there a polygonal curve of length at most k such that every point of the polygon has distance at most 1 from the polygonal curve?
Complexity: In $\exists\forall\mathbb{R}$, as shown by Volk [Vol23] and NP-hard, as shown by Arkin et al. [AFM00].
Comments: The problem is algebraically hard: Fekete et al. [FKP+23] show that an optimal tour even for a 2×2 square requires coordinates that are zeroes of an irreducible polynomial of degree 16. Volk [Vol23] shows his upper bound by using an intermediate problem variant in which the polygonal curve has at most k bends. Arkin et al. [AFM00] also showed NP-hardness of the *milling problem*, in which the mower may not cross the boundary of the polygon with holes.
Open Questions: Does the problem lie in $\exists\mathbb{R}$? This is true for the similar watchman route problem (CG-Open26). Fekete et al. [FKP+23] conjecture that the problem is $\exists\mathbb{R}$-complete.
Also see: Watchman route problem (CG-Open26), art gallery problem (CG16).

(CG-Open28) Sweeping Points
Given: Finite set $P \subseteq \mathbb{Q}^2$, $\ell \in \mathbb{Q}_{\geq 0}$.
Question: Can all the points be swept to the origin using sweeps of total length at most ℓ? In a single *sweep*, an infinite line is moved in parallel from one location to another location (the distance swept is the contribution of that sweep to the total length). Points encountered along the sweep get moved with the line.
Comments: Dumitrescu and Jiang [DJ11] attribute this problem to Paweł Żyliński; they study approximation algorithms, and ask whether it can be solved in polynomial time.
Open Questions: Can the shortest length solution always be achieved in a polynomial number of sweeps [DJ11]?

(CG-Open29) Hypercube
Given: Finite set of points $P \subseteq [0, 1]^d$, a natural number $k \in \mathbb{N}$.
Question: Is there a pointset $Q \subseteq [0, 1]^d$ of size k such that $P \subseteq \text{conv}(Q)$, where $\text{conv}(Q)$ is the convex hull of Q?
Complexity: The hypercube problem is NP-hard via a reduction from 3SAT, and lies in $\exists\mathbb{R}$ as shown by Kiefer and Wachter [KW14]. Kiefer and Wachter also observe that for fixed d, the problem lies in P.
Comments: The problem remains NP-hard if P is required to contain the origin, the *restricted hypercube problem*. Kiefer and Wachter introduced this problem to isolate the geometric difficulty at the root of the probabilitic automaton minimisation problem (ML-Open1). The hypercube problem is reminiscent of the nested polytope problem (CG72) as defined by [DHM19].

Open Questions: Kiefer and Wachter [KW14] observe that for $d = 2$ the set Q can always be assumed to be rational, and ask whether this remains true for higher dimensions.
Also see: Probabilitic automaton minimisation (ML-Open1), nested polytope problem (CG72).

(CG-Open30) k-Median of Polygonal Curves
Given: A family of polygonal curves $P_1, \ldots, P_n$ in $\mathbb{R}^d$, all of the same length, $c \in \mathbb{Q}$.
Question: Is there a polygonal curve P of length k with $\sum_{i\in[n]} d_F(P, P_i) \leq c$? Here, d_F denotes the Fréchet distance, see references for a definition.
Complexity: In $\exists\mathbb{R}$, as shown by Rohde [Roh22], and NP-hard by Buchin, Driemel, and Struijs [BDS20].
Comments: The Fréchet distance between two polygonal curves can be computed efficiently [AG92].
Also see: Fréchet distance between surfaces (CG-Open31).

(CG-Open31) Fréchet distance between Surfaces
Given: Two homeomorphic surfaces R and S together with immersions $\varphi_R : R \to \mathbb{R}^3$, and $\varphi_S : S \to \mathbb{R}^3$, $\delta \in \mathbb{Q}_{\geq 0}$.
Question: Is the Fréchet distance $\delta_F(R, S) \leq \delta$? Here $\delta_F(R, S)$ is the infimum over all homeomorphisms $f : R \to S$ of $\sup_{x\in R} \|\varphi_R(x) - \varphi_S(f(x))\|_2$.
Complexity: One has to assume that R, S, φ_R and φ_S have reasonable effective representations for this problem to fall into the domain of $\exists\mathbb{R}$. There are various relevant results, e.g. Godau's proof that the problem is NP-hard [God99], and Nayyeri and Xu's result that the problem lies in $\exists\mathbb{R}$ if one of the surfaces is a triangle [NX18].
Also see: k-Median of Polygonal Curves (CG-Open30), Hausdorff distance (A44).

(CG-Open32) Compatible Triangulations of Polygons with Holes
Given: Two vertex-labeled polygons P and Q with holes, an integer $k \geq 0$.
Question: Can P and Q be extended to vertex-labeled triangulations which are compatible by adding at most k points? Here two triangulations are *compatible* if they have the same rotation system, that is, the cyclic permutations of neighbors at each vertex is the same for both triangulations.
Complexity: In $\exists\mathbb{R}$ and NP-hard, as shown by Lubiw and Mondal [LM20], who also ask whether the problem is $\exists\mathbb{R}$-hard.
Comments: The additional points are known as *Steiner points*; the given polygons have to be compatible to start with for the problem to be solvable.
Open Questions: Lubiw and Mondal [LM20] ask whether the problem remains NP-hard for simple polygons (no holes).

(CG-Open33) Convexity Through Translation
Given: A family of point sets in $\mathbb{Q}^2$.
Question: Can the point sets be translated so that the resulting points are in convex position, that is, lie on the convex hull of the union of the translated point sets?
Complexity: In $\exists\mathbb{R}$ and NP-hard, as shown by Hoffmann et al. [HKR+13].

Comments: If the point sets can be scaled as well as translated, the problem is not even known to be NP-hard [HKR+13].
Also see: Convexity of semialgebraic set (A33), tiling the plane periodically (CG-Open34).

(CG-Open34) Tiling the Plane Periodically
Given: A finite set of simple polygons (with algebraic edge-lengths and angles), the *prototiles*. A *tiling* T of the plane is a family of non-overlapping *tiles*, isometric copies of the prototiles, whose union is the whole plane. The tiling T is *periodic* if there are two linearly independent translations that map T to T. A *fundamental domain* is a finite subset of T whose union is homeomorphic to a disk and which contains exactly one tile from each orbit, see [DL24] for full definitions.
Question: Do the prototiles allow a periodic tiling of the plane with a fundamental domain of area at most k?
Complexity: In $\exists\mathbb{R}$ as shown by Demaine and Langerman [DL24].
Open Questions: Dropping the area constraint on the fundamental domain leads to a computably enumerable problem, as observed in [DL24]. Is that problem undecidable?
Also see: Convexity through translation (CG-Open33)

(CG-Open35) Prescribed Area
Given: A plane graph G, an area assignment for G, where an *area assignment* for G assigns a positive real number to every face of G.
Question: Does G have a straight-line drawing realizing the area assignment, that is, each face having the area prescribed by the area assignment?
Complexity: In $\exists\mathbb{R}$. If some of the vertex locations can be fixed, the problem is $\exists\mathbb{R}$-complete, as shown by Dobbins et al. [DKMR23].
Comments: Dobbins et al. [DKMR23] conjecture that this problem is $\exists\mathbb{R}$-complete.
Also see: Prescribed area extension (CG61), area universality (CG-Open36), area universality for triples with partial area assignment (CG76).

(CG-Open36) Area Universality
Given: A plane graph G.
Question: An *area assignment* for G assigns a positive real number to every face of G. Can every area assignment be realized by a straight-line embedding of G? In that realization, every face must have the area assigned to it.
Complexity: Dobbins, Kleist, and Miltzow show that the problem lies in $\forall\exists\mathbb{R}$ and conjecture that it is $\forall\exists\mathbb{R}$-complete. They can show that a more restricted version, area universality for triples with partial area assignments (CG76), is $\forall\exists\mathbb{R}$-complete.
Also see: Area Universality for triples with partial area assignments (CG76), prescribed area (CG-Open35).

(CG-Open37) Segment Visibility Graph
Given: A graph G.
Question: Is G the visibility graph of a set of segments in the plane? The visibility graph of a set of segments represents each segment by a vertex, and there is an edge

between two vertices if the corresponding segments can see each other, that is, if there is a line segment connecting the two segments that does not intersect any other segment.
Complexity: In $\exists\mathbb{R}$.
Comments: There is also the endpoint segment visibility graph, which is the visibility graph of the endpoints of the segments, but the full segments are still the visibility obstacles.
Also see: Point visibility graph (CG36), visibility graph of polygon with holes (CG37), internal/external polygon visibility graph (CG38), obstacle number (CG-Open38).

(CG-Open38) Obstacle Number
Given: A graph G, an integer $k \geq 0$.
Question: Is G the visibility graph of a set of points (in general position) in the plane with at most k polygonal obstacles? Here two points can *see* each other if the straight-line segment connecting them does not intersect any obstacles.
Complexity: In $\exists\mathbb{R}$. This problem is not known to be NP-hard, though various variants and restrictions of it are, see the summary in [GOdMV18], and [BCG+24], which shows that the problem is NP-hard for a single, fixed obstacle.
Comments: Obstacle numbers were introduced by Alpert et al. [AKL10]. There also is a variant known as the *planar obstacle number* in which the visibility segments may not cross each other [GOdMV18].
Also see: (Point) visibility graph (CG36), segment visibility graph (CG-Open37).

(CG-Open39) Fiber Dimension
Given: A graph G, a natural number $d \in \mathbb{N}$.
Question: Does G have fiber dimension at most d? That is, is there a polytope $P \subseteq \mathbb{Q}^d$ and a symmetric set $M \subseteq \mathbb{Z}^d$ such that G is isomorphic to the graph on $V = P \cap \mathbb{Z}^d$ with an edge uv if and only if $u - v \in M$.
Complexity: We are not aware of any computational complexity results.
Comments: The fiber dimension of a graph was defined by Windisch [Win19]. Every graph has finite fiber dimension.

(CG-Open40) 4-dimensional Simplicial Polytope Realizability
Given: A combinatorial type of a polytope, given as a list of facets.
Question: Does there exist a 4-dimensional simplicial polytope of this combinatorial type?
Complexity: In $\exists\mathbb{R}$, known to be $\exists\mathbb{R}$-complete for arbitrary dimension (CG70).
Comments: Conjectured to be $\exists\mathbb{R}$-complete by Adiprasito and Padrol [AP17]. The complexity of deciding whether there exists a 4-dimensional simplicial and inscribed polytope (hence a 4-dimensional Delaunay triangulation) of the given combinatorial type is open as well.
Also see: Simplicial polytope realizability (CG70), 4-dimensional polytope realizability (CG69).

(CG-Open41) 4-dimensional Polytope Graph Realizability
Given: A graph G.
Question: Does there exist a 4-dimensional polytope with 1-skeleton G?
Comments: The input to known $\exists\mathbb{R}$-complete polytope realizability problems is the full combinatorial type of the polytope. Here the input is restricted to the faces of dimension 1. Note that *simple* polytopes are known to be completely determined by their graphs [Zie95, Sect. 3.5].
Also see: Treetope graph (CG-Open42), polytope realizability (CG68), 4-dimensional polytope realizability (CG69).

(CG-Open42) Treetope graph
Given: A graph G, an integer $k \geq 0$.
Question: Does there exist a k-treetope polytope P with 1-skeleton G? A k*-treetope* is a k-dimensional polytope with a special base facet F that intersects every face of dimension at least two in at least two points.
Complexity: In dimension four Eppstein showed that the question is polynomial-time solvable [Epp20]. The complexity of this problem is open for dimension five and above.
Comments: Eppstein [Epp20] states: "*Recognizing the graphs of pyramids over arbitrary 4-polytopes, and therefore also recognizing the graphs of 5-treetopes, is as difficult as recognizing the graphs of arbitrary 4-polytopes, which we expect to be complete for the existential theory of the reals.*".
Also see: 4-dimensional Polytope Graph Realizability (CG-Open41), polytope realizability (CG68), 4-dimensional polytope realizability (CG69).

5.4.9.4 Game Theory

(GT-Open1) Recursive Markov Chain Acceptance Probability
Given: A recursive Markov chain with associated monotone transition operator $P : \mathbb{R}^n_{\geq 0} \to \mathbb{R}^n_{\geq 0}$ and a probability vector $c \in \mathbb{Q}^b$ (for definitions, see [EY09]).
Question: Is there a fixed point of P below c, that is an x with $P(x) = x$ and $x \leq c$ (in all coordinates)?
Complexity: Etessami and Yannakakis [EY09] show that this (and some related problems) can be expressed in $\exists\mathbb{R}$, and is at least as hard as the sum of square root problem.
Also see: Büchi automaton accepting recursive Markov chain trajectory (GT-Open2).

(GT-Open2) Büchi Automaton Accepts Recursive Markov Chain Trajectory
Given: A Büchi automaton $\mathcal{A}$ and a recursive Markov chain $\mathcal{M}$ (for definitions, see [BLW13b]).
Question: Is there a trajectory of $\mathcal{M}$ that is accepted by $\mathcal{A}$?
Complexity: Building on work by Etessami and Yannakakis; Benedikt et al. [BLW13b] show that the problem lies in $\exists\mathbb{R}$.
Also see: Recursive Markov chain acceptance probability (GT-Open1)

5.4.9.5 Machine Learning and Probabilistic Reasoning

(ML-Open1) Probabilistic Automaton Minimisation
Given: A probabilistic automaton A, a natural number $n \in \mathbb{N}$.
Question: Does there exists a probabilistic automaton A' of size n so that A and A' are equivalent?
Complexity: The problem was shown to be NP-hard by Kiefer and Wachter [KW14] via a reduction from the restricted hypercube problem (CG-Open29), which is known to be NP-hard [KW14]. The problem is known to be contained in $\exists\mathbb{R}$ [KW14].
Comments: The restricted hypercube problem is similar to the known $\exists\mathbb{R}$-complete nested polytope problem (CG72), so there is hope to strengthen the result to $\exists\mathbb{R}$-completeness.
Also see: Hypercube problem (CG-Open29).

5.4.9.6 Markov Chains and Decision Processes

(MDP-Open1) Universal Safety for Probabilistic Finite State Automata
Given: A probabilistic finite state automata $\mathcal{A}$ [AGV18] and a safety polytope H given by its vertex representation.
Question: Is it true that for every starting distribution there exists a strategy such that $\mathcal{A}$ stays in the safety polytope H?
Complexity: Akshay, Genest, and Vyas show that the problem is in $\exists\mathbb{R}$ [AGV18].
Comments: The problem has many variants. One can ask for existential safety instead of universal safety, replace probabilistic finite state automata with Markov Decision Processes, or encode the safety polytope in different ways. None of these variants are known to be $\exists\mathbb{R}$-complete either [AGV18].

Acknowledgements We would like to thank the following people for their input and useful discussions: Mikkel Abrahamsen, Kristoffer Arnsfelt Hansen, Marcus Bläser, Peter Bürgisser, Balder ten Cate, Jeff Erickson, Sándor Fekete, Radoslav Fulek, Bill Gasarch, Xavier Goaoc, Christoph Hertrich, Jeremy Kirn, Jan Kynčl, Alessio Mansutti, Lucas Meijer, Themistoklis Melissourgos, Arnaud de Mesmay, Guillermo Alberto Perez, Konrad Swanepoel, Alexander Wolff.
We also owe thanks to the many anonymous referees who made copious useful suggestions.

References

[AAK02] Oswin Aichholzer, Franz Aurenhammer, and Hannes Krasser. Enumerating order types for small point sets with applications. *Order*, 19(3):265–281, 2002.
[AAM17] Mikkel Abrahamsen, Anna Adamaszek, and Tillmann Miltzow. Irrational guards are sometimes needed. In *33rd International Symposium on Computational Geometry*, volume 77 of *LIPIcs. Leibniz Int. Proc. Inform.*, pages Art. No. 3, 15. Schloss Dagstuhl. Leibniz-Zent. Inform., Wadern, 2017.

[AAM18] Mikkel Abrahamsen, Anna Adamaszek, and Tillmann Miltzow. The art gallery problem is ∃ℝ-complete. In *STOC'18—Proceedings of the 50th Annual ACM SIGACT Symposium on Theory of Computing*, pages 65–73. ACM, New York, 2018.

[AAM22] Mikkel Abrahamsen, Anna Adamaszek, and Tillmann Miltzow. The art gallery problem is ∃ℝ-complete. *J. ACM*, 69(1):Art. 4, 70, 2022.

[Aar11] Scott Aaronson. Projects aplenty. https://scottaaronson.blog/?m=201105, May 2011.

[Aar24] Scott Aaronson. Complexity zoo. https://complexityzoo.net/, 2024. [Online; accessed 22-January-2024].

[AB09] Sanjeev Arora and Boaz Barak. *Computational complexity. A modern approach.* Cambridge: Cambridge University Press, 2009.

[Abb08] Timothy Good Abbott. Generalizations of kempe's universality theorem. Master's thesis, Massachusetts Institute of Technology, Dept. of Electrical Engineering and Computer Science, 2008.

[ABB+22] Oswin Aichholzer, Manuel Borrazzo, Prosenjit Bose, Jean Cardinal, Fabrizio Frati, Pat Morin, and Birgit Vogtenhuber. Drawing graphs as spanners. *Discrete Comput. Geom.*, 68(3):774–795, 2022.

[Abe16] Zachary Abel. *On Folding and Unfolding with Linkages and Origami*. PhD thesis, Massachusetts Institute of Technology, Sep 2016.

[ABK+23] Patrizio Angelini, Michael A. Bekos, Julia Katheder, Michael Kaufmann, Maximilian Pfister, and Torsten Ueckerdt. Axis-parallel right angle crossing graphs. arXiv:2306.17073, 2023.

[ABKPM09] Eric Allender, Peter Bürgisser, Johan Kjeldgaard-Pedersen, and Peter Bro Miltersen. On the complexity of numerical analysis. *SIAM J. Comput.*, 38(5):1987–2006, 2008/09.

[ABM+23] Hugo A Akitaya, Maike Buchin, Majid Mirzanezhad, Leonie Ryvkin, and Carola Wenk. Realizability of free spaces of curves. arxiv:2311.07573, 2023.

[Abr22] Mikkel Abrahamsen. Covering polygons is even harder. In *2021 IEEE 62nd Annual Symposium on Foundations of Computer Science—FOCS 2021*, pages 375–386. IEEE Computer Soc., Los Alamitos, CA, 2022.

[Abr24] Mikkel Abrahamsen. personal communication, 2024.

[ACH+24] Patrizio Angelini, Sabine Cornelsen, Carolina Haase, Eleni Katsanou, Fabrizio Montecchiani, and Antonios Symvonis. Bipartite dichotomous ordinal graphs. In *40th European Workshop on Computational Geometry, Ioannina, Greece, March 13–15, 2024*, pages 17:1–8, 2024.

[ADD+16] Zachary Abel, Erik D. Demaine, Martin L. Demaine, Sarah Eisenstat, Jayson Lynch, and Tao B. Schardl. Who needs crossings? Hardness of plane graph rigidity. In *32nd International Symposium on Computational Geometry*, volume 51 of *LIPIcs. Leibniz Int. Proc. Inform.*, pages Art. No. 3, 15. Schloss Dagstuhl. Leibniz-Zent. Inform., Wadern, 2016.

[ADLDB+14] Patrizio Angelini, Giordano Da Lozzo, Marco Di Bartolomeo, Giuseppe Di Battista, Seok-Hee Hong, Maurizio Patrignani, and Vincenzo Roselli. Anchored drawings of planar graphs. In *Graph drawing*, volume 8871 of *Lecture Notes in Comput. Sci.*, pages 404–415. Springer, Heidelberg, 2014.

[AFM00] Esther M. Arkin, Sándor P. Fekete, and Joseph S. B. Mitchell. Approximation algorithms for lawn mowing and milling. *Comput. Geom.*, 17(1–2):25–50, 2000.

[AG92] Helmut Alt and Michael Godau. Computing the Fréchet distance between two polygonal curves. In *Eighth Annual ACM Symposium on Computational Geometry (Berlin, 1992)*, volume 5, pages 75–91. ACM, 1992.

[AGLK+20] Safwa Ameer, Matt Gibson-Lopez, Erik Krohn, Sean Soderman, and Qing Wang. Terrain Visibility Graphs: Persistence Is Not Enough. In Sergio Cabello and Danny Z. Chen, editors, *36th International Symposium on Computational Geometry (SoCG 2020)*, volume 164 of *Leibniz International Proceedings in Informatics*

(LIPIcs), pages 6:1–6:13, Dagstuhl, Germany, 2020. Schloss Dagstuhl – Leibniz-Zentrum für Informatik.

[AGLKW22] Safwa Ameer, Matt Gibson-Lopez, Erik Krohn, and Qing Wang. On the Visibility Graphs of Pseudo-Polygons: Recognition and Reconstruction. In Artur Czumaj and Qin Xin, editors, *18th Scandinavian Symposium and Workshops on Algorithm Theory (SWAT 2022)*, volume 227 of *Leibniz International Proceedings in Informatics (LIPIcs)*, pages 7:1–7:13, Dagstuhl, Germany, 2022. Schloss Dagstuhl – Leibniz-Zentrum für Informatik.

[AGV18] S. Akshay, Blaise Genest, and Nikhil Vyas. Distribution-based objectives for Markov decision processes. In Anuj Dawar and Erich Grädel, editors, *Proceedings of the 33rd Annual ACM/IEEE Symposium on Logic in Computer Science, LICS 2018, Oxford, UK, July 09-12, 2018*, pages 36–45. ACM, 2018.

[AH23] Inon Attias and Ishay Haviv. Local orthogonality dimension. *J. Graph Theory*, 104(2):387–419, 2023.

[AKL10] Hannah Alpert, Christina Koch, and Joshua D. Laison. Obstacle numbers of graphs. *Discrete Comput. Geom.*, 44(1):223–244, 2010.

[AKM21] Mikkel Abrahamsen, Linda Kleist, and Tillmann Miltzow. Training neural networks is $\exists\mathbb{R}$-complete. In *Advances in Neural Information Processing Systems 34: Annual Conference on Neural Information Processing Systems 2021, NeurIPS 2021, December 6-14, 2021, virtual*, pages 18293–18306, 2021.

[AKM23] Mikkel Abrahamsen, Linda Kleist, and Tillmann Miltzow. Geometric embeddability of complexes is $\exists\mathbb{R}$-complete. In *39th International Symposium on Computational Geometry*, volume 258 of *LIPIcs. Leibniz Int. Proc. Inform.*, pages Art. No. 1, 19. Schloss Dagstuhl. Leibniz-Zent. Inform., Wadern, 2023.

[AKPT16] Md. Jawaherul Alam, Stephen G. Kobourov, Sergey Pupyrev, and Jackson Toeniskoetter. Weak unit disk and interval representation of graphs. In *Graph-theoretic concepts in computer science*, volume 9224 of *Lecture Notes in Comput. Sci.*, pages 237–251. Springer, Berlin, 2016.

[ALD+20] Abu Reyan Ahmed, Felice De Luca, Sabin Devkota, Stephen G. Kobourov, and Mingwei Li. Graph drawing via gradient descent, $(gd)^2$. arXiv:2008.05584, 2020.

[AM19] Mikkel Abrahamsen and Tillmann Miltzow. Dynamic toolbox for ETRINV. arXiv:1912.08674, 2019.

[AMS20] Mikkel Abrahamsen, Tillmann Miltzow, and Nadja Seiferth. Framework for ER-completeness of two-dimensional packing problems. In *2020 IEEE 61st Annual Symposium on Foundations of Computer Science*, pages 1014–1021. IEEE Computer Soc., Los Alamitos, CA, 2020.

[AMS24] Mikkel Abrahamsen, Tillmann Miltzow, and Nadja Seiferth. Framework for $\exists\mathbb{R}$-Completeness of Two-Dimensional Packing Problems. *TheoretiCS*, Volume 3:1–78, April 2024.

[AOW11] Felix Arends, Joël Ouaknine, and Charles W. Wampler. On searching for small Kochen-Specker vector systems. In *Graph-theoretic concepts in computer science*, volume 6986 of *Lecture Notes in Comput. Sci.*, pages 23–34. Springer, Heidelberg, 2011.

[AP17] Karim A. Adiprasito and Arnau Padrol. The universality theorem for neighborly polytopes. *Combinatorica*, 37(2):129–136, 2017.

[APT15] Karim A. Adiprasito, Arnau Padrol, and Louis Theran. Universality theorems for inscribed polytopes and Delaunay triangulations. *Discrete Comput. Geom.*, 54(2):412–431, 2015.

[AW10] Kira Adaricheva and Marcel Wild. Realization of abstract convex geometries by point configurations. *European J. Combin.*, 31(1):379–400, 2010.

[Bas17] Saugata Basu. Algorithms in real algebraic geometry: a survey. In *Real algebraic geometry*, volume 51 of *Panor. Synthèses*, pages 107–153. Soc. Math. France, Paris, 2017.

[BBDJ23] Nicholas Bieker, Thomas Bläsius, Emil Dohse, and Paul Jungeblut. Recognizing unit disk graphs in hyperbolic geometry is $\exists\mathbb{R}$-complete. arXiv:2301.05550, 2023.
[BBK20] Hå vard Bakke Bjerkevik, Magnus Bakke Botnan, and Michael Kerber. Computing the interleaving distance is NP-hard. *Found. Comput. Math.*, 20(5):1237–1271, 2020.
[BBP20] Katharina Börsig, Ulrik Brandes, and Barna Pásztor. Stochastic gradient descent works really well for stress minimization. In David Auber and Pavel Valtr, editors, *Graph Drawing and Network Visualization - 28th International Symposium, GD 2020, Vancouver, BC, Canada, September 16-18, 2020, Revised Selected Papers*, volume 12590 of *Lecture Notes in Computer Science*, pages 18–25. Springer, 2020.
[BC06] Peter Bürgisser and Felipe Cucker. Counting complexity classes for numeric computations. II. Algebraic and semialgebraic sets. *J. Complexity*, 22(2):147–191, 2006.
[BC07] Maria Belk and Robert Connelly. Realizability of graphs. *Discrete Comput. Geom.*, 37(2):125–137, 2007.
[BC09] Peter Bürgisser and Felipe Cucker. Exotic quantifiers, complexity classes, and complete problems. *Found. Comput. Math.*, 9(2):135–170, 2009.
[BCG+24] Martin Balko, Steven Chaplick, Robert Ganian, Siddharth Gupta, Michael Hoffmann, Pavel Valtr, and Alexander Wolff. Bounding and Computing Obstacle Numbers of Graphs. *SIAM J. Discrete Math.*, 38(2):1537–1565, 2024.
[BCSS98] Lenore Blum, Felipe Cucker, Michael Shub, and Steve Smale. *Complexity and real computation*. Springer-Verlag, New York, 1998. With a foreword by Richard M. Karp.
[BCWP22] Dimitris Bertsimas, Ryan Cory-Wright, and Jean Pauphilet. Mixed-projection conic optimization: a new paradigm for modeling rank constraints. *Oper. Res.*, 70(6):3321–3344, 2022.
[BD93] Daniel Bienstock and Nathaniel Dean. Bounds for rectilinear crossing numbers. *J. Graph Theory*, 17(3):333–348, 1993.
[BD23] Nikhil Balaji and Samir Datta. USSR is in P/poly, 2023.
[BDC23] Olivier Bournez, Valentin Dardilhac, and Johanne Cohen. On the δ-decidability of decision problems for neural network questions. In *Continuity, Computability, Constructivity. From Logic to Algorithms. CCC'23*, Kyoto, Japan, 2023.
[BDLvdZ24] Markus Bläser, Julian Dörfler, Maciej Liskiewicz, and Benito van der Zander. The existential theory of the reals with summation operators. arXiv:2405.04697, 2024.
[BDS20] Kevin Buchin, Anne Driemel, and Martijn Struijs. On the hardness of computing an average curve. In *17th Scandinavian Symposium and Workshops on Algorithm Theory*, volume 162 of *LIPIcs. Leibniz Int. Proc. Inform.*, pages Art. No. 19, 19. Schloss Dagstuhl. Leibniz-Zent. Inform., Wadern, 2020.
[BEMK+23] Daniel Bertschinger, Nicolas El Maalouly, Linda Kleist, Tillmann Miltzow, and Simon Weber. The complexity of recognizing geometric hypergraphs. In *Graph drawing and network visualization. Part I*, volume 14465 of *Lecture Notes in Comput. Sci.*, pages 163–179. Springer, Cham, 2023.
[BEMM+22] Daniel Bertschinger, Nicolas El Maalouly, Tillmann Miltzow, Patrick Schnider, and Simon Weber. Topological art in simple galleries. In Karl Bringmann and Timothy M. Chan, editors, *5th Symposium on Simplicity in Algorithms, SOSA@SODA 2022, Virtual Conference, January 10-11, 2022*, pages 87–116. SIAM, 2022.
[BEMM+23] Daniel Bertschinger, Nicolas El Maalouly, Tillmann Miltzow, Patrick Schnider, and Simon Weber. Topological art in simple galleries. *Discrete Comput. Geom.*, page 1092–1130, 2023.
[BEW03] Prosenjit Bose, Hazel Everett, and Stephen Wismath. Properties of arrangement graphs. *Internat. J. Comput. Geom. Appl.*, 13(6):447–462, 2003.
[BFG+09] Ronen Basri, Pedro F. Felzenszwalb, Ross B. Girshick, Jacobs. David W., and Caroline J. Klivans. Visibility constraints on features of 3d objects. In *2009 IEEE Computer Society Conference on Computer Vision and Pattern Recognition (CVPR 2009), 20-25 June 2009, Miami, Florida, USA*, pages 1231–1238. IEEE Computer Society, 2009.

[BFKS23] Thomas Bläsius, Tobias Friedrich, Maximilian Katzmann, and Daniel Stephan. Strongly hyperbolic unit disk graphs. In Berenbrink. Petra, Patricia Bouyer, Anuj Dawar, and Mamadou Moustapha Kanté, editors, *40th International Symposium on Theoretical Aspects of Computer Science, STACS 2023, March 7-9, 2023, Hamburg, Germany*, volume 254 of *LIPIcs*, pages 13:1–13:17. Schloss Dagstuhl - Leibniz-Zentrum für Informatik, 2023.

[BFS99] Jonathan F. Buss, Gudmund S. Frandsen, and Jeffrey O. Shallit. The computational complexity of some problems of linear algebra. *J. Comput. System Sci.*, 58(3):572–596, 1999.

[BGdO90] Jürgen Bokowski and António Guedes de Oliveira. Simplicial convex 4-polytopes do not have the isotopy property. *Portugal. Math.*, 47(3):309–318, 1990.

[BGVdBW19] Robert Brijder, Floris Geerts, Jan Van den Bussche, and Timmy Weerwag. On the expressive power of query languages for matrices. *ACM Trans. Database Syst.*, 44(4): Art. 15, 31, 2019.

[BH21] Manon Blanc and Kristoffer Arnsfelt Hansen. Computational complexity of multi-player evolutionarily stable strategies. In *Computer science—theory and applications*, volume 12730 of *Lecture Notes in Comput. Sci.*, pages 1–17. Springer, Cham, 2021.

[BH22] Marie Louisa Tølbøll Berthelsen and Kristoffer Arnsfelt Hansen. On the computational complexity of decision problems about multi-player Nash equilibria. *Theory Comput. Syst.*, 66(3):519–545, 2022.

[BHJ+22] Daniel Bertschinger, Christoph Hertrich, Paul Jungeblut, Tillmann Miltzow, and Simon Weber. Training fully connected neural networks is $\exists\mathbb{R}$-complete. arXiv:2204.01368, 2022.

[BHM23] Vittorio Bilò, Kristoffer Arnsfelt Hansen, and Marios Mavronicolas. Computational complexity of decision problems about Nash equilibria in win-lose multi-player games. In *Algorithmic game theory*, volume 14238 of *Lecture Notes in Comput. Sci.*, pages 40–57. Springer, Cham, 2023.

[Bie91] Daniel Bienstock. Some provably hard crossing number problems. *Discrete Comput. Geom.*, 6(5):443–459, 1991.

[Bie20] Nicholas Bieker. Complexity of graph drawing problems in relation to the existential theory of the reals. Bachelor's thesis, Karlsruher Institut für Technologie, Karlsruhe, 2020. https://i11www.iti.kit.edu/_media/teaching/theses/ba-bieker-20.pdf (last accessed 12/17/2020).

[Bie22] Nicholas Bieker. Complexity of simple stretchability and related ER-complete problems in hyperbolic geometry. Master's thesis, Karlsruher Institut für Technologie, Karlsruhe, 2022. https://i11www.iti.kit.edu/_media/teaching/theses/ma-bieker-22.pdf (last accessed 3/7/2023).

[BIJL18] Markus Bläser, Christian Ikenmeyer, Gorav Jindal, and Vladimir Lysikov. Generalized matrix completion and algebraic natural proofs. In *STOC'18—Proceedings of the 50th Annual ACM SIGACT Symposium on Theory of Computing*, pages 1193–1206. ACM, New York, 2018.

[BIL+21] Markus Bläser, Christian Ikenmeyer, Vladimir Lysikov, Anurag Pandey, and Frank-Olaf Schreyer. On the orbit closure containment problem and slice rank of tensors. In *Proceedings of the 2021 ACM-SIAM Symposium on Discrete Algorithms (SODA)*, pages 2565–2584. [Society for Industrial and Applied Mathematics (SIAM)], Philadelphia, PA, 2021.

[BJ23] Boris Bukh and R. Amzi Jeffs. Planar convex codes are decidable. *SIAM J. Discret. Math.*, 37(2):951–963, 2023.

[BJ24] Peter Bürgisser and Gorav Jindal. On the hardness of PosSLP. In *Proceedings of the 2024 Annual ACM-SIAM Symposium on Discrete Algorithms (SODA)*, pages 1872–1886. SIAM, Philadelphia, PA, 2024.

[BJL16] Samuel Bald, Matthew P. Johnson, and Ou Liu. Approximating the maximum rectilinear crossing number. In *Computing and combinatorics*, volume 9797 of *Lecture Notes in Comput. Sci.*, pages 455–467. Springer, [Cham], 2016.

[BK96] Heinz Breu and David G. Kirkpatrick. On the complexity of recognizing intersection and touching graphs of disks. In *Graph drawing (Passau, 1995)*, volume 1027 of *Lecture Notes in Comput. Sci.*, pages 88–98. Springer, Berlin, 1996.

[BK98] Heinz Breu and David G. Kirkpatrick. Unit disk graph recognition is NP-hard. *Comput. Geom.*, 9(1-2):3–24, 1998.

[BK15] Amey Bhangale and Swastik Kopparty. The complexity of computing the minimum rank of a sign pattern matrix. arXiv:1503.04486, 2015.

[BKMS05] Jürgen Bokowski, Simon King, Susanne Mock, and Ileana Streinu. The topological representation of oriented matroids. *Discrete Comput. Geom.*, 33(4):645–668, 2005.

[BL76] Kellogg S. Booth and George S. Lueker. Testing for the consecutive ones property, interval graphs, and graph planarity using PQ-tree algorithms. *J. Comput. System Sci.*, 13(3):335–379, 1976.

[BLR15] Sándor Bozóki, Tsung-Lin Lee, and Lajos Rónyai. Seven mutually touching infinite cylinders. *Comput. Geom.*, 48(2):87–93, 2015.

[BLVS+99] Anders Björner, Michel Las Vergnas, Bernd Sturmfels, Neil White, and Günter M. Ziegler. *Oriented matroids*, volume 46 of *Encyclopedia of Mathematics and its Applications*. Cambridge University Press, Cambridge, second edition, 1999.

[BLW13a] Michael Benedikt, Rastislav Lenhardt, and James Worrell. LTL model checking of interval Markov chains. In *Tools and algorithms for the construction and analysis of systems. 19th international conference, TACAS 2013, held as part of the European joint conferences on theory and practice of software, ETAPS 2013, Rome, Italy, March 16–24, 2013. Proceedings*, pages 32–46. Berlin: Springer, 2013.

[BLW13b] Michael Benedikt, Rastislav Lenhardt, and James Worrell. Two variable vs. linear temporal logic in model checking and games. *Log. Methods Comput. Sci.*, 9(2):2:4, 37, 2013.

[BM13] Martijn Baartse and Klaus Meer. Topics in real and complex number complexity theory. In *Recent advances in real complexity and computation*, volume 604 of *Contemp. Math.*, pages 1–53. Amer. Math. Soc., Providence, RI, 2013.

[BM15] Martijn Baartse and Klaus Meer. The PCP theorem for NP over the reals. *Found. Comput. Math.*, 15(3):651–680, 2015.

[BM16] Vittorio Bilò and Marios Mavronicolas. A catalog of $\exists\mathbb{R}$-complete decision problems about Nash equilibria in multi-player games. In Nicolas Ollinger and Heribert Vollmer, editors, *33rd Symposium on Theoretical Aspects of Computer Science, STACS 2016, February 17-20, 2016, Orléans, France*, volume 47 of *LIPIcs*, pages 17:1–17:13. Schloss Dagstuhl - Leibniz-Zentrum für Informatik, 2016.

[BM20] Édouard Bonnet and Tillmann Miltzow. Parameterized hardness of art gallery problems. *ACM Trans. Algorithms*, 16(4):42:1–42:23, 2020.

[BM21] Vittorio Bilò and Marios Mavronicolas. $\exists\mathbb{R}$-complete decision problems about (symmetric) Nash equilibria in (symmetric) multi-player games. *ACM Trans. Econ. Comput.*, 9(3):Art. 14, 25, 2021.

[BMP05] Peter Brass, William Moser, and János Pach. *Research problems in discrete geometry*. Springer, New York, 2005.

[BMS01] Jürgen Bokowski, Susanne Mock, and Ileana Streinu. On the Folkman-Lawrence topological representation theorem for oriented matroids of rank 3. *European J. Combin.*, 22(5):601–615, 2001. Combinatorial geometries (Luminy, 1999).

[Boe21] Tobias Boege. Incidence geometry in the projective plane via almost-principal minors of symmetric matrices. arXiv:2103.02589, 2021.

[Boe22a] Tobias Boege. *The Gaussian conditional independence inference problem*. PhD thesis, Universität Magdeburg, 2022.

[Boe22b] Tobias Boege. On stable equivalence of semialgebraic sets. https://taboege.de/blog/2022/02/On-stable-equivalence-of-semialgebraic-sets/, 2022.

[BOKR86] Michael Ben-Or, Dexter Kozen, and John Reif. The complexity of elementary algebra and geometry. *J. Comput. System Sci.*, 32(2):251–264, 1986.

[BOZ18] Hossein Boomari, Mojtaba Ostovari, and Alireza Zarei. Recognizing visibility graphs of polygons with holes and internal-external visibility graphs of polygons. arXiv:1804.05105, 2018.

[BOZ21] Hossein Boomari, Mojtaba Ostovari, and Alireza Zarei. Recognizing visibility graphs of triangulated irregular networks. *Fund. Inform.*, 179(4):345–360, 2021.

[BPR06] Saugata Basu, Richard Pollack, and Marie-Françoise Roy. *Algorithms in real algebraic geometry*, volume 10 of *Algorithms and Computation in Mathematics*. Springer-Verlag, Berlin, second edition, 2006.

[BR10] Saugata Basu and Marie-Françoise Roy. Bounding the radii of balls meeting every connected component of semi-algebraic sets. *J. Symbolic Comput.*, 45(12):1270–1279, 2010.

[BR15] Michael A. Bekos and Chrysanthi N. Raftopoulou. On a conjecture of Lovász on circle-representations of simple 4-regular planar graphs. *J. Comput. Geom.*, 6(1):1–20, 2015.

[BRS17] Markus Bläser, B. V. Raghavendra Rao, and Jayalal Sarma. Testing polynomial equivalence by scaling matrices. In *Fundamentals of computation theory*, volume 10472 of *Lecture Notes in Comput. Sci.*, pages 111–122. Springer, Berlin, 2017.

[BSS89] Lenore Blum, Mike Shub, and Steve Smale. On a theory of computation and complexity over the real numbers: NP-completeness, recursive functions and universal machines. *Bull. Amer. Math. Soc. (N.S.)*, 21(1):1–46, 1989.

[BSV21] Vishwas Bhargava, Shubhangi Saraf, and Ilya Volkovich. Reconstruction algorithms for low-rank tensors and depth-3 multilinear circuits. In *STOC '21—Proceedings of the 53rd Annual ACM SIGACT Symposium on Theory of Computing*, pages 809–822. ACM, New York, 2021.

[BSZ24] Tobias Boege, Jesse Selover, and Maksym Zubkov. Sign patterns of principal minors of real symmetric matrices. arXiv:2407.17826, 2024.

[BvN36] Garrett Birkhoff and John von Neumann. The logic of quantum mechanics. *Ann. of Math. (2)*, 37(4):823–843, 1936.

[BZ10] Saugata Basu and Thierry Zell. Polynomial hierarchy, Betti numbers, and a real analogue of Toda's theorem. *Found. Comput. Math.*, 10(4):429–454, 2010.

[Can88a] John Canny. *The complexity of robot motion planning*, volume 1987 of *ACM Doctoral Dissertation Awards*. MIT Press, Cambridge, MA, 1988.

[Can88b] John Canny. Some algebraic and geometric computations in pspace. In *STOC '88: Proceedings of the twentieth annual ACM symposium on Theory of computing*, pages 460–469, New York, NY, USA, 1988. ACM.

[Car15] Jean Cardinal. Computational geometry column 62. *SIGACT News*, 46(4):69–78, 2015.

[Car24] Jean Cardinal. The complexity of intersection graphs of lines in space and circle orders. arXiv:2406.17504, 2024.

[CDR07] Sergio Cabello, Erik D. Demaine, and Günter Rote. Planar embeddings of graphs with specified edge lengths. *J. Graph Algorithms Appl.*, 11(1):259–276, 2007.

[CDT09] Xi Chen, Xiaotie Deng, and Shang-Hua Teng. Settling the complexity of computing two-player nash equilibria. *J. ACM*, 56(3):14:1–14:57, 2009.

[CdV90] Yves Colin de Verdière. Sur un nouvel invariant des graphes et un critère de planarité. *J. Combin. Theory Ser. B*, 50(1):11–21, 1990.

[CE95] Leizhen Cai and Hazel Everett. Visibility graphs of polygonal rings. In *Proceedings of the 7th Canadian Conference on Computational Geometry, Quebec City, Quebec, Canada, August 1995*, pages 157–161. Carleton University, Ottawa, Canada, 1995.

[CFL+23] Steven Chaplick, Krzysztof Fleszar, Fabian Lipp, Alexander Ravsky, Oleg Verbitsky, and Alexander Wolff. The complexity of drawing graphs on few lines and few planes. *J. Graph Algorithms Appl.*, 27(6):459–488, 2023.

[CFM+18] Jean Cardinal, Stefan Felsner, Tillmann Miltzow, Casey Tompkins, and Birgit Vogtenhuber. Intersection graphs of rays and grounded segments. *J. Graph Algorithms Appl.*, 22(2):273–295, 2018.

[CG97] Felipe Cucker and Dima Grigoriev. On the power of real Turing machines over binary inputs. *SIAM J. Comput.*, 26(1):243–254, 1997.

[CGM24] Martino Ciaperoni, Aristides Gionis, and Heikki Mannila. The hadamard decomposition problem. *Data Min. Knowl. Discov.*, page 2306–2347, 2024.

[CGS23] Suryajith Chillara, Coral Grichener, and Amir Shpilka. On hardness of testing equivalence to sparse polynomials under shifts. In *40th International Symposium on Theoretical Aspects of Computer Science*, volume 254 of *LIPIcs. Leibniz Int. Proc. Inform.*, pages Art. No. 22, 20. Schloss Dagstuhl. Leibniz-Zent. Inform., Wadern, 2023.

[CH17] Jean Cardinal and Udo Hoffmann. Recognition and complexity of point visibility graphs. *Discrete Comput. Geom.*, 57(1):164–178, 2017.

[Cho19] Ventsislav Chonev. Reachability in augmented interval Markov chains. In *Reachability problems*, volume 11674 of *Lecture Notes in Comput. Sci.*, pages 79–92. Springer, Cham, 2019.

[Chv75] V. Chvátal. A combinatorial theorem in plane geometry. *J. Combinatorial Theory Ser. B*, 18:39–41, 1975.

[CK95] Felipe Cucker and Pascal Koiran. Computing over the reals with addition and order: higher complexity classes. *J. Complexity*, 11(3):358–376, 1995.

[CK99] Olivier Chapuis and Pascal Koiran. Saturation and stability in the theory of computation over the reals. *Ann. Pure Appl. Logic*, 99(1-3):1–49, 1999.

[CK15] Jean Cardinal and Vincent Kusters. The complexity of simultaneous geometric graph embedding. *J. Graph Algorithms Appl.*, 19(1):259–272, 2015.

[CKM+17] Dmitry Chistikov, Stefan Kiefer, Ines Marušić, Mahsa Shirmohammadi, and James Worrell. Nonnegative matrix factorization requires irrationality. *SIAM J. Appl. Algebra Geom.*, 1(1):285–307, 2017.

[Cle20] Lorenzo Clemente. On the complexity of the universality and inclusion problems for unambiguous context-free grammars (technical report). arXiv:2006.05275, 2020.

[CM96] Felipe Cucker and Martín Matamala. On digital nondeterminism. *Math. Systems Theory*, 29(6):635–647, 1996.

[CN88] Wei-Pang Chin and Simeon Ntafos. Optimum watchman routes. *Inform. Process. Lett.*, 28(1):39–44, 1988.

[Col75] George E. Collins. Quantifier elimination for real closed fields by cylindrical algebraic decomposition. In *Automata theory and formal languages (Second GI Conf., Kaiserslautern, 1975)*, volume Vol. 33 of *Lecture Notes in Comput. Sci.*, pages 134–183. Springer, Berlin-New York, 1975.

[COW15] Ventsislav Chonev, Joël Ouaknine, and James Worrell. The polyhedron-hitting problem. In *Proceedings of the Twenty-Sixth Annual ACM-SIAM Symposium on Discrete Algorithms*, pages 940–956. SIAM, Philadelphia, PA, 2015.

[CR92] Felipe Cucker and Francesc Rosselló. On the complexity of some problems for the Blum, Shub & Smale model. In *LATIN '92 (São Paulo, 1992)*, volume 583 of *Lecture Notes in Comput. Sci.*, pages 117–129. Springer, Berlin, 1992.

[CR93] Joel E. Cohen and Uriel G. Rothblum. Nonnegative ranks, decompositions, and factorizations of nonnegative matrices. *Linear Algebra Appl.*, 190:149–168, 1993.

[CR94] Joseph C. Culberson and Robert A. Reckhow. Covering polygons is hard. *J. Algorithms*, 17(1):2–44, 1994.

[CS08] Vincent Conitzer and Tuomas Sandholm. New complexity results about nash equilibria. *Games Econ. Behav.*, 63(2):621–641, 2008.

[Cuc93] Felipe Cucker. On the complexity of quantifier elimination: the structural approach. *Comput. J.*, 36(5):400–408, 1993.

[CVC] CVC5. About cvc5. https://cvc5.github.io/.

[CW16] Markus Chimani and Tilo Wiedera. An ILP-based proof system for the crossing number problem. In *24th Annual European Symposium on Algorithms*, volume 57 of *LIPIcs. Leibniz Int. Proc. Inform.*, pages Art. No. 29, 13. Schloss Dagstuhl. Leibniz-Zent. Inform., Wadern, 2016.

[Dat03] Ruchira S. Datta. Universality of Nash equilibria. *Math. Oper. Res.*, 28(3):424–432, 2003.

[Dea02] Nathaniel Dean. Mathematical programming formulation of rectilinear crossing minimization. Technical Report 2002-12, DIMACS, USA, 2002.

[DFM20] Argyrios Deligkas, John Fearnley, and Themistoklis Melissourgos. Square-cut pizza sharing is PPA-complete. arXiv:2012.14236, 2020.

[DFM22] Argyrios Deligkas, John Fearnley, and Themistoklis Melissourgos. Pizza sharing is PPA-hard. In *Thirty-Sixth AAAI Conference on Artificial Intelligence, AAAI*, pages 4957–4965. AAAI Press, 2022.

[DFMS21] Argyrios Deligkas, John Fearnley, Themistoklis Melissourgos, and Paul G. Spirakis. Computing exact solutions of consensus halving and the Borsuk-Ulam theorem. *J. Comput. System Sci.*, 117:75–98, 2021.

[DFMS22] Argyrios Deligkas, John Fearnley, Themistoklis Melissourgos, and Paul G. Spirakis. Approximating the existential theory of the reals. *J. Comput. System Sci.*, 125:106–128, 2022.

[DGC99] Ernest Davis, Nicholas Mark Gotts, and Anthony G. Cohn. Constraint networks of topological relations and convexity. *Constraints*, 4(3):241–280, 1999.

[DGDH+15] Emilio Di Giacomo, Walter Didimo, Seok-Hee Hong, Michael Kaufmann, Stephen G. Kobourov, Giuseppe Liotta, Kazuo Misue, Antonios Symvonis, and Hsu-Chun Yen. Low ply graph drawing. In Nikolaos G. Bourbakis, George A. Tsihrintzis, and Maria Virvou, editors, *6th International Conference on Information, Intelligence, Systems and Applications, IISA 2015, Corfu, Greece, July 6-8, 2015*, pages 1–6. IEEE, 2015.

[DGM16] Stephane Durocher, Ellen Gethner, and Debajyoti Mondal. Thickness and colorability of geometric graphs. *Comput. Geom.*, 56:1–18, 2016.

[DGP09] Constantinos Daskalakis, Paul W. Goldberg, and Christos H. Papadimitriou. The complexity of computing a nash equilibrium. *SIAM J. Comput.*, 39(1):195–259, 2009.

[DGVVG99] Freddy Dumortier, Marc Gyssens, Luc Vandeurzen, and Dirk Van Gucht. On the decidability of semilinearity for semialgebraic sets and its implications for spatial databases. *J. Comput. System Sci.*, 58(3):535–571, 1999. Sixteenth ACM SIGACT-SIGMOD-SIGART Symposium on Principles of Database Systems (Tucson, AZ, 1997).

[DHK24] Simon Dohn, Kristoffer Arnsfelt Hansen, and Asger Klinkby. Improved hardness results for the clearing problem in financial networks with credit default swaps. arXiv:2409.18717, 2024.

[DHM03] Mike Develin, Stephen Hartke, and David Petrie Moulton. A general notion of visibility graphs. *Discrete Comput. Geom.*, 29(4):511–524, 2003.

[DHM19] Michael Gene Dobbins, Andreas F. Holmsen, and Tillmann Miltzow. A universality theorem for nested polytopes. arXiv:1908.02213, 2019.

[DJ11] Adrian Dumitrescu and Minghui Jiang. Sweeping points. *Algorithmica*, 60(3):703–717, 2011.

[DKL+24] Alexander Dobler, Stephen Kobourov, William J. Lenhart, Tamara Mchedlidze, Martin Nöllenburg, and Antonios Symvonis. Representing hypergraphs by point-line incidences. In *40th European Workshop on Computational Geometry, Ioannina, Greece, March 13-15, 2024*, pages 39:1–8, 2024.

[DKMR18] Michael Gene Dobbins, Linda Kleist, Tillmann Miltzow, and PawełRzążewski. $\forall\exists\mathbb{R}$-completeness and area-universality. In *Graph-theoretic concepts in computer science*, volume 11159 of *Lecture Notes in Comput. Sci.*, pages 164–175. Springer, Cham, 2018.

[DKMR23] Michael Gene Dobbins, Linda Kleist, Tillmann Miltzow, and PawełRzążewski. Completeness for the Complexity Class $\forall\exists\mathbb{R}$ and Area-Universality. *Discrete Comput. Geom.*, 70(1):154–188, 2023.

[DL24] Erik D. Demaine and Stefan Langerman. Tiling with three polygons is undecidable. arXiv:2409.11582, 2024.

[DLN+21] Julian D'Costa, Engel Lefaucheux, Eike Neumann, Joël Ouaknine, and James Worrell. On the complexity of the escape problem for linear dynamical systems over compact semialgebraic sets. In *46th International Symposium on Mathematical Foundations of Computer Science*, volume 202 of *LIPIcs. Leibniz Int. Proc. Inform.*, pages Art. No. 33, 21. Schloss Dagstuhl. Leibniz-Zent. Inform., Wadern, 2021.

[DMNW13] Stephane Durocher, Debajyoti Mondal, Rahnuma Islam Nishat, and Sue Whitesides. A note on minimum-segment drawings of planar graphs. *J. Graph Algorithms Appl.*, 17(3):301–328, 2013.

[DMO03] Erik D. Demaine, Joseph S. B. Mitchell, and Joseph O'Rourke. The open problems project. problem 69: Isoceles planar graph drawing, 2003.

[dMRST20] Arnaud de Mesmay, Yo'av Rieck, Eric Sedgwick, and Martin Tancer. Embeddability in $\mathbb{R}^3$ is NP-hard. *J. ACM*, 67(4):Art. 20, 29, 2020.

[DP11] Constantinos Daskalakis and Christos H. Papadimitriou. Continuous local search. In Dana Randall, editor, *Proceedings of the Twenty-Second Annual ACM-SIAM Symposium on Discrete Algorithms, SODA 2011, San Francisco, California, USA, January 23-25, 2011*, pages 790–804. SIAM, 2011.

[DT12] Adrian Dumitrescu and Csaba D. Tóth. Watchman tours for polygons with holes. *Comput. Geom.*, 45(7):326–333, 2012.

[Dub20] Jérémy Dubut. Bisimilarity of diagrams. In *Relational and algebraic methods in computer science*, volume 12062 of *Lecture Notes in Comput. Sci.*, pages 65–81. Springer, Cham, 2020.

[DvdZBL24] Julian Dörfler, Benito van der Zander, Markus Bläser, and Maciej Liskiewicz. Probabilistic and causal satisfiability: the impact of marginalization. arXiv:2405.07373, 2024.

[EBGJ+08] Alejandro Estrella-Balderrama, Elisabeth Gassner, Michael Jünger, Merijam Percan, Marcus Schaefer, and Michael Schulz. Simultaneous geometric graph embeddings. In *Graph drawing*, volume 4875 of *Lecture Notes in Comput. Sci.*, pages 280–290. Springer, Berlin, 2008.

[EHT65] Paul Erdös, Frank Harary, and William T Tutte. On the dimension of a graph. *Mathematika*, 12(2):118–122, 1965.

[Eid02] Stephan Eidenbenz. Inapproximability of finding maximum hidden sets on polygons and terrains. *Comput. Geom.*, 21(3):139–153, 2002.

[EJ85] Paul H. Edelman and Robert E. Jamison. The theory of convex geometries. *Geom. Dedicata*, 19(3):247–270, 1985.

[Eme94] T. Emerson. Relativizations of the $\mathrm{P} =? \mathrm{NP}$ question over the reals (and other ordered rings). *Theoret. Comput. Sci.*, 133(1):15–22, 1994.

[Epp20] David Eppstein. Treetopes and their graphs. *Discrete Comput. Geom.*, 64(2):259–289, 2020.

[EPR23] Kasper Engelen, Guillermo A. Pérez, and Shrisha Rao. Graph-based reductions for parametric and weighted mdps. In Étienne André and Jun Sun, editors, *Automated Technology for Verification and Analysis*, pages 137–157, Cham, 2023. Springer Nature Switzerland. arxiv:2305.05739.

[EPR25] Kasper Engelen, Guillermo A. Pérez, and Shrisha Rao. *Analyzing Value Functions of States in Parametric Markov Chains*, pages 145–165. Springer Nature Switzerland, Cham, 2025.

[Eri19] Jeff Erickson. Optimal curve straightening is $\exists\mathbb{R}$-complete. arXiv:1908.09400, 2019.

[Eri24] Jeff Erickson. personal communication, 2024.

[ERS+19] William Evans, Paweł Rzążewski, Noushin Saeedi, Chan-Su Shin, and Alexander Wolff. Representing graphs and hypergraphs by touching polygons in 3D. In *Graph drawing and network visualization*, volume 11904 of *Lecture Notes in Comput. Sci.*, pages 18–32. Springer, Cham, 2019.

[EvdHM20] Jeff Erickson, Ivor van der Hoog, and Tillmann Miltzow. Smoothing the gap between NP and ER. In *2020 IEEE 61st Annual Symposium on Foundations of Computer Science*, pages 1022–1033. IEEE Computer Soc., Los Alamitos, CA, 2020.

[EW90] Peter Eades and Nicholas C. Wormald. Fixed edge-length graph drawing is NP-hard. *Discrete Appl. Math.*, 28(2):111–134, 1990.

[EW96a] Peter Eades and Sue Whitesides. The logic engine and the realization problem for nearest neighbor graphs. *Theoret. Comput. Sci.*, 169(1):23–37, 1996.

[EW96b] Peter Eades and Sue Whitesides. The realization problem for Euclidean minimum spanning trees is NP-hard. *Algorithmica*, 16(1):60–82, 1996.

[EY09] Kousha Etessami and Mihalis Yannakakis. Recursive Markov chains, stochastic grammars, and monotone systems of nonlinear equations. *J. ACM*, 56(1):Art. 1, 66, 2009.

[EY10] Kousha Etessami and Mihalis Yannakakis. On the complexity of Nash equilibria and other fixed points. *SIAM J. Comput.*, 39(6):2531–2597, 2010.

[Fan20] Yuda Fan. Hidden points and hidden vertices: Hard visibility problems in classes of polygons. Master's thesis, ETH Zürich, Zürich, 2020. https://kurodakanbei.github.io/publications/MT.pdf (last accessed 5/23/2024).

[FHG23] Florian Frohn, Marcel Hark, and Jürgen Giesl. Termination of triangular polynomial loops. *Formal Methods in System Design*, page 70–132, 2023.

[FHH+93] Michael Formann, Torben Hagerup, James Haralambides, Michael Kaufmann, Frank Thomson Leighton, Antonios Symvonis, Emo Welzl, and Gerhard J. Woeginger. Drawing graphs in the plane with high resolution. *SIAM J. Comput.*, 22(5):1035–1052, 1993.

[FHKS16] Stephan Friedrichs, Michael Hemmer, James King, and Christiane Schmidt. The continuous 1.5D terrain guarding problem: discretization, optimal solutions, and PTAS. *J. Comput. Geom.*, 7(1):256–284, 2016.

[FHM90] Ronald Fagin, Joseph Y. Halpern, and Nimrod Megiddo. A logic for reasoning about probabilities. *Inform. and Comput.*, 87(1–2):78–128, 1990.

[FK99] Hervé Fournier and Pascal Koiran. Are lower bounds easier over the reals? In *STOC '98 (Dallas, TX)*, pages 507–513. ACM, New York, 1999.

[FKM+24] Henry Förster, Philipp Kindermann, Tillmann Miltzow, Irene Parada, Soeren Terziadis, and Birgit Vogtenhuber. Geometric thickness of multigraphs is $\exists\mathbb{R}$-complete. In José A. Soto and Andreas Wiese, editors, *LATIN 2024: Theoretical Informatics - 16th Latin American Symposium, Puerto Varas, Chile, March 18-22, 2024, Proceedings, Part I*, volume 14578 of *LNCS*, pages 336–349. Springer, 2024.

[FKO24] Henry Förster, Julia Katheder, and Giacomo Ortali. Outer-(ap)RAC graphs. arXiv:2411.17565, 2024.

[FKP+23] Sándor P. Fekete, Dominik Krupke, Michael Perk, Christian Rieck, and Christian Scheffer. The lawn mowing problem: From algebra to algorithms. In Inge Li Gørtz, Martin Farach-Colton, Simon J. Puglisi, and Grzegorz Herman, editors, *31st Annual European Symposium on Algorithms (ESA 2023)*, volume 274 of *Leibniz International Proceedings in Informatics (LIPIcs)*, pages 45:1–45:18, Dagstuhl, Germany, 2023. Schloss Dagstuhl – Leibniz-Zentrum für Informatik.

[FL78] Jon Folkman and Jim Lawrence. Oriented matroids. *J. Combin. Theory Ser. B*, 25(2):199–236, 1978.

[Fra10] Fabrizio Frati. A note on isosceles planar graph drawing. *Inform. Process. Lett.*, 110(12–13):507–509, 2010.

[FS20] Stefan Felsner and Manfred Scheucher. Arrangements of pseudocircles: on circularizability. *Discrete Comput. Geom.*, 64(3):776–813, 2020.

[FSTZ98] Charles M. Fiduccia, Edward R. Scheinerman, Ann Trenk, and Jennifer S. Zito. Dot product representations of graphs. *Discrete Math.*, 181(1–3):113–138, 1998.

[FT22] Radoslav Fulek and Csaba D. Tóth. Atomic embeddability, clustered planarity, and thickenability. *J. ACM*, 69(2):Art. 13, 34, 2022.

[Gaß09] Christine Gaßner. Oracles and relativizations of the P =? NP question for several structures. *J.UCS*, 15(6):1186–1205, 2009.

[GBC16] Ian Goodfellow, Yoshua Bengio, and Aaron Courville. *Deep learning*. Adaptive Computation and Machine Learning. MIT Press, Cambridge, MA, 2016.

[GGJ77] Michael R. Garey, Ron L. Graham, and David S. Johnson. The complexity of computing Steiner minimal trees. *SIAM J. Appl. Math.*, 32(4):835–859, 1977.

[GH22] Alexander Golovnev and Ishay Haviv. The (generalized) orthogonality dimension of (generalized) Kneser graphs: bounds and applications. *Theory Comput.*, 18:Paper No. 22, 22, 2022.

[Gho07] Subir Kumar Ghosh. *Visibility algorithms in the plane*. Cambridge University Press, Cambridge, 2007.

[GKK+10] Panos Giannopoulos, Rolf Klein, Christian Knauer, Martin Kutz, and Dániel Marx. Computing geometric minimum-dilation graphs is NP-hard. *Internat. J. Comput. Geom. Appl.*, 20(2):147–173, 2010.

[GKP13] Bruno Grenet, Pascal Koiran, and Natacha Portier. On the complexity of the multivariate resultant. *J. Complexity*, 29(2):142–157, 2013.

[GMVY17] Jugal Garg, Ruta Mehta, Vijay V. Vazirani, and Sadra Yazdanbod. Settling the complexity of Leontief and PLC exchange markets under exact and approximate equilibria. In *STOC'17—Proceedings of the 49th Annual ACM SIGACT Symposium on Theory of Computing*, pages 890–901. ACM, New York, 2017.

[GMVY18] Jugal Garg, Ruta Mehta, Vijay V. Vazirani, and Sadra Yazdanbod. $\exists\mathbb{R}$-completeness for decision versions of multi-player (symmetric) Nash equilibria. *ACM Trans. Econ. Comput.*, 6(1): Art. 1, 23, 2018.

[God99] Michael Godau. *On the complexity of measuring the similarity between geometric objects in higher dimensions*. Dissertation, Freie Universität Berlin, 1999.

[GOdMV18] John Gimbel, Patrice Ossona de Mendez, and Pavel Valtr. Obstacle numbers of planar graphs. In *Graph drawing and network visualization*, volume 10692 of *Lecture Notes in Comput. Sci.*, pages 67–80. Springer, Cham, 2018.

[GP80] Jacob E. Goodman and Richard Pollack. On the combinatorial classification of nondegenerate configurations in the plane. *J. Combin. Theory Ser. A*, 29(2):220–235, 1980.

[GP83] Jacob E. Goodman and Richard Pollack. Multidimensional sorting. *SIAM J. Comput.*, 12(3):484–507, 1983.

[GPS89] Jacob E. Goodman, Richard Pollack, and Bernd Sturmfels. Coordinate representation of order types requires exponential storage. In *STOC '89: Proceedings of the twenty-first annual ACM symposium on Theory of computing*, pages 405–410, New York, NY, USA, 1989. ACM.

[GPS90] Jacob E. Goodman, Richard Pollack, and Bernd Sturmfels. The intrinsic spread of a configuration in $\mathbf{R}^d$. *J. Amer. Math. Soc.*, 3(3):639–651, 1990.

[GPS20] Hugo Gimbert, Soumyajit Paul, and B. Srivathsan. A bridge between polynomial optimization and games with imperfect recall. In Amal El Fallah Seghrouchni, Gita Sukthankar, Bo An, and Neil Yorke-Smith, editors, *Proceedings of the 19th International Conference on Autonomous Agents and Multiagent Systems, AAMAS '20, Auckland, New Zealand, May 9–13, 2020*, pages 456–464. International Foundation for Autonomous Agents and Multiagent Systems, 2020.

[Gro24] Lennart Großkreutz. Hyperbolic sphericity. Bachelor's thesis, Karlsruher Institut für Technologie, Karlsruhe, 2024. https://scale.iti.kit.edu/_media/resources/theses/ba_lennart_grosskreutz.pdf (last accessed 12/1/2024).

[Grü72] Branko Grünbaum. *Arrangements and spreads*, volume No. 10 of *Conference Board of the Mathematical Sciences Regional Conference Series in Mathematics*. American Mathematical Society, Providence, RI, 1972.

[Grü03] Branko Grünbaum. *Convex polytopes*, volume 221 of *Graduate Texts in Mathematics*. Springer-Verlag, New York, second edition, 2003. Prepared and with a preface by Volker Kaibel, Victor Klee and Günter M. Ziegler.

[GSH89] Csaba P. Gabor, Kenneth J. Supowit, and Wen Lian Hsu. Recognizing circle graphs in polynomial time. *J. Assoc. Comput. Mach.*, 36(3):435–473, 1989.

[GV88] D. Yu. Grigor'ev and Nicolai N. Vorobjov, Jr. Solving systems of polynomial inequalities in subexponential time. *J. Symbolic Comput.*, 5(1-2):37–64, 1988.

[GV18] Nicolas Gillis and Stephen A. Vavasis. On the complexity of robust PCA and ℓ_1-norm low-rank matrix approximation. *Math. Oper. Res.*, 43(4):1072–1084, 2018.

[Han19] Kristoffer Arnsfelt Hansen. The real computational complexity of minmax value and equilibrium refinements in multi-player games. *Theory Comput. Syst.*, 63(7):1554–1571, 2019.

[Hås90] Johan Håstad. Tensor rank is NP-complete. *J. Algorithms*, 11(4):644–654, 1990.

[HHKV23] Teemu Hankala, Miika Hannula, Juha Kontinen, and Jonni Virtema. Complexity of neural network training and ETR: Extensions with effectively continuous functions. arXiv:2305.11833, 2023.

[HK01] Petr Hliněný and Jan Kratochvíl. Representing graphs by disks and balls (a survey of recognition-complexity results). *Discrete Math.*, 229(1–3):101–124, 2001. Combinatorics, graph theory, algorithms and applications.

[HK20] Alfredo Hubard and Roman Karasev. Bisecting measures with hyperplane arrangements. *Math. Proc. Cambridge Philos. Soc.*, 169(3):639–647, 2020.

[HKP11] Boris Horvat, Jan Kratochvíl, and Tomaž Pisanski. On the computational complexity of degenerate unit distance representations of graphs. In *Combinatorial algorithms*, volume 6460 of *Lecture Notes in Comput. Sci.*, pages 274–285. Springer, Heidelberg, 2011.

[HKR+13] Michael Hoffmann, Vincent Kusters, Günter Rote, Maria Saumell, and Rodrigo I. Silveira. Convex hull alignment through translation. In *Proceedings of the 25th Canadian Conference on Computational Geometry, CCCG 2013, Waterloo, Ontario, Canada, August 8–10, 2013.* Carleton University, Ottawa, Canada, 2013.

[HL13] Christopher J. Hillar and Lek-Heng Lim. Most tensor problems are NP-hard. *J. ACM*, 60(6):Art. 45, 39, 2013.

[HM18] Udo Hoffmann and Keno Merckx. A universality theorem for allowable sequences with applications. arXiv:1801.05992, 2018.

[HM21] Simon B. Hengeveld and Tillmann Miltzow. A Practical Algorithm with Performance Guarantees for the Art Gallery Problem. In Kevin Buchin and Éric Colin de Verdière, editors, *37th International Symposium on Computational Geometry (SoCG 2021)*, volume 189 of *Leibniz International Proceedings in Informatics (LIPIcs)*, pages 44:1–44:16, 2021.

[HMMP25] Thekla Hamm, Lucas Meijer, Tillmann Miltzow, and Subhasree Patro. Oracle separations for $\mathbb{R}$PH. arXiv:2502.09279, 2025.

[HMWW24] Michael Hoffmann, Tillmann Miltzow, Simon Weber, and Lasse Wulf. Recognition of Unit Segment and Polyline Graphs is $\exists\mathbb{R}$-complete. arXiv:2401.02172, 2024.

[Hof17] Udo Hoffmann. On the complexity of the planar slope number problem. *J. Graph Algorithms Appl.*, 21(2):183–193, 2017.

[Høg23] Kasper Høgh. *Complexity of Total Search Problems: Equilibria and Fair Division.* PhD thesis, Aarhus University, Denmark, 2023.

[Hon91] Hoon Hong. Comparison of several decision algorithms for the existential theory of the reals. Technical report, Johannes Kepler Univ., 1991.

[HS20] Kristoffer Arnsfelt Hansen and Steffan Christ Sølvsten. $\exists\mathbb{R}$-completeness of stationary Nash equilibria in perfect information stochastic games. In *45th International Symposium on Mathematical Foundations of Computer Science*, volume 170 of *LIPIcs. Leibniz Int. Proc. Inform.*, pages Art. No. 45, 15. Schloss Dagstuhl. Leibniz-Zent. Inform., Wadern, 2020.

[HSZ13] Christian Herrmann, Johanna Sokoli, and Martin Ziegler. Satisfiability of cross product terms is complete for real nondeterministic polytime Blum-Shub-Smale machines. In *Proceedings: Machines, Computations and Universality 2013*, volume 128 of *Electron. Proc. Theor. Comput. Sci. (EPTCS)*, pages 85–92. EPTCS, 2013.

[HVLnS17] César Hernández-Vélez, Jesús Leaños, and Gelasio Salazar. On the pseudolinear crossing number. *J. Graph Theory*, 84(3):297–310, 2017.

[HZ16] Christian Herrmann and Martin Ziegler. Computational complexity of quantum satisfiability. *J. ACM*, 63(2):Art. 19, 31, 2016.

[IIM24] Duligur Ibeling, Thomas Icard, and Mossé. Milan. On probabilistic and causal reasoning with summation operators. arXiv:2405.03069, 2024.

[IIMM23] Duligur Ibeling, Thomas Icard, Krzysztof Mierzewski, and Milan Mossé. Probing the quantitative–qualitative divide in probabilistic reasoning. *Annals of Pure and Applied Logic*, page 103339, 2023.

[JKM22] Paul Jungeblut, Linda Kleist, and Tillmann Miltzow. The complexity of the Hausdorff distance. In *38th International Symposium on Computational Geometry*, volume 224 of *LIPIcs. Leibniz Int. Proc. Inform.*, pages Art. No. 48, 17. Schloss Dagstuhl. Leibniz-Zent. Inform., Wadern, 2022.

[JKM23] Paul Jungeblut, Linda Kleist, and Tillmann Miltzow. The complexity of the Hausdorff distance. *Discrete Comput. Geom.*, page 177–213, 2023.

[JKPW21] Sebastian Junges, Joost-Pieter Katoen, Guillermo A. Pérez, and Tobias Winkler. The complexity of reachability in parametric Markov decision processes. *J. Comput. System Sci.*, 119:183–210, 2021.

[JP10] Gabriela Jeronimo and Daniel Perrucci. On the minimum of a positive polynomial over the standard simplex. *J. Symbolic Comput.*, 45(4):434–442, 2010.

[JT14] Peter Jonsson and Johan Thapper. Affine consistency and the complexity of semilinear constraints. In *Mathematical foundations of computer science 2014. Part II*, volume 8635 of *Lecture Notes in Comput. Sci.*, pages 420–431. Springer, Heidelberg, 2014.

[JT18] Peter Jonsson and Johan Thapper. Tractability conditions for numeric CSPs. *Theoret. Comput. Sci.*, 715:21–34, 2018.

[Jun23a] Paul Jungeblut. On the complexity of Lombardi graph drawing. In *Graph drawing and network visualization. Part I*, volume 14465 of *Lecture Notes in Comput. Sci.*, pages 180–194. Springer, Cham, 2023.

[Jun23b] Tim Janik Junginger. Robustness of the discrete real polynomial hierarchy. Bachelor's thesis, Karlsruher Institut für Technologie, Karlsruhe, 2023.

[Kai73] Paul C. Kainen. Thickness and coarseness of graphs. *Abh. Math. Sem. Univ. Hamburg*, 39:88–95, 1973.

[Kay12] Neeraj Kayal. Affine projections of polynomials [extended abstract]. In *STOC'12—Proceedings of the 2012 ACM Symposium on Theory of Computing*, pages 643–661. ACM, New York, 2012.

[KBŠ23] Damjana Kokol Bukovšek and Helena Šmigoc. Symmetric nonnegative matrix trifactorization. *Linear Algebra Appl.*, 665:36–60, 2023.

[KdMM23] Eunjung Kim, Arnaud de Mesmay, and Tillmann Miltzow. Representing matroids over the reals is $\exists\mathbb{R}$-complete. arXiv:2301.03221, 2023.

[Kem77] Alfred Bray Kempe. *How to Draw a Straight Line*. Classics in Mathematics Education. MacMillan, London, 1977.

[KK11] James King and Erik Krohn. Terrain guarding is NP-hard. *SIAM J. Comput.*, 40(5):1316–1339, 2011.

[KLPS17] Irina Kostitsyna, Maarten Löffler, Valentin Polishchuk, and Frank Staals. On the complexity of minimum-link path problems. *J. Comput. Geom.*, 8(2):80–108, 2017.

[KLR23] Alexander B. Kunin, Caitlin Lienkaemper, and Zvi Rosen. Oriented matroids and combinatorial neural codes. *Comb. Theory*, 3(1):Paper No. 14, 39, 2023.

[KM94] Jan Kratochvíl and Jiří Matoušek. Intersection graphs of segments. *J. Combin. Theory Ser. B*, 62(2):289–315, 1994.

[KM02] Michael Kapovich and John J. Millson. Universality theorems for configuration spaces of planar linkages. *Topology*, 41(6):1051–1107, 2002.

[KM12] Ross J. Kang and Tobias Müller. Sphere and dot product representations of graphs. *Discrete Comput. Geom.*, 47(3):548–568, 2012.

[KM14] Ross J. Kang and Tobias Müller. Arrangements of pseudocircles and circles. *Discrete Comput. Geom.*, 51(4):896–925, 2014.

[KM18] Katharina Klost and Wolfgang Mulzer. Recognizing generalized transmission graphs of line segments and circular sectors. In *LATIN 2018: Theoretical informatics*, volume 10807 of *Lecture Notes in Comput. Sci.*, pages 683–696. Springer, Cham, 2018.

[KNP15] Boris Klemz, Martin Nöllenburg, and Roman Prutkin. Recognizing weighted disk contact graphs. In *Graph drawing and network visualization*, volume 9411 of *Lecture Notes in Comput. Sci.*, pages 433–446. Springer, Cham, 2015.

[Knu92] Donald E. Knuth. *Axioms and hulls*, volume 606 of *Lecture Notes in Computer Science*. Springer-Verlag, Berlin, 1992.

[Koi92] Pascal Koiran. On the relations between dynamical systems and boolean circuits. Research Report LIP RR-1993-01, Laboratoire de l'informatique du parallélisme, December 1992.

[Koi94] Pascal Koiran. Computing over the reals with addition and order. *Theoret. Comput. Sci.*, 133(1):35–47, 1994.

[Koi96] Pascal Koiran. Hilbert's Nullstellensatz is in the polynomial hierarchy. *J. Complexity*, 12(4):273–286, 1996. Special issue for the Foundations of Computational Mathematics Conference (Rio de Janeiro, 1997).

[Koi97a] Pascal Koiran. Elimination of constants from machines over algebraically closed fields. *J. Complexity*, 13(1):65–82, 1997.

[Koi97b] Pascal Koiran. Randomized and deterministic algorithms for the dimension of algebraic varieties. In *38th Annual Symposium on Foundations of Computer Science, FOCS '97, Miami Beach, Florida, USA, October 19-22, 1997*, pages 36–45. IEEE Computer Society, 1997.

[Koi97c] Pascal Koiran. A weak version of the Blum, Shub, and Smale model. *J. Comput. System Sci.*, 54(1):177–189, 1997.

[Koi99] Pascal Koiran. The real dimension problem is $NP_{\mathbf{R}}$-complete. *J. Complexity*, 15(2):227–238, 1999.

[Koi00] Pascal Koiran. The complexity of local dimensions for constructible sets. In ., volume 16, pages 311–323. ., 2000. Real computation and complexity (Schloss Dagstuhl, 1998).

[KP97] Leonid Khachiyan and Lorant Porkolab. Computing integral points in convex semialgebraic sets. In *38th Annual Symposium on Foundations of Computer Science, FOCS '97, Miami Beach, Florida, USA, October 19-22, 1997*, pages 162–171. IEEE Computer Society, 1997.

[KP03] Volker Kaibel and Marc E. Pfetsch. Some algorithmic problems in polytope theory. In *Algebra, geometry, and software systems*, pages 23–47. Springer, Berlin, 2003.

[KP07] Jan Kratochvíl and Martin Pergel. Geometric intersection graphs: do short cycles help? (extended abstract). In *Computing and combinatorics*, volume 4598 of *Lecture Notes in Comput. Sci.*, pages 118–128. Springer, Berlin, 2007.

[KPW05] Jan Kára, Attila Pór, and David R. Wood. On the chromatic number of the visibility graph of a set of points in the plane. *Discrete Comput. Geom.*, 34(3):497–506, 2005.

[KPW23] Tirasan Khandhawit, Puttipong Pongtanapaisan, and Athibadee Wasun. Planar stick indices of some knotted graphs. arXiv:2312.06603, 2023.

[KR23] Benedikt Künzel and Jonathan Rollin. On the complexity of simultaneous geometric embedding for edge-disjoint graphs. arXiv:2312.09025, 2023.

[Kra91a] Jan Kratochvíl. String graphs. I. The number of critical nonstring graphs is infinite. *J. Combin. Theory Ser. B*, 52(1):53–66, 1991.

[Kra91b] Jan Kratochvíl. String graphs. II. Recognizing string graphs is NP-hard. *J. Combin. Theory Ser. B*, 52(1):67–78, 1991.

[Kre20] Gereon Kremer. *Cylindrical algebraic decomposition for nonlinear arithmetic problems*. PhD thesis, RWTH Aachen University, Germany, 2020.

[KRPS16] Roman N. Karasev, Edgardo Roldán-Pensado, and Pablo Soberón. Measure partitions using hyperplanes with fixed directions. *Israel J. Math.*, 212(2):705–728, 2016.

[KRW19] Myroslav Kryven, Alexander Ravsky, and Alexander Wolff. Drawing graphs on few circles and few spheres. *J. Graph Algorithms Appl.*, 23(2):371–391, 2019.

[KT19] Vojtěch Kaluža and Martin Tancer. Even maps, the Colin de Verdière number and representations of graphs. arXiv:1907.05055, 2019.

[KT20] Stefan Kiefer and Qiyi Tang. Comparing labelled Markov decision processes. In *40th IARCS Annual Conference on Foundations of Software Technology and Theoretical Computer Science*, volume 182 of *LIPIcs. Leibniz Int. Proc. Inform.*, pages Art. No. 49, 16. Schloss Dagstuhl. Leibniz-Zent. Inform., Wadern, 2020.

[KT24] Stefan Kiefer and Qiyi Tang. Minimising the probabilistic bisimilarity distance. In Rupak Majumdar and Alexandra Silva, editors, *35th International Conference on Concurrency Theory, CONCUR 2024, September 9-13, 2024, Calgary, Canada*, volume 311 of *LIPIcs*, pages 32:1–32:18. Schloss Dagstuhl - Leibniz-Zentrum für Informatik, 2024.

[KW82] David M. Kreps and Robert Wilson. Sequential equilibria. *Econometrica*, 50(4):863–894, 1982.

[KW05] Matthew Kitching and Sue Whitesides. The three dimensional logic engine. In *Graph drawing. 12th international symposium, GD 2004, New York, NY, September 29–October 2, 2004. Revised selected papers.*, pages 329–339. Berlin: Springer, 2005.

[KW14] Stefan Kiefer and Björn Wachter. Stability and complexity of minimising probabilistic automata. In *Automata, languages, and programming. Part II*, volume 8573 of *Lecture Notes in Comput. Sci.*, pages 268–279. Springer, Heidelberg, 2014.

[Kyn11] Jan Kynčl. Simple realizability of complete abstract topological graphs in P. *Discrete Comput. Geom.*, 45(3):383–399, 2011.

[Kyn20] Jan Kynčl. Simple realizability of complete abstract topological graphs simplified. *Discrete Comput. Geom.*, 64(1):1–27, 2020.

[KZ23] Jonathan Klawitter and Johannes Zink. Upward planar drawings with three and more slopes. *J. Graph Algorithms Appl.*, 27(2):119–145, 2023.

[Lad75] Richard E. Ladner. On the structure of polynomial time reducibility. *J. Assoc. Comput. Mach.*, 22:155–171, 1975.

[Lau09] Monique Laurent. Matrix completion problems. In Christodoulos A. Floudas and Panos M. Pardalos, editors, *Encyclopedia of Optimization*, pages 1967–1975. Springer US, Boston, MA, 2009.

[Law23] Rachel Lawrence. *Simplicial Reaction Networks and Dynamics on Graphs*. PhD thesis, EECS Department, University of California, Berkeley, Dec 2023.

[Lee13] Jae Hee Lee. *Qualitative reasoning about relative directions: computational complexity and practical algorithm*. PhD thesis, University of Bremen, 2013.

[Lee14] Jae Hee Lee. The complexity of reasoning with relative directions. In Torsten Schaub, Gerhard Friedrich, and Barry O'Sullivan, editors, *ECAI 2014 - 21st European Conference on Artificial Intelligence, 18–22 August 2014, Prague, Czech Republic - Including Prestigious Applications of Intelligent Systems (PAIS 2014)*, volume 263 of *Frontiers in Artificial Intelligence and Applications*, pages 507–512. IOS Press, 2014.

[Lie23] Caitlin Lienkaemper. Using oriented matroids to find low rank structure in presence of nonlinearity. arXiv:2312.17365, 2023.

[LLMM14] Leo Liberti, Carlile Lavor, Nelson Maculan, and Antonio Mucherino. Euclidean distance geometry and applications. *SIAM Rev.*, 56(1):3–69, 2014.

[LM20] Anna Lubiw and Debajyoti Mondal. On compatible triangulations with a minimum number of Steiner points. *Theoret. Comput. Sci.*, 835:97–107, 2020.

[LMM22] Anna Lubiw, Tillmann Miltzow, and Debajyoti Mondal. The complexity of drawing a graph in a polygonal region. *J. Graph Algorithms Appl.*, 26(4):421–446, 2022.

[LMT07] Ruggero Lanotte, Andrea Maggiolo-Schettini, and Angelo Troina. Parametric probabilistic transition systems for system design and analysis. *Formal Aspects Comput.*, 19(1):93–109, 2007.

[LO93] Peter J. Looges and Stephan Olariu. Optimal greedy algorithms for indifference graphs. *Comput. Math. Appl.*, 25(7):15–25, 1993.

[Lov79] László Lovász. On the Shannon capacity of a graph. *IEEE Trans. Inform. Theory*, 25(1):1–7, 1979.

[LS95] Yaw Ling Lin and Steven S. Skiena. Complexity aspects of visibility graphs. *Internat. J. Comput. Geom. Appl.*, 5(3):289–312, 1995.

[LSS03] Paul Lemke, Steven S. Skiena, and Warren D. Smith. Reconstructing sets from interpoint distances. In *Discrete and computational geometry*, volume 25 of *Algorithms Combin.*, pages 507–631. Springer, Berlin, 2003.

[Mac36] Saunders MacLane. Some Interpretations of Abstract Linear Dependence in Terms of Projective Geometry. *Amer. J. Math.*, 58(1):236–240, 1936.

[Mae91] Hiroshi Maehara. Distances in a rigid unit-distance graph in the plane. In ., volume 31, pages 193–200. Elsevier, 1991. First Canadian Conference on Computational Geometry (Montreal, PQ, 1989).

[Mak24] Michał Makowski. On the complexity of the conditional independence implication problem with bounded cardinalities. arXiv:2408.02550, 2024.

[Mat06] Yu. V. Matiyasevich. Hilbert's tenth problem: Diophantine equations in the twentieth century. In *Mathematical events of the twentieth century*, pages 185–213. Springer, Berlin, 2006.

[Mat14] Jiří Matoušek. Intersection graphs of segments and $\exists\mathbb{R}$. In J. Pach and G. Tóth, editors, *Courses in Discrete and Computational Geometry*, Bolyai Soc. Math. Studies, volume 31, Springer, 2026.

[Mém08] Facundo Mémoli. Gromov-hausdorff distances in euclidean spaces. In *IEEE Conference on Computer Vision and Pattern Recognition, CVPR Workshops 2008, Anchorage, AK, USA, 23-28 June, 2008*, pages 1–8. IEEE Computer Society, 2008.

[MII21] Milan Mossé, Duligur Ibeling, and Thomas Icard. Is causal reasoning harder than probabilistic reasoning? arXiv:2111.13936, 2021.

[Min24] Ilia Minkin. Defining the discrete real polynomial hierarchy with oracle machines. Bachelor's thesis, Karlsruher Institut für Technologie, Karlsruhe, 2024.

[MM97] Klaus Meer and Christian Michaux. A survey on real structural complexity theory. *Bull. Belg. Math. Soc. Simon Stevin*, 4(1):113–148, 1997.

[MM99] Terry A. McKee and F. R. McMorris. *Topics in intersection graph theory*. SIAM Monographs on Discrete Mathematics and Applications. Society for Industrial and Applied Mathematics (SIAM), Philadelphia, PA, 1999.

[MM10] Colin McDiarmid and Tobias Müller. The number of bits needed to represent a unit disk graph. In *Graph-theoretic concepts in computer science*, volume 6410 of *Lecture Notes in Comput. Sci.*, pages 315–323. Springer, Berlin, 2010.

[MM13] Colin McDiarmid and Tobias Müller. Integer realizations of disk and segment graphs. *J. Combin. Theory Ser. B*, 103(1):114–143, 2013.

[MM22] Lucas Meijer and Tillmann Miltzow. Sometimes two irrational guards are needed. arXiv:2212.01211, 2022.

[Mnë88] N. E. Mnëv. The universality theorems on the classification problem of configuration varieties and convex polytopes varieties. In *Topology and geometry—Rohlin Seminar*, volume 1346 of *Lecture Notes in Math.*, pages 527–543. Springer, Berlin, 1988.

[MR08] Wolfgang Mulzer and Günter Rote. Minimum-weight triangulation is NP-hard. *J. ACM*, 55(2):Art. 11, 29, 2008.

[MS92] Clyde Monma and Subhash Suri. Transitions in geometric minimum spanning trees. *Discrete Comput. Geom.*, 8(3):265–293, 1992. ACM Symposium on Computational Geometry (North Conway, NH, 1991).

[MS24] Tillmann Miltzow and Reinier F. Schmiermann. On Classifying Continuous Constraint Satisfaction Problems. *TheoretiCS*, Volume 3:1–54, 2024.

[MTW11] Jiří Matoušek, Martin Tancer, and Uli Wagner. Hardness of embedding simplicial complexes in $\mathbb{R}^d$. *J. Eur. Math. Soc. (JEMS)*, 13(2):259–295, 2011.

[Mut24] Oğuz Mutlu. On the computational complexity of doors. Bachelor's thesis, Karlsruher Institut für Technologie, Karlsruhe, 2024. https://scale.iti.kit.edu/_media/resources/theses/ba_oguz_mutlu.pdf (last accessed 11/30/2024).

[MvLvL13] Tobias Müller, Erik Jan van Leeuwen, and Jan van Leeuwen. Integer representations of convex polygon intersection graphs. *SIAM J. Discrete Math.*, 27(1):205–231, 2013.

[MW25] Klaus Meer and Adrian Wurm. Some structural complexity results for $\exists\mathbb{R}$. arXiv:2502.00680, 2025.

[Mye78] Roger B. Myerson. Refinements of the Nash equilibrium concept. *Internat. J. Game Theory*, 7(2):73–80, 1978.

[MZ09] Klaus Meer and Martin Ziegler. Real computational universality: the word problem for a class of groups with infinite presentation. *Found. Comput. Math.*, 9(5):599–609, 2009.

[NX18] Amir Nayyeri and Hanzhong Xu. On the decidability of the Fréchet distance between surfaces. In *Proceedings of the Twenty-Ninth Annual ACM-SIAM Symposium on Discrete Algorithms*, pages 1109–1120. SIAM, Philadelphia, PA, 2018.

[O'R82] Joseph O'Rourke. The complexity of computing minimum convex covers for polygons. In *20th Annual Allerton Conference on Communication, Control and Computing'*, pages 75–84. University of Illinois, 1982.

[O'R87] Joseph O'Rourke. *Art gallery theorems and algorithms*. International Series of Monographs on Computer Science. The Clarendon Press, Oxford University Press, New York, 1987.

[O'R18] Joseph O'Rourke. Visibility. In Jacob E. Goodman and Joseph O'Rourke, editors, *Handbook of Discrete and Computational Geometry, Third Edition*, pages 875–896. Chapman and Hall/CRC, 2018.

[ORW19] Yoshio Okamoto, Alexander Ravsky, and Alexander Wolff. Variants of the segment number of a graph. In *Graph drawing and network visualization*, volume 11904 of *Lecture Notes in Comput. Sci.*, pages 430–443. Springer, Cham, 2019.

[OS97] Joseph O'Rourke and Ileana Streinu. Vertex-edge pseudo-visibility graphs: characterization and recognition. In *SCG '97: Proceedings of the thirteenth annual symposium on Computational geometry*, pages 119–128, New York, NY, USA, 1997. ACM.

[OS98] Joseph O'Rourke and Ileana Streinu. The vertex-edge visibility graph of a polygon. *Comput. Geom.*, 10(2):105–120, 1998.

[OW14] Joël Ouaknine and James Worrell. Ultimate positivity is decidable for simple linear recurrence sequences. In *Automata, languages, and programming. Part II*, volume 8573 of *Lecture Notes in Comput. Sci.*, pages 330–341. Springer, Heidelberg, 2014.

[OW17] Joël Ouaknine and James Worrell. Ultimate positivity is decidable for simple linear recurrence sequences. arXiv:1309.1914, 2017.

[Pap77] Christos H. Papadimitriou. The Euclidean traveling salesman problem is NP-complete. *Theoret. Comput. Sci.*, 4(3):237–244, 1977.

[Pat06] Maurizio Patrignani. On extending a partial straight-line drawing. *Internat. J. Found. Comput. Sci.*, 17(5):1061–1069, 2006.

[Pet16] Dominik Peters. Recognising multidimensional euclidean preferences. arXiv:1602.08109, 2016.

[Pet17] Dominik Peters. Recognising multidimensional euclidean preferences. In Satinder Singh and Shaul Markovitch, editors, *Proceedings of the Thirty-First AAAI Conference on Artificial Intelligence, February 4-9, 2017, San Francisco, California, USA*, pages 642–648. AAAI Press, 2017.

[Pfe08] Florian Pfender. Visibility graphs of point sets in the plane. *Discrete Comput. Geom.*, 39(1-3):455–459, 2008.

[Pis] Tomaž Pisanski. Drawing 3-configurations of points and lines with straight lines. MathOverflow.

[PJ09] Grant Olney Passmore and Paul B. Jackson. Combined decision techniques for the existential theory of the reals. In *Intelligent computer mathematics. 16th symposium, Calculemus 2009, 8th international conference, MKM 2009, held as part of CICM 2009, Grand Bend, Canada, July 6–12, 2009. Proceedings*, pages 122–137. Berlin: Springer, 2009.

[Poo09] Bjorn Poonen. Characterizing integers among rational numbers with a universal-existential formula. *Amer. J. Math.*, 131(3):675–682, 2009.

[PS02] Hans Jürgen Prömel and Angelika Steger. *The Steiner tree problem*. Advanced Lectures in Mathematics. Friedr. Vieweg & Sohn, Braunschweig, 2002. A tour through graphs, algorithms, and complexity.

[PS07] Daniel Perrucci and Juan Sabia. Real roots of univariate polynomials and straight line programs. *J. Discrete Algorithms*, 5(3):471–478, 2007.

[PT14] Arnau Padrol and Louis Theran. Delaunay triangulations with disconnected realization spaces. In *Computational geometry (SoCG'14)*, pages 163–170. ACM, New York, 2014.

[Qua24] Valentin Andreas Quapil. Upward and upward-planar drawings with limited slopes. Bachelor's thesis, Karlsruher Institut für Technologie, Karlsruhe, 2024.

[Reda] Redlog. Applications. https://redlog.eu/references/.

[Redb] Redlog. Computing with logic. https://www.redlog.eu/.

[Ren88] James Renegar. A faster PSPACE algorithm for deciding the existential theory of the reals. In *29th Annual Symposium on Foundations of Computer Science, White Plains, New York, USA, 24-26 October 1988*, pages 291–295. IEEE Computer Society, 1988.

[Ren92a] James Renegar. On the computational complexity and geometry of the first-order theory of the reals. I. Introduction. Preliminaries. The geometry of semi-algebraic sets. The decision problem for the existential theory of the reals. *J. Symbolic Comput.*, 13(3):255–299, 1992.

[Ren92b] James Renegar. On the computational complexity of approximating solutions for real algebraic formulae. *SIAM J. Comput.*, 21(6):1008–1025, 1992.

[RG96] Jürgen Richter-Gebert. *Realization spaces of polytopes*, volume 1643 of *Lecture Notes in Mathematics*. Springer-Verlag, Berlin, 1996.

[RG99] Jürgen Richter-Gebert. The universality theorems for oriented matroids and polytopes. In *Advances in discrete and computational geometry (South Hadley, MA, 1996)*, volume 223 of *Contemp. Math.*, pages 269–292. Amer. Math. Soc., Providence, RI, 1999.

[RGZ95] Jürgen Richter-Gebert and Günter M. Ziegler. Realization spaces of 4-polytopes are universal. *Bull. Amer. Math. Soc. (N.S.)*, 32(4):403–412, 1995.

[Ric11] David A. Richter. How to draw a Tait-colorable graph. In *Graph drawing*, volume 6502 of *Lecture Notes in Comput. Sci.*, pages 353–364. Springer, Heidelberg, 2011.

[Rin55] Gerhard Ringel. Teilungen der Ebene durch Geraden oder topologische Geraden. *Math. Z.*, 64:79–102 (1956), 1955.

[Riv96] Igor Rivin. A characterization of ideal polyhedra in hyperbolic 3-space. *Ann. of Math. (2)*, 143(1):51–70, 1996.

[Roh22] Dennis Rohde. *k-Median Clustering of Spatial Data Sequences*. PhD thesis, Ruhr-Universität Bochum, 2022.

[RW19] Alexandre Rok and Bartosz Walczak. Outerstring graphs are χ-bounded. *SIAM J. Discrete Math.*, 33(4):2181–2199, 2019.

[Sau21] Lisa Sauermann. On the speed of algebraically defined graph classes. *Adv. Math.*, 380:Paper No. 107593, 55, 2021.

[Sax79] James B. Saxe. Embeddability of weighted graphs in k-space is strongly NP-hard. In *Seventeenth Annual Allerton Conference on Communication, Control, and Computing, Proceedings of the Conference held in Monticello, Ill., October 10–12, 1979*, pages xiv+1036, Urbana, 1979. University of Illinois Department of Electrical Engineering. Proceedings of the International School of Physics "Enrico Fermi", LXX*.

[SB18] Richard S. Sutton and Andrew G. Barto. *Reinforcement learning. An introduction.* Adapt. Comput. Mach. Learn. Cambridge, MA: MIT Press, 2nd edition edition, 2018.

[Sch99] Marcus Schaefer. Deciding the Vapnik-Červonenkis dimension is Σ_3^p-complete. *J. Comput. System Sci.*, 58(1):177–182, 1999.

[Sch10] Marcus Schaefer. Complexity of some geometric and topological problems. In *Graph drawing*, volume 5849 of *Lecture Notes in Comput. Sci.*, pages 334–344. Springer, Berlin, 2010.

[Sch13] Marcus Schaefer. Realizability of graphs and linkages. In *Thirty essays on geometric graph theory*, pages 461–482. Springer, New York, 2013.

[Sch14] Marcus Schaefer. Picking planar edges; or, drawing a graph with a planar subgraph. In *Graph drawing*, volume 8871 of *Lecture Notes in Comput. Sci.*, pages 13–24. Springer, Heidelberg, 2014.

[Sch15] André Schulz. Drawing graphs with few arcs. *J. Graph Algorithms Appl.*, 19(1):393–412, 2015.

[Sch21a] Marcus Schaefer. Complexity of geometric k-planarity for fixed k. *J. Graph Algorithms Appl.*, 25(1):29–41, 2021.

[Sch21b] Marcus Schaefer. On the complexity of some geometric problems with fixed parameters. *J. Graph Algorithms Appl.*, 25(1):195–218, 2021.

[Sch21c] Patrick Schnider. The complexity of sharing a pizza. In *32nd International Symposium on Algorithms and Computation*, volume 212 of *LIPIcs. Leibniz Int. Proc. Inform.*, pages Art. No. 13, 15. Schloss Dagstuhl. Leibniz-Zent. Inform., Wadern, 2021.

[Sch23a] Marcus Schaefer. The complexity of angular resolution. *J. Graph Algorithms Appl.*, 27(7):565–580, 2023.

[Sch23b] Marcus Schaefer. RAC-drawability is $\exists\mathbb{R}$-complete and related results. *Journal of Graph Algorithms and Applications*, 27(9):803–841, 2023.

[Sel75] Reinhard Selten. Reexamination of the perfectness concept for equilibrium points in extensive games. *Internat. J. Game Theory*, 4:25–55, 1975.

[Shi16a] Yaroslav Shitov. How hard is the tensor rank? arXiv:1611.01559, 2016.

[Shi16b] Yaroslav Shitov. A universality theorem for nonnegative matrix factorizations. arXiv:1606.09068, June 29 2016.

[Shi17a] Yaroslav Shitov. The complexity of positive semidefinite matrix factorization. *SIAM J. Optim.*, 27(3):1898–1909, 2017.

[Shi17b] Yaroslav Shitov. The nonnegative rank of a matrix: hard problems, easy solutions. *SIAM Rev.*, 59(4):794–800, 2017.

[Shi23] Yaroslav Shitov. Further $\exists\mathbb{R}$-complete problems with PSD matrix factorizations. *Found. Comput. Math.*, 23(4):1225–1248, 2023.

[Shi25a] Yaroslav Shitov. Further restricted versions of the nonnegative matrix factorization problem. *Linear Algebra Appl.*, 710:267–272, 2025.

[Shi25b] Yaroslav Shitov. Several remarks on tensor rank computation. *Pacific J. Math.*, 334(1):143–151, 2025.

[Sho91] Peter W. Shor. Stretchability of pseudolines is NP-hard. In *Applied geometry and discrete mathematics*, volume 4 of *DIMACS Ser. Discrete Math. Theoret. Comput. Sci.*, pages 531–554. Amer. Math. Soc., Providence, RI, 1991.

[SL15a] Steven Schockaert and Jae Hee Lee. Qualitative reasoning about directions in semantic spaces. In Qiang Yang and Michael J. Wooldridge, editors, *Proceedings of the Twenty-Fourth International Joint Conference on Artificial Intelligence, IJCAI*

2015, Buenos Aires, Argentina, July 25-31, 2015, pages 3207–3213. AAAI Press, 2015.
[SL15b] Steven Schockaert and Sanjiang Li. Realizing RCC8 networks using convex regions. *Artificial Intelligence*, 218:74–105, 2015.
[SM73] Larry J. Stockmeyer and Albert R. Meyer. Word problems requiring exponential time: preliminary report. In *Fifth Annual ACM Symposium on Theory of Computing (Austin, Tex., 1973)*, pages 1–9. Association for Computing Machinery, New York, 1973.
[Spi94] Jeremy Spinrad. Recognition of circle graphs. *J. Algorithms*, 16(2):264–282, 1994.
[SŠ17] Marcus Schaefer and Daniel Štefankovič. Fixed points, Nash equilibria, and the existential theory of the reals. *Theory Comput. Syst.*, 60(2):172–193, 2017.
[SŠ18] Marcus Schaefer and Daniel Štefankovič. The complexity of tensor rank. *Theory Comput. Syst.*, 62(5):1161–1174, 2018.
[SŠ23] Marcus Schaefer and Daniel Štefankovič. Beyond the existential theory of the reals. *Theory Comput. Syst.*, page 195–226, 2023.
[SŠ24] Marcus Schaefer and Daniel Štefankovič. The complexity of tensor rank. arXiv:1612.04338, 2024.
[SŠ25] Marcus Schaefer and Daniel Štefankovič. Beyond the existential theory of the reals. arXiv:2210.00571, 2025.
[SSŠ03] Marcus Schaefer, Eric Sedgwick, and Daniel Štefankovič. Recognizing string graphs in NP. *J. Comput. System Sci.*, 67(2):365–380, 2003. Special issue on STOC'02 (Montreal, QC).
[ST22] Jack Stade and Jamie Tucker-Foltz. Topological universality of the art gallery problem. arXiv:2202.11076, 2022.
[Sta23] Jack Stade. The point-boundary art gallery problem is $\exists\mathbb{R}$-hard. arXiv:2210.12817, 2023.
[Str05] Ileana Streinu. Non-stretchable pseudo-visibility graphs. *Comput. Geom.*, 31(3):195–206, 2005.
[Stu87] Bernd Sturmfels. On the decidability of Diophantine problems in combinatorial geometry. *Bull. Amer. Math. Soc. (N.S.)*, 17(1):121–124, 1987.
[Stu05] Milan Studený. *Probabilistic conditional independence structures*. Information Science and Statistics. Springer, London, 2005.
[SW23] Patrick Schnider and Simon Weber. On the complexity of recognizing nerves of convex sets. arXiv:2302.13276, 2023.
[Tan10] Martin Tancer. d-collapsibility is NP-complete for $d \geq 4$. *Chic. J. Theoret. Comput. Sci.*, pages Article 3, 32, 2010.
[Tar48] Alfred Tarski. *A Decision Method for Elementary Algebra and Geometry*. The Rand Corporation, Santa Monica, CA, 1948.
[tCKO13] Balder ten Cate, Phokion G. Kolaitis, and Walied Othman. Data exchange with arithmetic operations. In Giovanna Guerrini and Norman W. Paton, editors, *Joint 2013 EDBT/ICDT Conferences, EDBT '13 Proceedings, Genoa, Italy, March 18-22, 2013*, pages 537–548. ACM, 2013.
[TGS95] Paul J. Tanenbaum, Michael T. Goodrich, and Edward R. Scheinerman. Characterization and recognition of point-halfspace and related orders (preliminary version). In *Graph drawing (Princeton, NJ, 1994)*, volume 894 of *Lecture Notes in Comput. Sci.*, pages 234–245. Springer, Berlin, 1995.
[Tho74] L. B. Thomas. Rank factorization of nonnegative matrices. *SIAM Review*, 16(3):393–394, 1974.
[Tri90] Eberhard Triesch. A note on a theorem of Blum, Shub, and Smale. *J. Complexity*, 6(2):166–169, 1990.
[Vah23] Mihir Vahanwala. Robust positivity problems for linear recurrence sequences. arXiv:2305.04870, 2023.
[Vak06] Ravi Vakil. Murphy's law in algebraic geometry: badly-behaved deformation spaces. *Invent. Math.*, 164(3):569–590, 2006.

[Val79] Leslie G. Valiant. Completeness classes in algebra. In *Conference Record of the Eleventh Annual ACM Symposium on Theory of Computing (Atlanta, Ga., 1979)*, pages 249–261. ACM, New York, 1979.

[Vav09] Stephen A. Vavasis. On the complexity of nonnegative matrix factorization. *SIAM J. Optim.*, 20(3):1364–1377, 2009.

[vD84] Eric E. C. van Damme. A relation between perfect equilibria in extensive form games and proper equilibria in normal form games. *Internat. J. Game Theory*, 13(1):1–13, 1984.

[vdHLS99] Hein van der Holst, Lázló Lovász, and Alexander Schrijver. The Colin de Verdière graph parameter. In *Graph theory and combinatorial biology (Balatonlelle, 1996)*, volume 7 of *Bolyai Soc. Math. Stud.*, pages 29–85. János Bolyai Math. Soc., Budapest, 1999.

[vdZBL23] Benito van der Zander, Markus Bläser, and Maciej Liskiewicz. The hardness of reasoning about probabilities and causality. arXiv:2305.09508, 2023.

[Ver23] Emil Verkama. Repairing the universality theorem for 4-polytopes. Master's thesis, Aalto University School of Science, 2023.

[Vol23] Mirco Volk. Komplexitätsanalyse des lawn mowing problems und verwandter probleme. Bachelor's thesis, Karlsuhre Institute of Technology, 2023.

[Vor84] Nikolai N. Vorob'ev. Estimates of real roots of a system of algebraic equations. *Zap. Nauchn. Sem. Leningrad. Otdel. Mat. Inst. Steklov. (LOMI)*, 137:7–19, 1984. Theory of the complexity of computations, II.

[Weg67] Gerd Wegner. *Eigenschaften der Nerven homologisch-einfacher Familien im $\mathbb{R}^n$*. PhD thesis, Universität Göttingen, 1967.

[Wik24] Wikipedia contributors. Existential theory of the reals — Wikipedia, the free encyclopedia. https://en.wikipedia.org/wiki/Existential_theory_of_the_reals, 2024. [Online; accessed 22-January-2024].

[Wil12] Stedman McDee Wilson. *Embeddings of Polytopes and Polyhedral Complexes*. PhD thesis, University of California, Los Angeles, 2012.

[Win19] Tobias Windisch. The fiber dimension of a graph. *Discrete Math.*, 342(1):168–177, 2019.

[WL10] Diedrich Wolter and Jae Hee Lee. Qualitative reasoning with directional relations. *Artificial Intelligence*, 174(18):1498–1507, 2010.

[Wur23] Adrian Wurm. Complexity of reachability problems in neural networks. In *Reachability problems*, volume 14235 of *Lecture Notes in Comput. Sci.*, pages 15–27. Springer, Cham, 2023.

[Wur24] Adrian Wurm. Complexity of Reachability and Verification Problems in Neural Networks. working paper or preprint, September 2024.

[Yap00] Chee Keng Yap. *Fundamental problems of algorithmic algebra*. Oxford University Press, New York, 2000.

[Yem79] Yechiam Yemini. Some theoretical aspects of position-location problems. In *20th Annual Symposium on Foundations of Computer Science (San Juan, Puerto Rico, 1979)*, pages 1–8. IEEE, New York, 1979.

[Z3] Z3. An efficient smt solver. https://www.microsoft.com/en-us/research/project/z3-3/.

[Zha92] Xiao-Dong Zhang. Complexity of neural network learning in the real number model. In *Workshop on Physics and Computation*, pages 146–150. IEEE, 1992.

[Zie95] Günter M. Ziegler. *Lectures on polytopes*, volume 152 of *Graduate Texts in Mathematics*. Springer-Verlag, New York, 1995.

[Zie08] Günter M. Ziegler. Nonrational configurations, polytopes, and surfaces. *Math. Intelligencer*, 30(3):36–42, 2008.

[ZMO24] Fabian Zaiser, Andrzej S. Murawski, and C.-H. Luke Ong. Guaranteed bounds on posterior distributions of discrete probabilistic programs with loops. arXiv:2411.10393, 2024.

Index

J. Pach and G. Tóth (eds.), *Courses in Discrete and Computational Geometry*, Bolyai Society Mathematical Studies 31,
https://doi.org/10.1007/978-3-032-10503-5

The manufacturer's authorised representative in the EU is Springer Nature Customer Service Centre GmbH, Europaplatz 3, 69115 Heidelberg, Germany. If you have any concerns regarding our products, please contact ProductSafety@springernature.com

Printed and bound by CPI Group (UK) Ltd, Croydon, CR0 4YY
05/10/2026
02228057-0006